FILE ORGANIZATION AND PROCESSING

FILE ORGANIZATION AND PROCESSING

Alan L. Tharp

North Carolina State University

John Wiley & Sons

WILEY New York • Chichester • Brisbane • Toronto • Singapore

FILE ORGANIZATION AND PROCESSING

Library of Congress Cataloging in Publication Data

Tharp, Alan L.
  File organization and processing.

  Includes bibliographical references and index.
  1. File organization (Computer science)   I. Title.
QA76.9.F5.T48   1988   005.74   87-14977
ISBN 0-471-60521-2

Printed in the United States of America

10 9 8 7 6 5 4

Printed and bound by Malloy Lithographing, Inc.

To
Kay and *Andrew*

Contents

Preface

One of the commonest mistakes is selecting the wrong hammer for the job; a hammer that is too large or too small or too heavy or too light will almost always lead to problems[1]

Without question, ours is an information society. We have an almost insatiable need for information. Computers provide one tool, a hardware one, for coping with our information needs. In conjunction with the computer, however, we also need software tools—methods for organizing and structuring information—to be able to accomplish many information-oriented tasks effectively and efficiently and, in some instances, to be able to accomplish them at all.

This text introduces the many useful and powerful data structures for representing information physically. (This is in contrast to a database management system, which represents information with logical structures.) Because most of our information requirements are large, the focus of this book is on large quantities of information and hence the title, *File Organization and Processing.* The primary audience for the book is computer science majors in the junior to beginning graduate level range who are taking a course in file structures. A secondary audience exists for those people who are involved or interested in software development and did not take a course in file structures during their formal education.

From one perspective, this text and the associated course could alternatively be called "Advanced Data Structures," since it is a continuation of the data structures course that is basic to every computer science curriculum. That initial course in data structures is one in which the students learn the basic "tools" of the trade. Arrays, stacks, queues, linked lists, and trees used by computer scientists may be compared with the basic tools such as the hammer, saw, drill, and screwdriver used by carpenters. Just as carpenters in their early training must learn about each tool of their trade, so too must computer scientists. Continuing with the carpentry analogy, in the course for which this text is intended, the students will learn the "secrets" of the trade to transform them into **craftsmen.** They will learn the equivalent of there being not one but

[1]Gene Austin, *Do-it-yourself,* Philadelphia Inquirer, January 26, 1986

nine kinds of "hammers," five types of "saws," and two kinds of "screwdrivers." Even though an elementary data structure may get the job done, a specialized data structure may do it much better. Just as there are many weekend "handymen" who think there is one all-purpose handy, dandy tool, usually a hammer, to solve all problems, so also are there too many computer software developers who think there is just one data structure for all purposes. Whatever their problem, they fit it into their favorite data structure. A goal of this book is to discourage that mentality among our future software developers. We are interested in preparing *true craftsmen* of computer science.

Just as new tools are continually being developed for the carpentry trade, so too are new data structures being discovered for use in software development. One feature that distinguishes this text from most existing ones is its emphasis on recent data structures. This emphasis is intended to assist in disseminating information about them so that they may be introduced into practice more rapidly.

Providing instruction about file organization and processing techniques is certainly the primary goal of this text, but a secondary, though nonetheless important, goal is providing an opportunity for the student to exercise and refine his *problem solving skills*. Many of the individual data structures considered in the text will lose their relevance over time in this field of rapidly changing technology. Learning about the specific data structures is then *not* as important over the long term as developing a facility to approach a novel situation and choose or synthesize a data structure appropriate to that situation. The *process* is as important as the goal.

The content and aims of this text parallel those of the CS5 course in the ACM's CURRICULUM '78. Nevertheless, because many of the file structures build upon the basic data structures of CS7, it is my belief that it is preferable if the file structures course *follows* the data structures course. This text therefore assumes a prerequisite of a basic data structures course.

This text has a practical rather than a theoretical orientation; it does not require the reader to be familiar with complex mathematics to understand the file structure concepts. Although the computational complexity of an algorithm is often noted when it is appropriate to the discussion, the derivation of the complexity is usually beyond the purposes of this text. The intent of this text is to assist the student in understanding the advanced "tools of the trade" and in applying them rather than in analyzing them. A primary goal of this text is to present the algorithms in such a straightforward manner that readers' efforts may be directed to a profound understanding of the structures and to improving, comparing, contrasting, and applying them, rather than directing their efforts to deciphering the basic notions. It is to elucidate rather than obfuscate. The audience is people who are unfamiliar with file structures.

We have all had the experience in which we are introduced to something new, maybe a camera or a computer system, which is *seemingly* complex until we understand it and then it appears simple. We are forever changed by our understanding of the mechanisms. We are in awe of a magic trick until we understand how it works. This text is aimed primarily at those who do not yet understand the mechanisms of file structures. We want to let readers in on the secrets and change them forever when it comes to understanding file structures.

How should algorithms for processing the file structures be represented? A conscious decision was made not to use a popular programming language such as Pascal. Instead, two types of representations were chosen:

- An outline in English for the more complicated algorithms. This is the approach used in stepwise refinement.
- A pseudocode for the simpler algorithms, and segments of the more complicated ones. This pseudocode is similar to the code of almost any block structured language but without the refinements necessary to make it into a programming language.

These representations are described in more detail in Chapter 1.

Unlike the algorithms in a basic data structures course, the algorithms for file processing are more complex because they involve more powerful and complicated data structures. Looking at page upon page of program listings may not be the best way to *understand* an algorithm (the trees for the forest syndrome). If the algorithm is short, maybe; if it is long, probably not. A high-level description in English that is shorter and more easily read should assist in comprehension. This decision on algorithm presentation agrees with the notion of "chunking" of information to aid human comprehension. Psychologists have observed that people can effectively handle only about seven units of information at a time.[1] If too much information about an algorithm is given to the reader at one time, it may inhibit comprehension rather than aid it. This text is *not* intended to be a "cookbook" in which readers merely choose the appropriate code and do not need to *understand* what they are doing.

A second reason for not choosing Pascal as the algorithm description language is that Standard Pascal does not have many of the structures needed for processing external files.

Third, a student who reaches the intermediate and upper levels of a computer science program is beginning to learn other programming languages such as C, Ada, Modula 2, Snobol, Icon, or REXX. Restricting the text to a single language would make it less desirable for someone wanting to program in another general-purpose programming language. The file structures described in this text transcend a particular programming language.

Still another reason for not choosing a programming language as the algorithm description language is that at some point in his career, a student will need to write a program from a description in a specification language. An important skill to develop is to be able to implement an algorithm readily given a description of it. In practice, the computer science graduate usually will not be given a task in which everything is already defined in terms of a programming language.

What will the reader gain from studying this text? Among the intended benefits are:

- An understanding of the basic mechanisms of each data structure discussed (i.e., how it works). And, given data, the reader should know enough about a particular file structure to store the data into it.
- An understanding of the advantages and disadvantages of each method. There is not yet one all-purpose data structure; so an important skill is to know when to use a data structure and when not to.

[1]Miller, George A., "The Magical Number Seven, Plus or Minus Two: Some Limits on Our Capacity for Processing Information," *Psychological Reviews,* 1956, pp.81–97.

● An ability to select the appropriate data structure for a given realistic application. Often, so much emphasis is placed on how each structure works that the student is not exercised in the equally important skill of what to use where. What is the ultimate reason to study file structures? It is to be able to use them in new situations. Only knowing a method is not enough. If one knows a method well, but *doesn't* know when to use it, it is really of little practical value. The final chapter on Applying File Structures is intended to underscore this goal.

How else will the text assist students in achieving these goals? First, the discussions are intended to be straightforward. The plan is to allow readers to learn the new information quickly and easily rather than to impress them (or maybe depress them) with how much they do not already know. The discussion is intended to be inviting rather than intimidating. Figures and illustrations are included to assist the reader in overcoming the initial hurdles to understanding. Important definitions are set off from the text so that they may be easily identified and referenced. Important techniques and principles are highlighted. As readers progress through the text, they will be surprised at how readily they are able to understand the new topics because of understanding the previous ones. They will be better able to notice improvements, strengths, and weaknesses of the individual file structures as they progress through the text. The emphasis is on what they **can** do rather than on what they **can't**.

Second, the text includes many examples. We learn a lot by observing; it is usually easier to watch someone do something, for example, replace a light bulb or log in to a new computer system, than to read about how to do it. For that reason, after introducing and describing each file structure in general, the text illustrates its functioning with example data so that the reader can "watch" it operate. One example may be worth many words.

Third, exercises appear at the end of each section and answers to selected ones appear in the appendix. The exercises were chosen to assist the reader in:

● Understanding basic concepts.
● Applying a basic concept to a unique situation.
● Combining several basic concepts to solve a novel problem.
● Recognizing idiosyncracies of a technique.

Finally, a list of key terms appears at the end of each section. It can serve as a measure of how well the reader understands the material and what topics may need to be restudied. These terms are not intended to be memorized but rather are ones that readers need to "understand" so that what follows is comprehensible. They are to *use* them.

Because the text contains slightly more material than would normally be covered in one course, and because many of the sections are independent of one another, instructors may tailor the course content to meet the needs and abilities of their students. For example, some may want to include the in-depth coverage of direct file organizations while others may wish to spend more time on sorting techniques.

ACKNOWLEDGMENTS

I want to thank the numerous people who helped with this project. First I am indebted to the many students over the years who expressed an enthusiasm for the subject and encouraged me to put this material into book form. I also appreciate the many students who studied from drafts of the manuscript and offered suggestions for improving the presentation.

I also wish to express my gratitude to the people who reviewed the manuscript: Hussein Abdul-Wahab of North Carolina State University; Henry Etlinger of Rochester Institute of Technology; Nancy Green of the University of Pennsylvania; Charles Hall of Rex Hospital; Mary Jane Lee of California State University at Sacramento; Jeff Mincy of Thinking Machines Corporation; Willard Ruchte of Wake County Schools; Eric Wang of Silver Springs, Md; and Lynn R. Ziegler of Michigan Technological University.

Alan L. Tharp

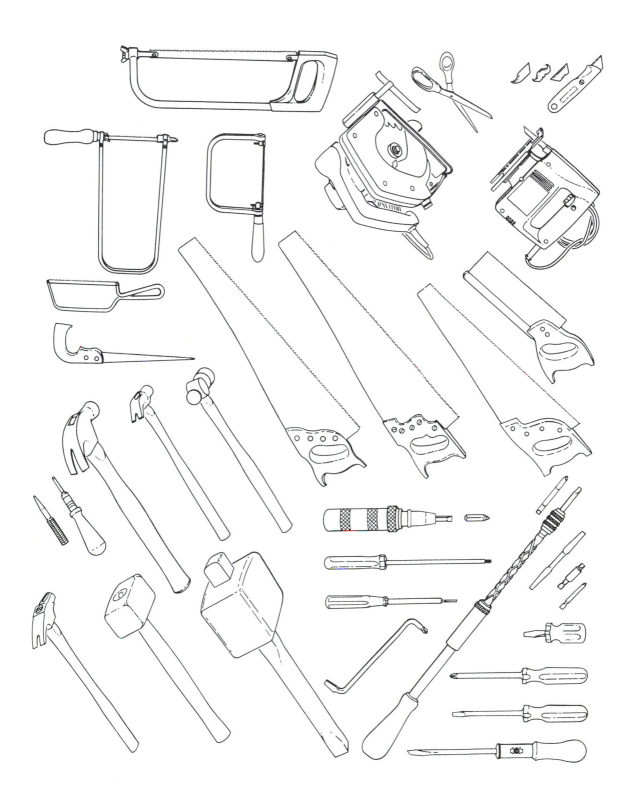

Chapter 1

Introduction

PURPOSE

Over the past 30 years, computers have dramatically changed society and the way that we do things. Extrapolate from these accomplishments and imagine what things might be like 30 years hence. It is probably impossible to envision more than a few years into the future, but you certainly can think of many new and appealing uses of computers. Very likely, you will help develop and/or apply many of them. All these uses of a computer, past, present and future, utilize **information** in some respect. The appropriate processing of information is what *makes* the application—whether it be an expert diagnostic system from artificial intelligence or an airline marketing analysis scheme from management information systems. Essentially every software system makes use of information—operating systems, language processors, information systems, graphic systems, etc. No matter what you ultimately plan to do in computer or information science, you will work with systems that use information.

> To build the vital software systems of the future, we need to be able to structure and manipulate information effectively and efficiently.

Devising such information structures is the topic of this text. It continues the discussion of how to handle data that was begun with the study of data structures. The primary differences are that in this instance (1) the quantities of data that we deal with are larger, (2) the number of structures is greater, and (3) their uses are more specialized. These larger quantities of data are referred to as files, but we will talk more about definitions later. To be able to build the systems of the future, you need to understand the information structuring and processing components well. Even in using such software systems, it will be helpful if you understand something about how they are constructed, just as it is helpful for you to know something about computer architecture even though you may never build a computer.

The history of computer science is strewn with software systems that were not successful; one prominent underlying theme for these failures has been an improper choice of a data or file structure or an inability to discover a proper one. Many **billions** of dollars have been wasted because these systems were not able to accomplish what they were intended to.

- Fast response time is often a stipulation for a software system. There are many examples of a system doing everything that it was intended to do but not fast enough. When we want information, we want it **now**.
- Frequently, the performance constraint on a software system is dictated by its use in a real time application. Factory automation projects, for example, often have real time constraints. In a real time application, the data must be processed within a fixed amount of time, usually a fraction of a second, to enable the application system to function properly. If the data can't be processed quickly enough, the system is of little or no value. Many projects have been started with the belief that the proper data structures could be found during the course of the development. Discovering them was not always achievable.
- A common constraint on developing a software system is having a limited amount of time in which to complete it. In computer science, technology changes so rapidly that if a development group does not build a system quickly enough, it may be obsolete when completed or a competitor may build a similar system sooner.
- A software system may have a memory constraint. If it cannot be written to fit within the available memory, it is not achievable. This constraint is more quantitative than the others, for the consequences of exceeding it are immediately obvious: everything stops. Either you have enough memory or you don't.
- Usability is another requirement. Many unsuccessful computer systems were designed from an "ease of implementation" perspective rather than an "ease of use" one. A system, however, is implemented once but used often. If the ultimate user, one who is not sophisticated in computer science, finds the system unusable, it will be scrapped. Such systems are legion.

Many times, commercial systems have not even been brought to market because they were not fast enough, not small enough, or not capable enough. One influence for many of these failures has been the developers' lack of understanding of file and data structures. With a better understanding of them, perhaps the failure rate of software systems will be reduced.

Selecting the "right" data or file structure will enable:

- Some applications to be developed that would otherwise be **impossible.**
- Some applications to be developed that would otherwise be too **slow.**
- Some applications to be developed that would otherwise be too **large.**

ALGORITHMS

Associated with most structures for organizing data are algorithms for processing them. Unlike many of the algorithms that you have considered in earlier courses, the algorithms that we consider, although not overly complex, contain multiple facets.

Algorithm 1.1
UNDERSTANDING A FILE ORGANIZATION

I. Read the introduction and the general description of a file organization together with its associated processing.

II. Study the example of how data is processed through the file organization.

III. Ponder the advantages, disadvantages, extensions, and alternatives to the file organization.

IV. For each key term,

    A. If its meaning is not known, restudy that portion of the text defining or discussing it.

    B. If its meaning is known, continue.

V. While unattempted exercises remain at the end of the section and one's understanding of a file organization or its associated processing is incomplete,

    A. Work an unattempted exercise.

    B. While the answer is not correct.

        1. Obtain additional information.

        2. Redo the exercise.

Explaining each facet in absolute detail might obscure the overall thrust of the algorithm. For that reason, we are using two formats for describing algorithms. One format describes the algorithm in an outline form using English. This format is similar to what you construct when you develop an algorithm using stepwise refinement. The other format is a block-structured pseudocode that can be readily converted into a programming language such as Pascal. Most of the algorithms will be presented in the English outline form. If the reader has an understanding of the technique, he should be able to proceed from this outline form to the pseudocode stage with a reasonable amount of effort by refining the individual components of the algorithm. The higher level English description requires that the reader understand the algorithm before programming it. The pseudocode format will be used to explain compact algorithms, for example, the sorting algorithms or critical portions of larger algorithms.

    An example of the outline format appears in Algorithm 1.1. As you can recognize, this algorithm suggests how to use this text to understand a new file organization. Notice that it gives you a general impression of what needs to be done without describing each detail. You will of course need to be able to resolve some of these details, such as how to determine if an answer is correct, whether an exercise has been attempted, or what to do if the exercises are depleted before you understand a topic. But your previous learning experiences have provided you with the means for handling these particulars so they do not need to detract from your understanding of the overall process.

    Algorithm 1.2 uses the pseudocode format to define an algorithm for taking an exam. The control structures of **for, while, if . . . then, repeat . . . until**, etc., function as they do in most block-structured languages. If you are unfamiliar with them or need

Algorithm 1.2
STRATEGY FOR TAKING AN EXAM

---

proc   take__exam

     /*   This is a comment; this procedure describes a strategy
          for taking an exam.   */

```
1       take(deep__breath)
2       for i = first to last do
3               scan(question[i])
4               if answer__to__question = "straightforward" then
5                       do
6                               read__thoroughly(question[i])
7                               answer[i] := your__best__solution
8                       end
9       end
10      while current__time < end__of__the__period
11              for i = first to last do
12                      if answer[i] = "empty" then
13                              do
14                                      read__thoroughly(question[i])
15                                      answer[i] := your__best__solution
16                              end
17              end
18              for i = first to last do
19                      check(answer[i])
20              end
21      end
22      turn__in(answer)
23      repeat
24                      worry__about!(answer)
25      until exam__is__returned
end     take__exam
```

---

a review, consult, for example, [1, pp. 9–15]. The **end** statement terminating a procedure definition acts as a **return** statement. ":=" signifies an assignment statement; x[ ] is an array and y( ) is a procedure call. As you can observe, the pseudocode format contains more detail than the outline format and hence takes more statements to describe the same algorithm.

METRICS

In the beginning of this chapter, we noted that it was essential that the file organizations and processing techniques be *efficient and effective*. But how does one know when or if he has achieved these goals? How do you know whether a piece of furniture will fit into a space within a room? You need a metric; you need a measuring device, for example, a yardstick or a rule. The same is true when evaluating file structures. We need metrics to determine whether a structure will meet the desired goal. Unfortunately, many of the metrics that we use for measuring file structures are not as accurate as ones we have for measuring distances. You have likely seen these software measures in your study of data structures. We will review them briefly. Note that many of them overlap.

- *Simplicity.* This rule, which has existed for some time, is often referred to as Occam's razor (or rule) which gives credit to the fourteenth century philosopher who formulated it so succinctly. If one has alternatives, and all other things are equal, he should choose the simplest one. Although this seems obvious, it is sometimes overlooked when in haste.
- *Reliability.* We all like things that function properly and we are all too familiar with the costs when something malfunctions. In choosing a file structure or processing technique, we want to be sure that it will work properly under *all* circumstances. We cannot always guarantee reliability or program correctness, but we need to make our selections such that the reliability is commensurate with the application.
- *Programmability.* Can one program the processing algorithm in a reasonable amount of time? If one cannot, another technique may be preferable.
- *Maintainability.* Once the technique is programmed, can it be modified or updated easily? If it is so complex that it is difficult to modify, another procedure may be better.
- *Storage requirements or space complexity.* When we consider storage media later in this chapter, we will say more about storage limitations and their importance in selecting a technique.
- *Computational or time complexity.* Performance is one of the most prominent constraints on a software system, and the lack of it is a primary cause for one's failure. The computational complexity of an algorithm estimates how much processing time will be needed when the algorithm is applied to a large quantity of data.

The computational complexity together with the storage requirements or space complexity are the most quantitative of the measures and the ones that we will concentrate on.

Since we will be using computational complexity as a frequent measure, let's review it quickly. We will *not* be deriving any but the most simple computational complexities; therefore, it is not essential that you completely comprehend the mathematics of it *if* you understand what to do with the results. Using computational complexity when choosing a file organization or processing technique moves computer science into the realm of a science rather than an art.

After stating the formal definition of what we mean by computational complexity, we will discuss what it means informally. Formally, the computation time of an algorithm is proportional to a function,

$$f(n) = O(g(n))$$

if and only if there exists a constant $c \geq 0$ and a constant $n_0 \geq 0$ such that

$$|f(n)| \leq c \ |g(n)| \qquad \text{for all } n \geq n_0$$

where n specifies the number of inputs to the algorithm. $f(n) = O(g(n))$ is read "f of n equals big-oh of g of n." What this means is that for a large number of inputs ($\geq n_0$), the computing time for an algorithm is bounded by a function $g(n)$ times a constant c. This constant c, a constant of proportionality, is inconsequential in determining the computation time when n_0 is very large; hence it is usually ignored when giving the complexity. Therefore, two algorithms with complexity functions of n^2 and $10n^2 + 1$ are in the same problem class and both have computational complexities of $O(n^2)$. The computing time for an algorithm is then proportional to, or of the order of magnitude of $g(n)$, that is, $O(g(n))$. Figure 1.1 graphs several common functions demonstrating what happens as n increases.

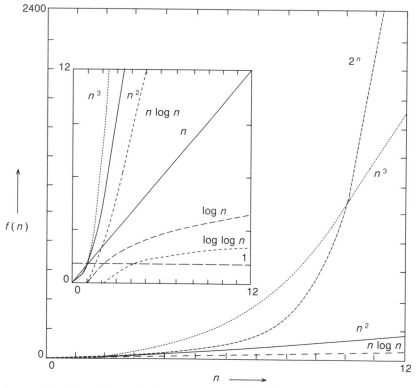

Figure 1.1 Rate of growth of common complexity functions. (*Source:* Edward M. Reingold and Wilfred J. Hansen, *Data Structures in Pascal*, copyright 1986 by Edward M. Reingold and Wilfred J. Hansen, reprinted by permission of Little, Brown and Company.)

Classes of Asymptotic Behavior (in Increasing Time)

- $O(1)$. The cost of using this algorithm is independent of the number of inputs. Any function that is $O(c)$ is $O(1)$. An algorithm of $O(1)$ is said to have **constant complexity**.
- $O(\log_2 \log_2 n)$. An algorithm with $\log_2 \log_2 n$ **complexity** has a performance bounded by $k \log_2 \log_2 n$ where k is a constant.
- $O(\log_2 n)$. An algorithm with **logarithmic complexity** is bounded by $k \log_2 n$.
- $O(n)$. An algorithm with **linear complexity** has a cost of execution no more than k times n.
- $O(n \log_2 n)$. An algorithm with $n \log_2 n$ **complexity** has a performance bound of k times $n \log_2 n$.
- $O(n^2)$. An algorithm with **quadratic complexity**.
- $O(n^3)$. An algorithm with **cubic complexity**.
- $O(c^n)$ where $c > 1$. An algorithm with **exponential complexity**.

By analyzing algorithms through the use of computational complexity, one can determine which algorithm is more efficient. For two algorithms such that all other factors are equal, one will usually choose the algorithm with the lower complexity, the more efficient one. An $O(n)$ algorithm would usually be chosen over an $O(n^2)$ algorithm.

All the metrics need to be considered when making the final selection of a technique. It is rare that a technique is the best among all the measures.

THE mod FUNCTION

Most of you have encountered the **mod** function previously, perhaps you have even programmed with it since it is built into many programming languages. We will make extensive use of this function in the remainder of the text so it is important that everyone understand it. The **mod** function, which has two parameters x and y, is useful for mapping a number x into a fixed range of values, 0 through $y - 1$. y is called the modulo, which explains where the function gets its name, for example, x modulo y. For positive integers, x **mod** y is merely the remainder after evenly dividing x by y. For example,

$$4 \text{ } \mathbf{mod} \text{ } 3 \text{ is } 1 \qquad 35 \text{ } \mathbf{mod} \text{ } 11 \text{ is } 2 \text{ and} \qquad 27 \text{ } \mathbf{mod} \text{ } 7 \text{ is } 6$$

The **mod** function has a formal definition of

$$x \text{ } \mathbf{mod} \text{ } y \equiv x - y \left\lfloor \frac{x}{y} \right\rfloor \qquad \text{if } y \neq 0, \text{ else } x \text{ } \mathbf{mod} \text{ } 0 \equiv x$$

where $\lfloor z \rfloor$ represents the floor(z) or the greatest integer $\leq z$.

You may envision the **mod** function as a "big wheel" that rolls along the line of numbers. The circumference of the "wheel" is divided into y slots of a length equivalent to the distance between the integers on the number line. The slots are identified by the integers 0 through $y - 1$. As the "wheel" rolls along, it maps numbers from the number line into its slots. The identifier for the slot that a number is mapped into is the result of moding that number with y. Figure 1.2 depicts this arrangement. Since there are three slots on the "wheel" in this example, the computations are for modulo 3 arithmetic. From it, you can easily see why 4 **mod** 3 is 1, why 8 **mod** 3 is 2, and why computing with a negative value for x requires the use of the formal definition, for example, -1 **mod** 3 is 2. We will be dealing exclusively with positive integers; so it should be easy for you to compute with the function.

DATABASE MANAGEMENT SYSTEMS

As noted previously, this text will consider the organization and processing of information, usually in large quantities. A database management system is a special-purpose software system that is specifically designed for the purpose of storing and manipulating information. The question then naturally arises as to what the relationship is between the techniques of file organization and processing and those of a database management system (DBMS).

Differences

The primary difference between these two approaches is that the file organization and processing structures are used to manage information on a *physical* level whereas a

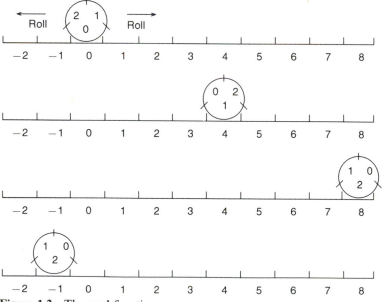

Figure 1.2 The **mod** function.

DBMS is used to manage information on a *logical* level. What are the meanings of physical level and logical level? A physical structuring is the manner in which the information is *actually* stored internally within a computer system. The physical structuring is a computer-oriented format in which usually many different constructs are used to store the information. A logical structuring is an *abstraction* of how the information is actually stored. It is how the user envisions that the information is arranged although it may physically be stored in a much different fashion. It is a format that is aimed at the human user and usually contains relatively few constructs. The physical level is then a lower level for representing information than the logical level. There are many advantages for representing information at a logical level, which explains the popularity of DBMSs [2]. A DBMS transforms the information from a logical representation into a physical one.

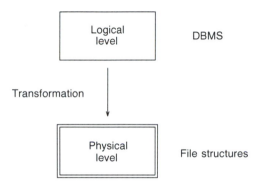

When does one use the principles and techniques described in this text and when does one use a DBMS? DBMSs are used primarily to develop **application** software such as an automated preregistration system, whereas the file organization and processing techniques are used primarily to build **system** software such as an operating system, language processor, or yes, even a DBMS. Anyone who builds system software then needs to be quite knowledgeable of file structures. Any DBMS course that delves into the implementation of a DBMS needs a file organization course as a prerequisite. In the past, file organization techniques were taught either as an appendix to a data structures course or as a preliminary to a DBMS course. However, with the great quantity of new information in both database management and file organization, it is no longer possible to discuss all the basic information on both topics in a single course. And there is not enough time at the end of a data structures course to discuss all the important file structures. That is one reason that it is important to have a distinct course on file organization techniques.

But what if you are more interested in application software? Should you still be knowledgeable of file structures? If you will be primarily a user of a DBMS rather than a builder of one, will it be important for you to understand file structures? As mentioned earlier, you may be a user of computers who never plans to build a computer, but it is important for you to know something about computer architecture. By knowing about the underlying structures and internals of a DBMS, you will be able to understand and use it more efficiently and effectively. You will better understand why certain operations of a DBMS system require so much time to complete, for example, a relational JOIN. You will know what it means to have the DBMS create an index for a data attribute and what the ramifications of doing it are. One quality that will separate a DBMS user who is a computer science graduate from one who is not will be the computer science graduate's greater knowledge. DBMSs are now simple enough and available enough that one not need to be a computer science graduate to apply one. But to really be a master, some knowledge of how it works is still necessary.

Even if you expect to be designing or building application software exclusively, you may not always use a DBMS. A DBMS incurs significant system overhead. It saves people time but requires considerable computer time. There may be situations in which the task that you are working on has constraints, for example, real time ones, which may prohibit the use of a DBMS to solve the problem. Or it may be a relatively simple problem in which calling a DBMS would take too much time. It is the old case of trying to crack a peanut with a sledge hammer. A DBMS may be overkill. In such a case it might be better to develop the application using file organization techniques.

KEY TERMS

Algorithm format	**mod** function
English outline	$N \log n$ complexity
Pseudocode	Occam's razor
Application software	Physical level
Constant complexity	Programmability
Control structures	Quadratic complexity
Algorithms	Real time
Cubic complexity	Reliability
Database management systems	Response time
Exponential complexity	Simplicity
Floor	Space complexity
Linear complexity	Algorithm analysis
Logarithmic complexity	System software
Logical level	Technology change
Maintainability	Time complexity
Metrics	Algorithm analysis
Algorithm analysis	Usability

EXERCISES

1. List three reasons why software systems in the past have failed.

2. Write a procedure in an English outline format for deciding what courses to take for the next term.

3. Write a procedure in pseudocode for preparing for an exam.

4. Which algorithm is preferable? One with exponential complexity or one with quadratic complexity? Why?

5. What is 39 **mod** 11?

6. List three differences between physical and logical representations for information.

STORAGE MEDIA

The reason for reviewing storage media is that they affect the decision of how best to structure and process information. Just as the time required to access or locate something of yours may influence what you do, so too does it influence how information is handled within a software system. If you had considered studying your notes for a course, but you find that you do not have them with you, you may decide to study something else. Unless, of course, it is the day of an exam for that course; then you may decide that it is worth a trip to recover the notes.

With a software system, the *access time* is the length of time required to locate a piece of information. If information is stored on a device with high-speed access, meaning that relatively little time is needed to locate an item of information, one way of doing things may be appropriate; whereas with a storage medium with slower access, a different way may be preferable. But why not store *all* information in a high-speed storage medium?

That is a question that I have been asked frequently for many years. The resolution suggested by the question certainly sounds appealing because it would make many of the organization and processing decisions simple. In conjunction with that question, I have often heard the statement, "high-speed memory prices are getting cheaper," which is undeniably true. That statement, though, implies that all information can be stored in high-speed memory so there is no need to learn about the various file structures that we will soon examine. Often it has been suggested that if we had only X amount of high-speed memory, we could do anything. The interesting thing is that the value for X has constantly increased. What seemed as if it would be more than sufficient 10 years ago, is now considered less than adequate. What has happened is that our need for information has increased at least as rapidly as the decrease in the cost and the increase in the capacity of high-speed memory.

Even though capacities have increased and costs have come down, there are still limitations on the amount of high-speed memory that may be attached to a computer; the constraints may be physical or economic. The addressing scheme of the computer system may restrict the amount of high-speed memory to 640 kilobytes for a micro or 16 megabytes for a larger computer. In addition, budgetary constraints may come into play. Even though high-speed memory is cheaper, it is still not free. How many of you who own a microcomputer have the maximum amount of memory attached to it? You may want the maximum amount but you are just not able to afford it at the present. *Cost* and *capacity* are important considerations in addition to access time. The restrictions that they impose coupled with our almost insatiable thirst for storing information will probably, at least for the foreseeable future, make it impossible to store *all* the information that we want in primary memory. There will still be a need for a variety of storage media *and* techniques for structuring and processing information.

We will briefly review several storage media. For greater detail, refer to an introductory computer science text [e.g., 3 or 4] or one on computer organization [e.g., 5]. Since storage media has different (1) access times, (2) storage capacities, and (3) costs, they form a hierarchy.

Storage Hierarchy

The media for storing information within a computer system is usually divided into two main types:

Primary Memory

Auxiliary Memory

Primary memory is characterized by having faster access, costing more per bit, and having a smaller capacity. Auxiliary memory has attributes that are the converse; it has slower access, costs less, and has a larger capacity. Let's consider each type in turn.

Primary Memory   The most common implementation of high-speed primary memory uses semiconductor technology. A semiconductor memory chip is a very large scale integration (VLSI) of transistors and other electronic components. A typical access time for semiconductor memory is 5×10^{-7} seconds. Other technologies used for primary memory include charge-coupled devices and bubble memories. Older computers used ferrite core technology. In addition to fast access, high cost, and smaller capacity (since several megabytes cannot be classed as small), primary memory has the characteristic that any address may be located from any other in a fixed amount of time. This fixed time property makes primary memory desirable.

Auxiliary Memory   In considering the auxiliary storage media, we will follow a progression of decreasing unit storage costs, but with increasing storage capacities and access times, that is,

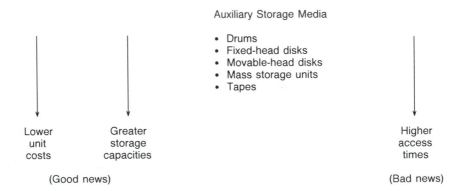

Auxiliary Storage Media

- Drums
- Fixed-head disks
- Movable-head disks
- Mass storage units
- Tapes

Lower unit costs   Greater storage capacities   Higher access times

(Good news)   (Bad news)

Drums   The drum, sketched in Figure 1.3, consists of a cylindrically shaped drum (hence the name) containing information recorded as magnetic spots on parallel tracks around the circumference of the cylinder.

> A **track** is a unit for storing and referencing information.

In a drum, a track is a circular band of information around the surface of the drum. Each track has a read/write head associated with it, so a rotation of the drum is the only motion needed to find a specific location. The rotational delay in locating a specific address is referred to as the latency. The representative access time for a drum is 10^{-2} seconds. Just as with primary memory, the drum and all the other auxiliary storage devices that we consider except for the tape may move from one storage location to any other. However, *unlike* primary memory, the amount of time that it takes to complete a move is variable and computing what it is may often be complex [see 6]. The time for moving is a function of the device plus the location of the current and subsequent addresses. In comparison with primary memory, a drum has a slower access time but also lower unit costs and greater storage capacities. Drums are used primarily with mainframe computer systems.

Fixed-Head Disks   A disk drive, often shortened to "disk," is a device that writes and reads information to and from recording platters that resemble phonograph records except that the tracks form concentric circles instead of a spiral. Of course, the recording technologies are also different. The density of information on a disk recording surface is greater on the inner tracks than on the outer ones because the same amount of information is stored on each track (see Figure 1.4). A disk contains two recording surfaces per platter, one on the top and the other on the bottom, except for the very top and the very bottom platters of the device, as shown in Figure 1.5. The distinctive feature of a fixed-head disk is that *each* track has a read/write head associated with it, so just as with a drum, the only delay in accessing a specific location is a rotation. The access time is therefore also the same as that of the drum, that is, 10^{-2} seconds. Because of the hardware associated with each track, the recording surfaces are not removable.

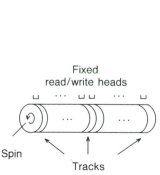

Figure 1.3   A drum storage device.

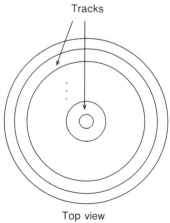

Figure 1.4   A disk recording surface.

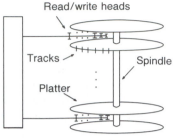

Figure 1.5 A fixed-head disk.

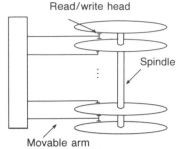

Figure 1.6 A movable-head disk.

Movable-Head Disks   A movable-head disk is similar to a fixed-head disk except that it has only a single read/write head per recording surface. The read/write heads are situated on an access arm that can be moved so that the read/write heads can be positioned at any track on the recording surface. All the read/write heads move together and thus access tracks in the same vertical plane. The time required to position the heads at the desired track is called the seek time. To find a specific location, both a rotational and an access arm movement may be necessary. Therefore, access time is usually greater than with a fixed-head disk, typically 10^{-1} seconds, but the cost per bit of storage is lower because less hardware is necessary. See Figure 1.6. Since the access arm may be retracted such that the read/write heads are not between the disk platters, the storage disks are removable; the removable component is referred to as a disk pack. The movable-head disk is the most common type of auxiliary storage medium; it is usually what we envision when we refer to auxiliary storage in subsequent discussions. Even disk storage is a scarce resource in many applications [7].

Table 1.1 contains timing and capacity values for several IBM disks. The number of tracks in parentheses tells how many additional backup tracks exist per recording surface; these backup tracks are used to store information should a primary one become damaged. The larger the number designating the disk, the more recent it is.

A floppy disk is a type of movable-head disk drive. It contains only a single platter with one or two read/write heads. The number of heads determines whether the removable storage medium, called a diskette, is one-sided or two-sided. The $5\frac{1}{4}$-inch diskette used with the IBM PC/AT class of computers has 96 tracks per inch (tpi) and after formatting is capable of storing 1.2 megabytes of data. The $5\frac{1}{4}$-inch diskettes introduced earlier have fewer tracks per inch and thus lower storage capacities. A

TABLE 1.1 CHARACTERISTICS OF SEVERAL IBM MOVABLE-HEAD DISKS

Disk	3330	3350	3380
Tracks per recording surface	404 (7)	555 (5)	885 (1)
Bytes per track	13,030	19,069	47,476
Recording surfaces	19	30	30
Access arm motion average (ms)	30	25	16
Rotational delay average (ms)	8.4	8.4	8.3
Data transfer (Mbytes/s)	0.806	1.198	3

Winchester or hard-disk drive also has movable heads and has versions in which the recording medium is both removable and not. The access time for a floppy-disk drive is typically 10^{-1} seconds whereas for a hard disk it is 10^{-2} seconds.

The recent technology of optical disks provides greater storage capacities at lower costs than their magnetic counterparts. An optical disk drive stores information by using a laser to burn holes into one of the layers of the recording medium. On reading, each hole allows light to pass through it so that a binary zero can be distinguished from a binary one. A small form of an optical disk, called a CD-ROM (compact disk—read only memory), can store up to 550 megabytes of data, which is equivalent to 150,000 printed pages or over 450 high-density (96 tpi) diskettes, at a cost of only a few dollars. It is suitable for storing reference materials, for example, dictionaries, encyclopedias, and catalogs. Its disadvantage, of course, is suggested in its name; it is a *read only* medium.

Mass Storage Devices and Tapes A mass storage device is a hybrid of disk and tape technology. A tape is a serial access storage medium. That means that the only addresses that a tape may move to from the current one are the two adjacent ones— either forward or reverse. No other positions are possible without scanning all the intervening positions. Obviously then, accessing a specific position is relatively slow with a tape. But a tape is an inexpensive storage media used primarily for archiving or transporting information. One reason that we are mentioning the mass storage device is that it uses a principle that we will see again in the text, that is: **divide and conquer**. If we have a tape rolled out lengthwise, we may have to search the entire tape to find the record we want. But what if we take a scissors to the tape and cut it into smaller pieces? Then, if we have an index or a means of knowing which piece of tape contains the record that we are seeking, we need to search, at most, only a single piece of tape; limiting the search to one segment of the information improves access time. The concept is simple, that is:

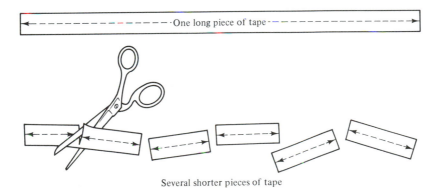

Several shorter pieces of tape

A mass storage device does just that. It uses tape as the recording medium because of its relatively low cost but divides the tape into smaller units. These smaller pieces of tape are wrapped around a canister the size of a soft drink can. These canisters are stored in what might be considered a large honeycombed vending machine. When you need to access a specific record, the appropriate canister is retrieved and copied to a disk in the event that it may be accessed again in the immediate future. Retrieving a canister and copying the information takes much less time than performing a sequential search of a tape. A mass storage device can store over 200 **billion** characters of information. Figure 1.7 illustrates the honeycombed storage area of a mass storage device. Figure 1.8 shows the relative sizes of a mass storage unit's canister and a disk pack with equivalent storage.

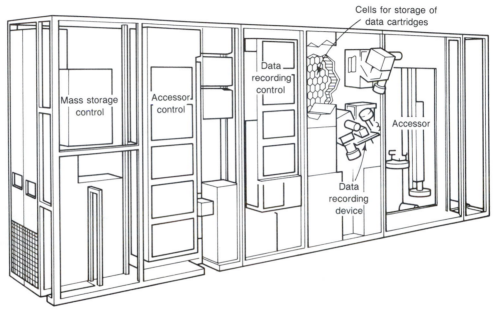

Figure 1.7   Honeycombed storage area for a mass storage device. (Courtesy of International Business Machines Corporation.)

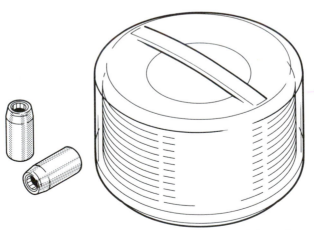

Figure 1.8   Two mass storage cartridges and their equivalent capacity in a disk pack. (Courtesy of International Business Machines Corporation.)

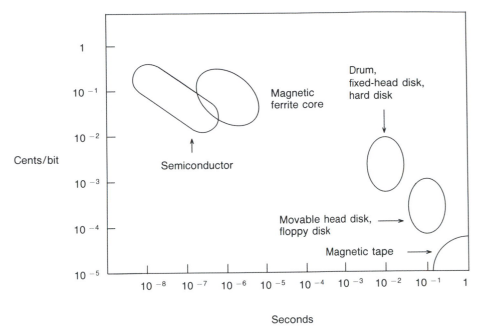

Figure 1.9 Access time vs. costs. (Adapted from Figure 5-14 in [3] with permission.)

Impact of Storage Hierarchy

The important concept to grasp from this brief review is that we need to use a hierarchy of storage media in which the individual devices have different properties. Figure 1.9 illustrates the relative unit costs and access times for the storage media that we have discussed. It is important for you to understand the differences in the relative access times—to understand beyond merely noting what the numbers are, that is, to comprehend what impact the differences have. Let's look more closely at the two most common storage devices: primary semiconductor memory and auxiliary storage in the form of a movable-head disk. The access times vary from 5×10^{-7} seconds for semiconductor memory to 10^{-1} seconds for a movable-head disk. That means that 5 seconds worth of accessing semiconductor memory would be equivalent to 1 million seconds on a movable-head disk, or 5 seconds vs. $\approx 11\frac{1}{2}$ *days*. That is a big difference. And that is why it is so important to choose the proper data structure. Let's look at these values differently. What if access time were money (which in a way it is)?

> Then what would cost 5 cents in primary memory accesses would cost $10,000 in auxiliary storage accesses.

And that is a significant difference! We don't give much thought to our 5-cent purchases but most of us give considerable thought to our infrequent $10,000 purchases. An access to auxiliary storage is *not* almost the same as an access to primary memory, just as a stick of chewing gum is not almost the same as a new automobile. (Even a stick of chewing gum costs more than 5 cents now.)

Until you have something to compare against, it is difficult to comprehend these relative differences in more than just numbers. In a recent class on file organization, I gave an assignment in which the students were asked to process 50 records. Most of the students' previous experience with storing information had been with primary memory. Unfortunately, the minicomputer that the students were to use for the programming assignment did not allow them to store the records directly in auxiliary storage. But a few students who owned their own micros asked if they could do the assignment on their own machines. I said "yes" if they stored the records on auxiliary storage (floppy disks) rather than in primary memory. One of the students came back and said that he couldn't believe it, even though I had made the previous remarks to him prior to his embarking on the project. He kept checking his algorithms to see if they were correct and they seemed to be. But it took him 14 hours of total run time to get his answers; whereas the total elapsed time for those who used the primary memory in the time-shared minicomputer was only a few minutes. Where you store information *does* make a difference!

Track Formats

One of two addressing schemes is used to format information stored on a track of a drum or a disk: either block addressability or sector addressability. With block addressability, the storage areas may be a variable size whereas with sector addressing, all the areas are a fixed size.

Block Addressability Block addressability makes use of the concept of a cylinder of information. We have previously considered a recording surface that is a horizontal subsetting of the information on a disk. A *cylinder*, which is a vertical subsetting of information, is all the information that can be accessed with one positioning of the access arm on a movable-head disk. All the read/write heads move in parallel, and, at a given time, the same track value identifies each head's position on a recording surface.

For improved retrieval performance, it is actually better to store information from a file in a cylinder, that is, on the same track numbers but on different recording surfaces, than on adjacent tracks on the same recording surface. By storing the information in a cylinder, we eliminate the access arm movement from one track to another. Changing the read/write head can be done electronically which takes little time. Figure 1.10 illustrates the concept of a cylinder.

With the definition of a cylinder, we can now consider Figure 1.11 which represents how information is stored using block addressability; this is the layout for the auxiliary storage devices used with the larger IBM and compatible computers. Within the block addressability scheme there are two basic formats: count-data and count-key-data. The difference depends upon whether a separate key or identifier area is associated with each block of information. The index marker denotes the beginning of the track and the home address area identifies it. A track is designated by a combination of its horizontal and vertical positions within a disk. The head number tells which recording surface within a cylinder and the cylinder number denotes which cylinder. Each block of information has a count and/or key area associated with it that gives information used by the system. Within the count area for each block of information, there is a field that gives the record number for that block within the track. Record R0 is another system area. All the information areas, both those used by the system and those employed by the user have *gaps* between them (to allow time for equipment functions to take place). The size of these gaps, called interrecord gaps, varies with the device, the location of the gap, and the length of the preceding area. The considerable storage needed for the system functions in the form of gaps and nondata areas is then not available to the user. Since much of the system overhead requirements are on a per data block basis, meaning that a certain amount of overhead is needed for each data area,

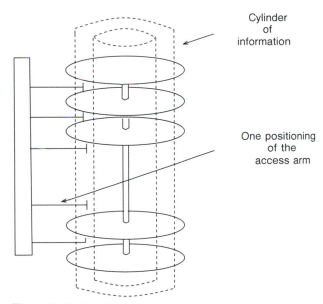

Cylinder
of
information

One positioning
of the
access arm

Figure 1.10 Vertical subsetting with a cylinder.

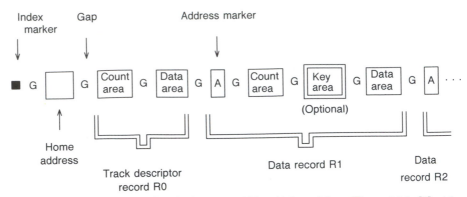

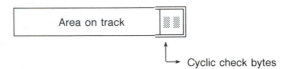

Figure 1.11 Track layouts for block addressability. (Adapted from Figure 3.3 in [8] with permission.)

the larger the data block, the greater the amount of information that can be stored per track.

Each area on a track contains two cyclic check bytes which are used in checking for transmission errors.

They perform the same function as parity bits but require much less storage. With parity bits, each character or byte of information has one parity bit associated with it. The value of the parity bit is determined by the values of the other bits comprising the byte. Some computer systems use even parity and others use odd. For odd parity, the total *count* of one bits, including the parity bit, must be an odd number. For an even parity, the total must be an even number. For the bit pattern

$$11010010$$

the parity bit would be a "1" so that the total number of one bits would be five which is an odd number. If even parity were used, the parity bit would be a "0." After a transmission of data, if the proper number of "1"s does not exist in a byte, it is retransmitted. Since many people are either never exposed to the concept of parity bits or forget about them because they are rarely mentioned, they think that they are free (something like "buy eight and get one free"). The parity bits, however, cost just as much as the other bits. Each byte of information requires one. That then is a $\frac{1}{8}$ or 12.5 percent overhead or increase in storage costs. Since cyclic check bytes usually take considerably less space than parity bits, they save expense. A cyclic check byte area of only 2 bytes is associated with each block of data no matter what size it is. For a 200-byte information area, only 2 bytes are needed for error checking with cyclic check bytes versus 200 bits (25 bytes) for parity bits. The overhead when using cyclic check bits in this example would be only 1 percent rather than 12.5 percent for parity bits.

To produce the value for the cyclic check bytes, a function is applied to the contents of the associated information area, that is,

$$f(\text{information area}) \rightarrow \text{cyclic check bytes}$$

Upon transmission of an area, the function is applied to the noncyclic check byte portion of the area to check if that result is equivalent to the contents of the cyclic check byte area. If they differ, a retransmission of the area is made in an attempt to correct the error. The field of study that develops efficient error checking functions is referred to as algebraic error correction or coding. Many universities offer entire courses on how to develop such functions. We will encounter one type of such a function in a subsequent chapter.

The IBM student text on direct-access storage devices [8] provides additional information on the purpose of the fields in the various areas of a track when using block addressability.

Sector Addressability Figure 1.12 pictures a sector formatting of a disk. A track is divided into equal-sized storage units called sectors. A sector is addressed by its track and sector numbers. It is typically 128, 256, or 512 bytes. The gaps between the sectors allow the system to perform certain functions such as processing the current record and preparing for the next record. Another similarity to block formatting is that each sector has two cyclic check bytes associated with it for error checking. In addition to the gaps between sectors, each track has a preamble and postamble storage area reserved for use by the computer system. As a result of the system storage requirements, much of the actual storage is unavailable to the user. Of the 1.6 megabytes of actual storage on a PC/AT class diskette, the user may only use 1.2 megabytes. Floppy disks and hard disks used with micro systems use a sector formatting. A hard-sectored diskette always

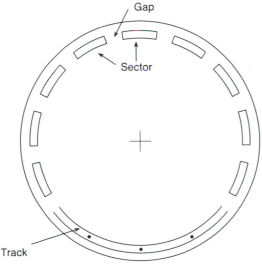

Figure 1.12 Sector addressability.

has sectors of the same fixed size. With a soft-sectored diskette, the size of the sectors may be specified by the system's control program when the disk is formatted.

SUMMARY

We have now completed the preliminaries to the study of file structures. You have been introduced to the reasons for studying them together with the formats and evaluation measures of the algorithms for processing them. In addition, you have been reacquainted with or introduced to the **mod** function, storage media, and database management systems. You now have the background to better understand and appreciate the how and why of the various file structures that we will consider.

KEY TERMS

Auxiliary memory
Block addressability
CD-ROM
Cyclic check bytes
     Block addressability
Cylinder of information
Disk pack
Disk recording surface
Diskette
Drum
Fixed-head disk
Floppy disk
Hard disk
Interrecord gaps
Latency
Mass storage devices
Movable-head disks

Parity bits
Platters
Primary memory
Sector addressability
Seek time
Semiconductor memory
Storage hierarchy
Storage media
     Access time
     Capacity
     Cost
Tape
Track
Track formats
Transmission errors
Winchester disk

EXERCISES

1. What is the storage capacity in bytes of a cylinder on an IBM 3350 disk?

2. Without performing any calculations, determine which has the greater storage capacity in a cylinder, a 3350 or a 3380 disk drive. Why?

3. **(a)** On a 3350 disk drive, is the storage capacity greater for a platter or for a cylinder? **(b)** for a recording surface or a platter? **(c)** for a track or a cylinder?

4. What is the total storage capacity of an IBM 3330 disk?

5. Give an explanation for the relative data transfer rates for the disks in Table 1.1.

6. What would be the equivalent time on a movable-head disk for 1 second's worth of accessing primary semiconductor memory?

7. What percentage of overhead storage would be saved by using cyclic check bytes rather than parity bits with a 1000-byte data area?

8. What is the storage capacity of a diskette that is double-sided with 40 tracks per side, 16 sectors per track, and 256 bytes per sector?

9. **(a)** What are the advantages of having a sector size of 512 bytes vs. 256 bytes? **(b)** the disadvantages?

REFERENCES

1. Horowitz, Ellis, and Sartaj Sahni, *Fundamentals of Data Structures*, Computer Science Press, Rockville, MD, 1982.

2. Date, C. J., *An Introduction to Database Management Systems*, 4th Edition, Addison-Wesley, Reading, MA, 1986.

3. Pick, James B., *Computer Systems in Business*, PWS Publishers, Boston, MA, 1986.

4. Shelly, Gary B., and Thomas J. Cashman, *Computer Fundamentals with Application Software*, Boyd & Fraser Publishing Company, Boston, MA, 1986.

5. Gorsline, G. W., *Computer Organization: Hardware/Software*, Prentice-Hall, Englewood Cliffs, NJ, 1986.

6. Peterson, W. Wesley, and Art Lew, *File Design and Programming*, John Wiley & Sons, New York, 1986.

7. Stamps, David, "DASD: Too Much Ain't Enough," *Datamation*, December 15, 1985, pp. 22–28.

8. *Introduction to IBM Direct-Access Storage Devices and Organization Methods*, IBM, Poughkeepsie, NY, 1978, GC20–1649.

PART ONE

PRIMARY FILE ORGANIZATIONS

PREVIEW

The topic of this text is the organization and processing of information, usually records stored in a file. With the preliminaries of the previous chapter behind us, we are now prepared to embark on the study of data structures appropriate for organizing:

- Large quantities of records.
- Information that may have special space or performance requirements.
- Information that may need to be processed in different ways for different applications.
- Information that enables new tasks to be accomplished.

We are interested in locating not only a data structure that will meet our needs, but also one that will do it efficiently in terms of both space and performance.

This part presents the three primary file organizations of (1) sequential, (2) direct, and (3) indexed sequential, which form the basis for many applications. It is important to make a distinction between how we store or *organize* information and how we process or *access* it. The distinction may be illustrated by:

Organization	Access
Sequential	Sequential
Indexed sequential	Sequential and direct
Direct	Direct

Sequential access refers to accessing *multiple* records, often an entire file, and usually according to a predefined *order*, whereas direct access, also called random access, refers to locating a *single* record. To have an effective organization, we need to match the type of organization with the type of intended access. We need to fit the organization to the activity. One organization may be ideal for one type of access but completely inappropriate for another use. It is essential then that we match the use with the structure. Just as we wouldn't use a hammer to transform a board into two smaller ones, we shouldn't use an inappropriate data organization to structure information for an application. A sequential organization, as the name suggests, is suitable for sequential access and a direct organization is fitting for direct access. Indexed sequential organization is primarily for applications that require *both* types of access.

Chapter 2

Sequential File Organization

BACKGROUND

Definitions

First, let us assume that we have units of information called *records* which contain data about individual entities. A record may be decomposed into more atomic units called *fields* which contain specific values for attributes describing the entity, for example,

record	field $_1$	field $_2$	. . .	field $_n$

Information about an employee, for example, might be stored in the following way:

employee record

employee name	number	title	dept.	manager	salary	other fields

with the field names of "employee name," "number," "title," etc. The *primary key* is a field, or a composite of several fields, which uniquely distinguishes a record from all others; all the remaining fields are the *secondary keys* or attributes. For example, either the "employee name" or "number" field might be the primary key while the "title" field, "department" field, and others are secondary keys.

> A **file** consists of records of the same format.

Figure 2.1 contains an example file of records of employee data. Each row in the figure represents a record; the vertical lines separate the fields of the record. The sizes of the fields within a record may vary. For fixed-length records, the type that we will consider mostly, all the records within a file have the same length. With variable-length records, as the name implies, all the records within a file do not need to be the same length. Most files that we will deal with will have a potentially large number of records; that is why it is essential to match the organization with the activity, for a slight inefficiency

Employee name	Number	Title	Department	Manager	Salary	
Scott Matthews	0123	Programmer	Accounting	Melissa Jones	0026000	. . .
Tammy Boger	0240	Analyst	Data proc.	Morris Lancaster	0032000	. . .
Morris Lancaster	0067	Manager	Data proc.	Charles Hall	0045000	. . .
Larry Cookman	0189	Programmer	Security	John Bittle	0025000	. . .
David Caudle	0423	Analyst	Accounting	Melissa Jones	0047000	. . .
Richard Lehner	0847	Instructor	Training	Tom Nelson	0031000	. . .
Robert Blom	1002	Programmer	Accounting	George Steel	0027000	. . .
Nancy White	0417	Programmer	Operations	Curt Alexander	0028000	. . .
Randy Wheeler	0293	Analyst	Data proc.	William Crocker	0031500	. . .
John Bittle	0367	Manager	Security	Robert Wilson	0043000	. . .

Figure 2.1   File of records.

in locating a single record becomes a significant inefficiency when magnified many times.

Organization

Sequential organization is probably already familiar to you from your previous experience with computers, so we will not dwell on it. In a sequential organization, the $i + 1$st element of a file is stored contiguous to and immediately after the ith element, for example,

1	2	3	. . .	i	i + 1	i + 2	. . .	n − 1	n

Access

As the name implies, sequential organization is well suited for sequential access, for example, in applications in which many or all the records will be processed, such as with a mailing list for the members of an organization. It is a simple process to move from one record in the file to the next by incrementing the address of the current record by the record size. Most computers can perform this operation rapidly. A sequential array is an example of sequentially organized information. All the elements of the array can be processed easily with a loop construct that varies the subscript for each iteration of the loop.

We can also process a single record directly, with little effort, *if* we know the subscript. But what if the subscript isn't the primary identifier of a record, for example, in the employee file of Figure 2.1 with name as the primary key? We would then need to perform a sequential or linear search, record by record, to locate the desired record.

> A **sequential search** processes the records of a file in their order of occurrence until it either locates the desired record or processes all the records.

We then *can* do direct access on a sequentially organized file but it is not efficient. In fact, in a file of n records, $n/2$ probes, where a **probe** is an access to a distinct location, are needed to locate the record of interest. To retrieve the record for Robert Blom in the example employee file in Figure 2.1 requires seven probes. How many probes are required for an unsuccessful retrieval? In that case, we need to probe the entire file of n records. For a large n, which is usually the situation, such a processing procedure would produce unacceptable performance. In the employee file, if we wish to locate the record for David Christofaro, we need to probe all ten locations before determining that such a record is not in the file. Although its performance is poor for large n, the simplicity of a sequential search makes it appropriate when n is small.

What can be done to improve the retrieval performance? We can do extra processing on the list before retrieving on it. For example, we could *sort* the records in the file to obtain a linear ordering based upon the key values of the records, either alphabetical or numerical. (We will discuss sorting procedures in detail in Chapter 13.) If the records in the file are ordered such that

$$\text{key}_1 < \text{key}_2 < \cdots < \text{key}_i < \cdots < \text{key}_n$$

then only $n/2$ records need to be examined on a record by record search for the desired record. For when we go beyond the position in the file where the desired record would normally appear in the sorted order, we end the search. To search for David Christofaro's record in a sorted employee file requires only five probes rather than the ten needed with the unsorted file. Sorting the file illustrates a technique that we will see repeated throughout the remaining chapters of this text. That is: Spend extra time storing records so that they may be retrieved more efficiently. Even $n/2$ gives unsatisfactory performance when n is large, so we need a still better organization of the records.

BINARY SEARCH

If we have an ordered file of records, we can further reduce the number of probes needed to retrieve a single record by applying a binary search technique.

> A **binary search** compares the key of the sought record with the *middle* record of the file. Then either the sought record has been located or half of the file has been eliminated from further consideration. In the latter case, the process of comparing against the middle record is continued on the remaining records.

If the $\text{key}_\text{sought} < \text{key}_\text{middle}$, we have eliminated the upper portion of the file including the record compared against. If the $\text{key}_\text{sought} > \text{key}_\text{middle}$, we have eliminated the lower portion of the file including the record compared against. In repeating the process of comparing against the middle record, we will eventually locate the desired record or determine that it is not in the file when no records remain under consideration. The

Algorithm 2.1
BINARY SEARCH

proc binary_search

/*   The n records of the file are ordered in increasing order of the keys.   */

```
1       LOWER := 1
2       UPPER : = n
3       while  LOWER ≤ UPPER do
4              MIDDLE := ⌊ (LOWER + UPPER) / 2 ⌋
5              if key[sought] = key[MIDDLE]
6                     then terminate successfully.
7                     else if key[sought] > key[MIDDLE]
8                            then LOWER := MIDDLE + 1
9                            else UPPER := MIDDLE − 1
10      end
11      terminate unsuccessfully.
     end binary_search
```

complete algorithm for a binary search appears in Algorithm 2.1. Since the records of the file are stored sequentially, it is a straightforward procedure to perform the address calculations needed to locate the next record for comparison.

Example

Let's consider an example of applying a binary search to a file of records ordered numerically. In this example, we are searching for the record with key 39, where ∧ indicates the current record being compared against and the brackets delimit the records still under consideration; [is LOWER and] is UPPER.

[13,	16,	18,	27,	28,∧	29,	38,	39,	53]
13,	16,	18,	27,	28,	[29,	38,∧	39,	53]
13,	16,	18,	27,	28,	29,	38,	[39,∧	53]

In this example, only three comparisons were needed vs. the eight that would have been required if we were using a sequential search of the file. In general, the computational complexity of the binary search algorithm for a successful search is $O(\log_2 n)$ vs. $O(n/2) \equiv O(n)$ for the sequential search of an ordered file. (The constant of proportionality of $\frac{1}{2}$ *is* significant if we are comparing a sequential search of an ordered file vs. an unordered one.) The complexity for an unsuccessful binary search is also $O(\log_2 n)$. Because of its performance characteristics, the binary search is sometimes referred to as a "logarithmic search." Another name is "bisection." In the previous example, the search was successful, whereas in the following, it is unsuccessful. If we attempt to locate a record with the key 17, we observe these actions.

[13,	16,	18,	27,	28,	29,	38,	39,	53]
[13,	16,	18,	27],	28,	29,	38,	39,	53
13,	16,	[18,	27],	28,	29,	38,	39,	53
13,	16],	[18,	27,	28,	29,	38,	39,	53

Discussion

As noted in the previous section, a binary search provides better performance than a sequential search of an ordered file since $O(\log_2 n)$ is less than $O(n)$ for large values of n. The primary reason though for discussing the binary search is that it is a well-known, classic technique that is applicable in many situations. Nevertheless, many people overuse it. **Caution:** As n becomes large, $\log_2 n$ becomes significant in terms of the number of comparisons needed in a search. Therefore, the binary search is a viable technique only for small values of n. Even with n equal to only 128, up to seven comparisons may be necessary to locate an individual item. We want to be able to retrieve a record in a single access. We can only do that with a sequentially organized file in those rare cases in which the key of the record is also its address, for example, the subscript in an array. In more typical situations, as n increases, we get further from the goal of a single access by using the binary search algorithm.

The preprocessing cost of ordering a file and the continuing cost of maintaining the order must also be considered when contemplating the use of the binary search algorithm.

INTERPOLATION SEARCH

Discussion

Do we as humans use a binary search technique when we look up information in a sorted list? If you wish to find the phone number for Phyllis Bishop, do you begin your search at the middle of the telephone book? And then do you move to a position one-fourth of the way into the book, etc? Most likely not. The difference in the two strategies is that a binary search does not use the key of the sought record to *approximate* its relative position within the remainder of the file. You would probably begin a search for Phyllis Bishop's telephone number near the front of the book, and then you would determine how many pages to turn for your next comparison based upon the closeness of your first comparison to the sought item.

> An **interpolation search** chooses the next position for a comparison based upon the estimated position of the sought key relative to the remainder of the file to be searched.

Instead of computing a MIDDLE position as with the binary search algorithm, an interpolation search chooses the next position, assuming numeric keys, based upon

$$NEXT := \left\lceil LOWER + \frac{key[sought] - key[LOWER]}{key[UPPER] - key[LOWER]} (UPPER - LOWER) \right\rceil$$

where $\lceil z \rceil$ represents the ceiling (z) or the smallest integer $\geq$ z. If the sought record has a key that would normally appear near the front of the list, the first position checked will be near the front. Likewise, if the key of the sought record would normally appear near the end of the list, the first position checked will be near the end.

Although the worst case computational complexity for an interpolation search is $O(n)$, the average complexity is $O(\log_2 \log_2 n)$. Its performance improves as the distribution of the keys becomes more uniform. A binary search is preferable to an interpolation search when the data is stored in primary memory because the additional calculations needed for the interpolation search cancel any savings gained from fewer probes. However, when the data is stored in auxiliary memory and the key distribution is uniform, an interpolation search is preferable because an access of auxiliary storage is an order of magnitude greater than the time required for the additional calculations. For the same reasons as given for the binary search, the interpolation search is of limited use because as the file grows, its average number of probes moves further from the goal of one probe. The main reason for considering it is that it makes effective use of the value of the sought key to accelerate the search.

SELF-ORGANIZING SEQUENTIAL SEARCH

We observed that the average performance for retrieving a record when using a sequential search on an unordered file, $n/2$, is unacceptable for large n. We then looked at the effects of ordering the list. It improved the performance for an unsuccessful linear search and also made the binary search possible, which improved the performance for both successful and unsuccessful searches. The disadvantage of ordering a file, however, is the cost incurred in doing so. In the previous searches, we assumed that all records in the file would be accessed equally often. If this assumption is not true, we can improve the retrieval performance for a sequential search by organizing the file to take advantage of the frequency of access for the records. In that case, we would naturally desire the most frequently accessed records to appear at the beginning of the file.

> As it retrieves records, a **self-organizing sequential search** modifies their order for the purpose of moving the most frequently retrieved records to the beginning of the file.

Many algorithms and variants exist for performing a self-organizing sequential search [1, 2]. Most of them are quite simple and only add a few lines of code to that which would already be needed for a sequential search. Three popular algorithms are Move—to—front, Transpose, and Count.

Move__to__front

In the Move__to__front algorithm, when the sought record is located, it is moved to the front position of the file and all the intervening records are moved back one position to make room for the move. Because of this extensive amount of repositioning of the records, a linked implementation is preferable even though it takes more storage. This algorithm has the potential of making big mistakes if a record is accessed, moved to the front of the file, and then rarely if ever accessed again. Such a move only has the effect of *increasing* the retrieval time of all the other records in the file. The Move__to__ front algorithm does handle locality of accesses well; **locality** means that a record that has recently been accessed is more likely to be accessed again in the near future. The locality of reference phenomenon is observed when using actual data. For example, if the data records contained information about your program files, you are likely to use or access one of the files frequently over a short period of time, for example, while you are working on a programming assignment and then not access it again for a while. All the self-organizing sequential search methods assume some degree of locality of accesses. The Move__to__front algorithm is essentially the same as the LRU (least recently used) paging algorithm used by operating systems to determine which data page to remove from primary memory to provide space for an incoming page. The LRU algorithm, however, is concerned with the item at the end of the list rather than at the beginning. **Recommendation:** Move__to__front is appropriate when space is not limited and locality of access is important.

An Example   Although the Move__to__front algorithm is straightforward, an example follows to illustrate how the file is reorganized over time. The file consists of 26 records containing the letters of the alphabet. To make the example clearer, the records appear in alphabetical order initially; in practice, if the file were already ordered, a binary search would be preferable. The records are accessed in the order of "fileorganization"

	a	b	c	d	e	f	g	h	i	j	k	l	m	n	o	p	q	r	s	t	u	v	w	x	y	z
t							. . .																			
i	l	i	f	a	b	c	d	e	g	h	j	k	m	n	o	p	q	r	s	t	u	v	w	x	y	z
m							. . .																			
e	a	g	r	o	e	l	i	f	b	c	d	h	j	k	m	n	p	q	s	t	u	v	w	x	y	z
							. . .																			
↓	n	o	i	t	a	z	g	r	e	l	f	b	c	d	h	j	k	m	p	q	s	u	v	w	x	y

Even with this brief example, you can see that the more frequently occurring letters are beginning to move to the front of the file. Most of the vowels, for example, now appear in the first half of the file. You also saw the disadvantage of the method when the "z" was moved to the front of the file.

Transpose

The Transpose algorithm interchanges the sought record with its immediate predecessor unless, of course, the sought record is in the first position. With this approach, a record needs to be accessed many times before it is moved to the front of the list. It converges to a near steady state more slowly than the Move__to__front algorithm but it is more stable, that is, it does not make big mistakes in reorganizing the file. If a record is not to be accessed again, the retrieval performance for only one other record is degraded. The algorithm can be readily implemented using sequential storage and thus does not need the additional space required for a linked implementation, such as with Move__to__front. **Recommendation:** It should be used when space is at a premium.

Count

The Count algorithm keeps a count of the number of accesses of each record. The record is then moved in the file to a position in front of all the records with fewer accesses. The file is then always ordered in a decreasing order of frequency of access. The disadvantages of the Count algorithm are that it requires extra storage to keep the count of the accesses and it does not handle the locality of access phenomenon well. **Recommendation**: Because of its storage requirements, use it only when the counts are needed for another purpose.

Discussion

Although much analysis and experimentation has been performed on these algorithms, there is no clear choice of which one to use in all cases because their performances are data dependent. By following the recommendations given with the description of the algorithms, you can make a good choice. If you have an application in which a sequential search is appropriate, but you need just a little more speed, you may want to consider a self-organizing sequential search.

In the next chapter, we will consider several organizations intended primarily for direct access which will provide a single access or an average number of accesses close to one.

KEY TERMS

Attribute
Binary search
Bisection
Ceiling
Count
Direct access
External search
Field
File
Internal search
Interpolation search
Linear search
Locality of access

Logarithmic search
Move__to__front
Primary key
Probe
Record
　Fixed-length
　Variable-length
Secondary key
Sequential access
Sequential file organization
Sequential search
　Self-organizing
Transpose

EXERCISES

1. **(a)** How many probes are necessary to retrieve "October" from the list of months in calendar order using a sequential search? **(b)** How many probes are necessary if the months are stored in alphabetical order? **(c)** How many probes are needed if a binary search is used?

2. **(a)** How many probes are necessary to retrieve the record for John Bittle in Figure 2.1 using a sequential search? **(b)** How many probes are necessary if the records in the employee file are stored in alphabetical order of the employees' last names? **(c)** How many probes are needed if a binary search is used?

3. **(a)** How many probes are necessary to retrieve the record for Morris Lancaster using a binary search of the data in Figure 2.1 sorted by employee number? **(b)** with an interpolation search?

4. When is a binary search appropriate?

5. Write a complete algorithm for an interpolation search. The test for executing the **while** loop should be different from that used with the binary search.

6. Is a sequential, binary, or interpolation search preferable for locating the record with key 7 in the file of ordered data consisting of the records with the keys

$$1,2,3,4,5,6,7,10000?$$

Explain.

7. Repeat Problem 6 with the data

$$1,2,3,4,5,6,7,8.$$

8. Apply the Transpose algorithm to the original data file and accessed data that was used with the Move_to_front example. **(a)** Give the final ordering of the records. **(b)** What differences do you note in comparison with the Move_to_front results?

9. Apply the Count algorithm to the original data file and accessed data that was used with the Move_to_front example. **(a)** Give the final ordering of the records. **(b)** What differences do you note in comparison with the Move_to_front and Transpose results?

10. Why is the Move_to_front algorithm not appropriate with limited space?

11. In the Count algorithm, what is an advantage of moving a record in front of all the records with fewer accesses rather than in front of all records with the *same* or fewer accesses?

REFERENCES

1. Bentley, Jon L., and C. C. McGeoch, "Amortized Analysis of Self-Organizing Sequential Search Heuristics," *CACM*, vol. 28, no. 4 (April 1985), pp. 404–411.
2. Hester, James, H., and Daniel S. Hirschberg, "Self-Organizing Linear Search," *Computing Surveys*, vol. 17, no. 3 (September 1985), pp. 295–311.

Chapter 3

Direct File Organization

LOCATING INFORMATION

One thing that distinguishes our society from earlier ones is our need for information *and* the impatience we have in obtaining it. We desire rapid access to large volumes of data. These two goals are often contradictory; people can do it but computers cannot. One well-known technique for simulating the associative human storage and retrieval processes is quaintly referred to as hashing. Even though hashing has been used for some time, it is continually being improved upon and applied to new situations. Since hashing has such widespread applicability, it is important to acquaint those unfamiliar with the concept, to update the others, and to familiarize all with its many applications. Before considering hashing though, let's look at two other ways to organize a file for direct access.

The Key Is a Unique Address

Ideally, when we want to retrieve the record associated with a certain primary key, we want to go *directly* to the address where the record is stored. This would be possible if the key *were* also an address. For an application in which the record contained employee information, if the employee's nine-digit social security number were the record's identifying number (or primary key of the record), then 10^9, or 1 billion table locations, would be necessary as illustrated in Figure 3.1. In this case, we have excellent retrieval time—one probe per record retrieved, but what is the disadvantage? Obviously a great deal of space must be reserved for the table—one location for each *possible* social security number. Unless a company had a billion employee records, much if not most of the space would remain unused. This technique would be acceptable in situations in which the employee numbers could be assigned sequentially, 1, 2, 3, 4, ... But after employees began leaving the company, gaps would begin to appear between current employee numbers and the space would begin to be wasted just as in the case of the social security numbers.

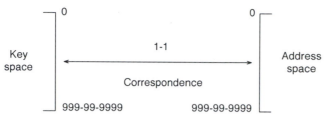

Figure 3.1 1–1 correspondence between keys and addresses.

The Key Converts to a Unique Address

A procedure that is one step removed from having the primary key *be* the storage address is to use an algorithm to convert the primary key field (possibly a composite) into a unique storage address. If the subscript of an array element is considered to be its unique identifier, then a straightforward formula exists to determine its address when the records are stored in contiguous locations:

$$\text{Location } A[i] = \alpha + m \times s$$

where A = an array of any dimension
i = typical element
α = location of first element in array
m = number of elements preceding ith element according to a standard ordering such as lexicographic order
s = size (in address units) of an element of the array

By applying the formula, only a single probe is necessary to locate an element of the array. Using a formula such as this is a common way in which arrays are implemented in most programming systems. See a data structures text, for example, [1,2], for details on how to determine m.

Another example of converting information into a unique address is an airline reservation system. If an airline, which numbers its flights from 1 to 999, wants to keep reservations for an entire year in which the days are numbered from 1 to 366, then the flight number and day of the year could be *concatenated* to determine the location for the record containing the reservations for that flight on that particular day.

$$\text{Location} = \text{flight number} \parallel \text{day\_of\_the\_year}$$

where $\parallel$ denotes concatenation. The addresses would range from 001001 to 999366. Does the order of the concatenation have an effect on the efficiency of the procedure? We need only about one-third as much storage if we form the location by

$$\text{Location} = \text{day\_of\_the\_year} \parallel \text{flight number}$$

Why is that? It is preferable to place the wider range last so that we do not have large gaps in the key values, for example, key values ???367 through ???999 would not exist but nevertheless space would have to be reserved for them. (? matches any digit.) The range of the key values with the alternative formation method is 001001 to 366999. Notice that the range of address values is therefore about one-third; so much less space needs to be reserved. A simple matter such as the order of concatenation can have an enormous effect on the amount of space needed for storing the records. Since the airline probably would not have 999 flights, many potential flight numbers would still go unused.

The Key Converts to a Probable Address

The reason that considerable space is wasted in many situations using the two previous techniques is that the range of acceptable values of keys is *much greater* than the actual number of keys, so many gaps exist. If we remove most of the empty spaces in the address space, we then have a key space larger than the address space; an example in which social security numbers are the primary keys is illustrated in Figure 3.2.

Note. We have lost the 1–1 correspondence between keys and addresses since we now have more key values than address values.

We need a function then to map the wider range of key values into the narrower range of address values. Such functions, referred to as **hashing functions**, will necessarily map multiple key values into a single address value.

Hash(key) → probable address

The output of a hashing function generates a **probable address** since we have lost the guarantee of a unique address for every key. This initial probable address for locating a record in a table is also known as the **home address** for the record. Just as its namesake, the food dish, is a "mixing" of ingredients, the hashing function "mixes up" the key value to obtain an address value. Any function that maps the keys into the range of addresses is acceptable but it is desirable if the function also

- Evenly distributes the keys among the addresses.
- Executes efficiently.

The latter guideline is intended to keep the retrieval time to a minimum. The first one is meant to reduce the number of *collisions*. A **collision** occurs when two *distinct*

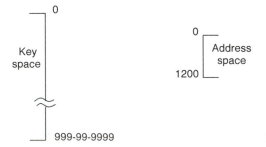

Figure 3.2 Compressed address space.

keys map to the same address. Hashing is then composed of two aspects:

- The function.
- The collision resolution method.

A collision resolution mechanism is necessary when the number of records that map to a given location exceed its capacity. For most of these discussions, we assume an address capacity of one record.

HASHING FUNCTIONS

Key mod *N*

One of the most popular hashing functions is

$$f(\text{key}) = \text{key } \textbf{mod } N$$

where N is the table size. For positive integers the **mod** function returns the remainder of dividing the *key* by N, for example, 27 **mod** 8 is 3. An advantage of this function is that it returns only values in the range of the address space (0 to $N-1$).

Key mod *P*

A variation of the previous function is

$$f(\text{key}) = \text{key } \textbf{mod } P$$

where P is the smallest prime number $\geq N$, the table size. P is then the new table size. Some collision resolution methods require this hashing function.

Truncation

Another class of hashing functions is **truncation** or substringing. If we have social security numbers as the primary keys for example,

123-45-6789

and wish to map them into a table of 1000 addresses, three digits (not necessarily contiguous) could be used as the address. The function would select which three digits to use. It does *not* have to be truncation in the usual sense of the word in which the rightmost digits are dropped. The digits may come from *any* part of the key. Would it be reasonable to truncate the social security number key after the first three digits?

It depends upon the application. If we are applying this function to employees who work at a single plant, it would not be a good choice. Why? Since social security numbers are assigned by geographic regions, most of the workers would have the same leftmost two or three digits, hence the addresses would not be evenly distributed. For example, in North Carolina, most of the social security numbers would map into the addresses between 220 and 240 (using only 21 of the 1000 possibilities). A better choice would be to use the three rightmost digits.

Folding

A third distinct class of hashing functions uses **folding** (which derives its name from a process similar to folding a piece of paper). Two kinds of folding exist:

- Folding by boundary.
- Folding by shifting.

If we have the nine-digit key

1	2	3	4	5	6	7	8	9

written on a large piece of paper, we could fold it on the dashed *boundaries* into three parts. The digits would then be superimposed as

```
    3  2  1
    4  5  6
+   9  8  7
_____
```

If the superimposed codes were added together (without carry), we would then obtain the probable address of 654. Since this example hashing function uses nine-digit keys, it could be used to convert social security numbers into three-digit addresses.

Folding by shifting is actually similar to tearing the paper at the boundaries and then sliding the parts one over another until all are superimposed. For this example, the segments would be

```
    1  2  3
    4  5  6
+   7  8  9
_____
```

which would yield the different address of 258. For both methods we could also have added with carry and then used only the three rightmost digits. The results with the example would then have been (1)764 and (1)368, respectively.

Squaring

Squaring a key and then substringing or truncating a portion of the result for the address is yet another way of "mixing up" a key value to evenly distribute it throughout the possible addresses.

Radix Conversion

To scatter key values into the address space with **radix conversion**, the key is considered to be in a base other than 10 and is then converted into a number in base 10, for example, the key

$$1\ 2\ 3\ 4$$

in base 11 would become

$$(1 \times 11^3) + (2 \times 11^2) + (3 \times 11^1) + (4 \times 11^0) = 1331 + 242 + 33 + 4$$
$$= 1610$$

when converted to base 10. Substringing or truncation could then be used to obtain the desired number of digits.

Polynomial Hashing

The function used to produce the cyclic check bytes for checking for transmission errors in data, that is,

$$f(\text{information area}) \rightarrow \text{cyclic check bytes}$$

is in fact a hashing function. The type of function typically used for this purpose is referred to as a polynomial hashing function since the key is divided by a polynomial [3]. Although for this application, this type of function is used to transform a data area into two cyclic check bytes, it could also be used for converting a key into an address. Its study, however, is beyond the scope of this text.

Alphabetic Keys

You probably observed that the key values used as arguments to all the example hashing functions were *numeric*. All the hashing functions described in this section do indeed require integer arguments. But does that imply that the key field of records organized for direct access must be numeric? Does that suggest that fields containing *alphabetic* information such as a person's name are excluded from consideration as a primary key? Is that the reason that so many computer applications assign numbers to people and other entities prior to storing information about them? Fortunately, the answer to the first two questions is a resounding "no." It should also be the same for the third question but occasionally ignorance of how computers operate or of the capabilities available from an implementation language causes numeric keys to be assigned when it is unnecessary. Numeric keys *may* be needed in some situations to obtain unique values for the primary keys but it should not be done merely to allow a key to be used as input to a hashing function. Why? Because alphabetic or alphanumeric key values can be input to a hashing function if the values are *interpreted* as integers. You know that all information is represented internally in a computer as bits, 1s and 0s: integers, real numbers, characters, everything. Therefore, in order to use an alphabetic field as an argument for a hashing function, it is only necessary to have it, or a portion of it, interpreted as an integer. The bit string must be interpreted as some data type anyway, so it is just as simple for it to be interpreted as an integer as anything else. When printed or displayed in other situations, the interpretation could still be character. It would be interpreted as an integer only when used in conjunction with a hashing function. Many high-level programming languages provide facilities for accomplishing this. In Pascal, the construct of **variant records** may be used. When a variant record is used, a field is referred to by multiple names. In the declaration for the record, an association is made as to what data type or interpretation goes with each name. In this way a field can be character when it is printed and an integer when it is hashed. (See a Pascal programming text for details on how to accomplish this.) Other languages provide similar facilities, for example, in PL-I, the UNSPEC function is used to interpret a character string as an integer. If, for some reason, the high-level language that you are working with does not provide such a facility, you may be able to write an external assembly language routine to accomplish the same result. If all else fails, you can always resort to the method of associating a numeric digit with each letter of your alphabet. In all cases, then, you can have alphabetic or alphanumeric key fields. With that said, we will continue to use integer key values in the examples so that we don't complicate the examples by explaining the integer equivalent of an alphanumeric key. In your future use of direct-access files, however, do not unnecessarily restrict yourself to integer keys.

Collisions

The reason that it is useful to be aware of a variety of hashing functions is that for a given set of data, one hashing function may distribute the keys more evenly over the address space than another. One of the hashing function types described previously should distribute your data evenly over the address space. If that is not possible, you may have to devise your own scheme; one strategy for finding such a hashing function would be to combine simpler functions. A hashing function that has a large number of collisions or synonyms is said to exhibit **primary clustering**. The fewer the number of collisions, the fewer the times that it will be necessary to look elsewhere for the desired record and that in turn will keep the average number of probes or accesses of storage near one. The reason that we are so interested in minimizing the number of retrieval probes is that in most commercial applications, the number of records is so large that they need to be stored on auxiliary storage. Secondary memory is at least an order of magnitude slower than primary memory. Many accesses to auxiliary storage would then slow a program considerably. It is better to have a slightly more expensive hashing function if it reduces the number of collisions.

One method for reducing collisions is to change hashing functions. Another method is to reduce the packing factor. The **packing factor** of a file is the ratio of the number of items stored in the file to the capacity of the file, that is,

$$\text{Packing factor} = \frac{\text{number of records stored}}{\text{total number of storage locations}}$$

It is a measure of the storage utilization. Two other names for it are the packing density and the load factor. As the packing factor increases, the likelihood of a collision increases. Just as with automobiles, as their density or packing factor increases, the possibility of a collision increases. If more of the space is occupied, there is a greater chance that the key of a record to be inserted will collide with a key of a record already in the table. The disadvantage of decreasing the packing factor to reduce the number of collisions is that a lower packing factor takes more space to store the same number of records. Building more highways would probably reduce the number of automobile collisions, but that may be too costly a solution. The predicament in choosing a packing factor is then an example of a time-space tradeoff. If we decrease the packing factor, we need more storage, but doing so reduces the chance of a collision which improves performance. The relationship between collisions and storage may be expressed in a graph of the form

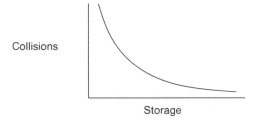

Collisions typically increase rapidly when the packing factor goes beyond about 90 percent. Usually we choose a packing factor that is somewhere between the extremes of storage and performance.

COLLISION RESOLUTION

Changing the hashing function or decreasing the packing factor may *reduce* the number of collisions but it will usually not *eliminate* them. Therefore, we need a procedure to position a synonym at another location when a collision does occur. What we are attempting to accomplish with a collision resolution method is to place a synonym record in a location that requires a minimum number of additional probes from its home address. Recall that a **probe** is an access to a distinct location.

One way to accomplish this task is to *point* to the location of the synonym record. If multiple synonyms occur for a particular home address, we form a chain of synonym records. If the records A, B, and C are synonyms with a home address of r, a synonym chain may appear as in Figure 3.3a. The link field of location r points to location s, which is where the first synonym, B, is stored. The Λ or NULL pointer in the link field of t indicates the end of the synonym chain.

A disadvantage of using pointers is that storage is needed for the link fields *in addition to that needed for the record*. For a few records, that amount of space may not seem like much, but for thousands of records, that additional space may be more than we can afford, either physically or monetarily. We may therefore need a mechanism for locating a synonym that does not involve the use of actual links. How can we achieve this? We can use *implied* links by applying a *convention*, or set of rules for deciding where to go next. Then instead of looking up the next search address in a link field, we *compute* the next search address by applying the set of rules. A simple convention is to look at the next location in memory. If it is occupied, we then examine the following location. An empty or Λ location indicates the end of the synonym chain. If the records A, B, and C are synonyms with a home address of r, then a synonym chain in which the successor location is computed may appear as in Figure 3.3b. In this example, the record X is not a synonym of A, B, and C. The disadvantage of not using links is that we may need to make more probes in locating the record of interest. For example, four probes would be needed to locate record C in Figure 3.3b whereas only three are needed in Figure 3.3a.

In subsequent sections of this chapter, we consider several mechanisms for resolving collisions. These methods may be grouped according to the mechanism used to locate the next address to search and the attribute of whether a record once stored can be relocated; the classifications are

- Collision resolution with links.
- Collision resolution without links.
  - Static positioning of records.
  - Dynamic positioning of records.
- Collision resolution with pseudolinks.

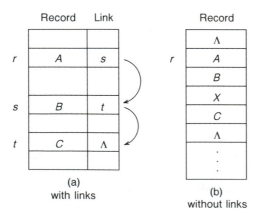

(a) with links

(b) without links

Figure 3.3 Synonym chains: with and without links.

One reason that we will study several collision resolution methods is that we often have different constraints in different situations. For a specific circumstance, one method may be preferable to others. A second and perhaps more important reason for considering several collision resolution techniques is that it will illustrate the concept of having a variety of "tools" for performing a given task so that we may *choose* the most appropriate one. Collision resolution in and of itself may not be important but the exposure to a wide variety of alternative data structures and the experience in *thinking* about a problem in different ways is.

In Chapter 10, we will consider other ways of organizing a direct-access file.

KEY TERMS

Address
    Home
    Probable
Address space
Auxiliary storage
Collision resolution
Cyclic check bytes
    Hashing
Direct file organization
Folding by boundary
Folding by shifting
Hashing
Hashing function
    Alphanumeric key
    Folding
    Key mod N
    Key mod P
    Polynomial
    Radix conversion
    Squaring
    Truncation
Key
    as unique address

Key conversion
    Probable address
    Unique address
Key space
Load factor
Mod function
Packing density
Packing factor
Primary clustering
Primary key
    Alphabetic
Storage utilization
Substringing
Synonym
Synonym chain
Tools
UNSPEC
    PL-I
Variant record
    Pascal

EXERCISES

1. What is the primary disadvantage of using the key as the actual storage address?

2. What are the two most important characteristics of a hashing function?

3. What are the components of a hashing procedure?

4. **(a)** Convert the decimal key 987654321 into a three-digit decimal address using a folding by boundary hashing function. **(b)** Repeat using folding by shifting.

5. Repeat Problem 4 using an add with carry.

6. Use radix conversion and left truncation to transform the key 4321 into a three-digit decimal address; assume that the key is in base 13.

7. Would a truncation or a radix conversion hashing function typically execute faster? Explain.

8. Comment on the advisability of devising a hashing function using truncation for converting the two-digit and three-digit numbers of an interstate highway in the United States into an address space of a size within 10 percent of the total number of such highways.

9. Would a folding by boundary hashing function be easier to implement than one using folding by shifting? Explain.

10. What other three-digit address is possible when folding the example nine-digit key of 123456789 by boundary?

11. Assume that several keys in a file with alphabetic keys have identical *prefixes*. If the bit pattern for a key is interpreted as an integer in the hashing process, what effect would this distribution of keys versus a uniform distribution have on the number of collisions using a key **mod** P hashing function?

12. Repeat Problem 11 but assume that several of the keys have identical *suffixes*.

COALESCED HASHING

Coalesced hashing is a collision resolution method that uses pointers to connect the elements of a synonym chain. For an overview of its process for inserting and linking, let's use the example of Figure 3.3a. In it, record A is inserted first and placed at its home address at location r. In the process of inserting record B, it also hashes to location r, which results in a collision requiring that record B be stored elsewhere. Space is available at location s, and B is inserted there; the link field at r is then set to point to s. The insertion of a synonym is the same as the insertion into a linked list. During the insertion of record C, another collision occurs at location r. Record C is inserted at location t and then linked into the end of the existing synonym chain. On retrieval, the elements of the synonym chain are searched until the desired record is located or until the end of the chain, indicated by a Λ link, is reached.

Coalesced hashing obtains its name from what occurs when we attempt to insert a record with a home address that is already occupied by a record from a chain with a *different* home address. This situation would occur, for example, if we attempted to insert a record with a home address of s into the table of Figure 3.3a. What occurs is that the two chains with records having different home addresses **coalesce** or grow together. In the example in Figure 3.4, the records with keys X, D, and Y were inserted in the given order into the table in Figure 3.3a. A, B, C, and D form one set of synonyms and X and Y form another set. In Figure 3.4a, when we attempt to insert the record X, it collides with the record B which is *not* a synonym of X since B has a home address of r and X has a home address of s. As a result, the two chains coalesce, where ▓ indicates the portion of the chain in which coalescing has occurred, | represents the insertions on the synonym chain with r as its home address, and **|** represents the insertions on the chain with s as its home address. In order to access records at the end of a coalesced chain, we have to move over the records of more than one chain. Since this will obviously result in more probes than would be required for noncoalesced chains, we want to minimize the amount of coalescing. In Figure 3.4a, when X is inserted into the table with coalescing, it must be inserted at the *end* of the chain that

Algorithm 3.1
COALESCED HASHING

I. Hash the key of the record to be inserted to obtain the home address, or the probable address for storing the record.

II. If the home address is empty, insert the record at that location, else if the record is a duplicate, terminate with a "duplicate record" message, else

    A. Examine the records on the probe chain to check for a duplicate and to locate the end of the chain, signified by a Λ link.

    B. Find the bottommost empty location in the table. If none is found, terminate with a "full table" message.

    C. Insert the record into the identified empty location and set the link field of the record at the end of the chain to point to the location of the newly inserted record.

---

it is coalescing with. Instead of needing only one probe to retrieve X, three are needed. The greater the coalescing, the longer the probe chains will be, and as a result, retrieval performance will be degraded. When record **D** is now added, it must be inserted at the *end* of the coalesced chains; we must move over record X from the other chain then to locate **D**. With the example of Figure 3.4a in which coalescing exists, we need a total of 19 probes to retrieve each record once: 1 for **A**, 2 for **B**, 3 for **C**, 3 for X, 5 for **D**, and 5 for Y. This total compares with only 13 probes for retrieving the same records in Figure 3.4b when the two synonym chains are separated. You can see then that coalescing is something to be minimized to improve retrieval performance; later we consider variants of the standard algorithm that are intended to reduce the amount of coalescing.

Coalesced hashing originated with Williams [4] and is also referred to as *direct chaining*. The mechanisms of the basic process are given in Algorithm 3.1.

The insertion of a new record is made at the bottommost (highest address) empty location as a matter of convention. In searching for an empty location, an available space pointer is continually decremented from its current position until either an empty

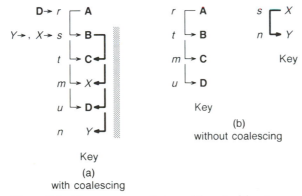

Figure 3.4 Two synonym chains: with and without coalescing.

location is found or it points to an address less than the minimum address bound of the table ($<$ zero in the subsequent example). The latter condition indicates that the table is full. The available space pointer is initially set to a location one record length beyond the last storage location in the table.

An Example

For the collision resolution examples of this chapter we use the hashing function of

$$\text{Hash(key)} = \text{key } \textbf{mod } 11$$

where 11 is the prime number table size. For comparison purposes, we will also use the same initial subset of data consisting of records with the keys of

$$27, 18, 29, 28, 39, 13, \text{ and } 16$$

The file will then have 11 entries with addresses of 0 through 10. We hash the key of the first record to be inserted, Hash(27) = 5 for a probable address. Since this is the first record to be inserted, there is no chance for a collision and it is inserted into location 5. The next record with a key of 18 hashes to a probable address of 7. Again there is no collision. But when we attempt to insert the record with key 29 we *do* have a collision since its home address of 7 is occupied by the record with the key of 18. Where do we then place the record with key 29? The available space pointer, *R,* is decremented from one record length beyond the bottom of the file; we therefore check if location 10 is available. Since it is, we insert 29 into location 10 and set the link field in the home address location of 7 to point to location 10. The file then appears as

	Record	Link
0		Λ
1		Λ
2		Λ
3		Λ
4		Λ
5	27	Λ
6		Λ
7	18	10
8		Λ
9		Λ
$R\rightarrow$ 10	29	Λ

We next insert the record with key 28, which hashes to the unoccupied location 6. Key 39 collides at location 6 with 28 and must then be inserted at another location. After decrementing *R* by one, it points to location 9, which is unoccupied. 39 is inserted there and the link field at location 6 is set to 9. The file then appears as

with links

	Record	Link
0		Λ
1		Λ
2		Λ
3		Λ
4		Λ
5	27	Λ
6	28	9
7	18	10
8		Λ
R→ 9	39	Λ
10	29	Λ

The record with key 13 is inserted without complication into location 2. When inserting 16, a collision occurs at location 5. After decrementing R, it points to location 8, which is unoccupied, so the record with key 16 may be inserted there. The hashed file then becomes

	Record	Link
0		Λ
1		Λ
2	13	Λ
3		Λ
4		Λ
5	27	8
6	28	9
7	18	10
R→ 8	16	Λ
9	39	Λ
10	29	Λ

At this stage we have inserted the common set of records that we will be using throughout the discussions of collision resolution. To retrieve each record once would require an average of 1.4 probes. However, we have *no* coalescing; we have three separate chains with home addresses of 5, 6 and 7. If we add two additional records with keys of 42 and 17 and home addresses of 9 and 6, respectively, we will have coalescing of the chains with these two home addresses. R will be decremented repeatedly until it points to an unoccupied location, in this case, location 4, or until it has a value less than zero which indicates that the table is full. 42 will then be inserted at location 4 and 17 subsequently at location 3 to yield the final table as shown in Figure 3.5. Notice how many probes are now required to retrieve the record with key 17. We need to make four probes including a probe at location 4 which contains a record *not* on the same probe chain as 17, that is, not having the same home address. Because of the coalescing of the chains with home addresses of 6 and 9, we need to make one extra probe in locating 17, that of an item not on the same probe chain. The average number of probes to retrieve each record in Figure 3.5 once is about 1.8 with coalescing, whereas it would have been 1.6 if coalescing had not occurred. If we add records to both of the probe chains, the performance of each chain would be degraded because of the interference of the other chain, that is, probing locations containing records from the other chain. We therefore want to minimize the amount of coalescing.

Discussion

The packing factor of the final table in the example is $\frac{9}{11}$ or 82 percent. One method of reducing the coalescing is to reduce the packing factor, because the lower it is the less chance that there will be a collision of a record from one chain with that of another chain. We will also consider several variants of coalesced hashing intended to reduce coalescing.

How does the retrieval process terminate for an unsuccessful search, that is, what happens when we seek a record that does not exist in the table? We must search the entire probe chain of records beginning at the home address produced by hashing the key of the desired record. We know that we have reached the end of the probe chain when we encounter the Λ value in the link field. Obviously then, an unsuccessful search takes more time than a successful one since we need to examine more records.

Notice that the number of retrieval probes is a function of the *order of insertion* of the records into a file. A record inserted early in the process will be placed near the beginning of the probe chain and will therefore require fewer retrieval probes than a record inserted later in the process and appearing deeper in the probe chain. So if the frequencies are known, it would be advisable to place the most frequently accessed records early in the insertion process.

Deleting a record from a table in which coalesced hashing is being used as the

	Record	Link
0		Λ
1		Λ
2	13	Λ
$R \rightarrow$ 3	17	Λ
4	42	3
5	27	8
6	28	9
7	18	10
8	16	Λ
9	39	4
10	29	Λ

Figure 3.5 Insertion using coalesced hashing.

collision resolution method is complicated by the fact that multiple probe chains may
have coalesced. As a result, we cannot merely remove the record as if we were deleting
an element from a singly linked list. For if we did, we might lose the remainder of a
probe chain that had coalesced at that location as a result of its being its home address.
If coalescing has occurred, a simple deletion procedure is to move a record later in the
probe chain into the position of the deleted record. The relocated record should be the
last element on the probe chain having the home address of the location of the deleted
record. This relocation process is repeated at the vacated location, and in turn at
subsequent vacated locations until no other records later in the chain need to be moved.
At that stage, the prior-to-moving predecessor of the most recently moved record is
relinked to the moved record's prior-to-moving successor. In that way, we ensure that
any coalesced chains are not lost and that the moved record(s) may still be retrieved.
A more complex scheme could be used to reposition records in the coalesced chains
to reduce the amount of coalescing. Determining if coalescing has occurred in a probe
chain may be a time-consuming process, so it may be preferable to assume that it has
and make the adjustments accordingly. If we know, however, that a coalescing of chains
has not occurred, then we do not need to move any records; we can just remove the
deleted record from a singly linked list. No other records in the probe chain would need
to be moved; only the link of the predecessor element needs to be set to the address
of the successor element.

To reuse the space of the deleted record, it should be reinitialized to null and the
available space pointer, R, should be reset to the maximum of (1) its current address
and (2) the vacated address plus one. For example, to delete the record with key 39
(at location 9) from Figure 3.5 requires that

- Because coalescing has occurred, we move the last element in the chain with
  a home address of 9, that is, the record with key 42 into location 9 and adjust
  the links and table entries appropriately.
- R is set to location 5 (the vacated address plus one). The resulting table is then:

		Record	Link
	0		Λ
	1		Λ
	2	13	Λ
	3	17	Λ
	4		Λ
R →	5	27	8
	6	28	9
	7	18	10
	8	16	Λ
	9	42	3
	10	29	Λ

with links

Variants

Many suggestions have been made for reducing the coalescing of probe chains and thereby lowering the number of retrieval probes which in turn improves performance. The variants may be classified in three ways.

- The table organization (whether or not a separate overflow area is used).
- The manner of linking a colliding item into a chain.
- The manner of choosing unoccupied locations.

Coalescing may be reduced by modifying the table organization. Instead of allocating the entire table space for *both* overflow records *and* home address records, the table is divided into a primary area and an overflow area, such as

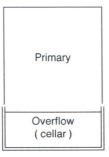

where the *primary* area is the address space that the hash function maps into. The *overflow* or cellar area contains only overflow records. The **address factor** is the ratio of the primary area to the total table size, that is,

$$\text{Address factor} = \frac{\text{primary area}}{\text{total table size}}$$

For a fixed amount of storage, as the address factor decreases, the cellar size increases, which reduces the coalescing but because the primary area becomes smaller, it increases the number of collisions. More collisions mean more items requiring multiple retrieval probes. Vitter [5] determined that an address factor of 0.86 yields nearly optimal retrieval performance for most load factors. To be most effective, though, the cellar should use additional storage.

The original algorithm given in Algorithm 3.1 is called late insertion standard coalesced hashing (LISCH) since new records are inserted at the *end* of a probe chain. The "standard" in the name refers to the lack of a cellar. The variant of Algorithm 3.1 that uses a cellar is called LICH, late insertion coalesced hashing. If we insert the example records, that is, those with the keys

27, 18, 29, 28, 39, 13, 16, 42, and 17

with links

into a table of seven primary locations and four cellar locations using the key **mod** 7
hashing function and LICH for collision resolution, we obtain

	Record	Link
0	28	8
1	29	Λ
2	16	Λ
3	17	Λ
4	18	10
5		Λ
6	27	9
7		Λ
$R\rightarrow$ 8	42	Λ
9	13	Λ
10	39	Λ

Cellar (rows 7–10)

The average number of probes needed to retrieve each record in this table once is 1.3
vs. the 1.8 for the table in Figure 3.5 which used LISCH for insertion. In general, for
a 90 percent packing factor, using a cellar will reduce the number of probes by about
6 percent compared with LISCH [6].

A second technique is to vary the position in which new records are inserted into
a probe chain. Chen and Vitter [7] discuss several variants of this type. A variant called
early insertion standard coalesced hashing (EISCH) inserts a new record into a position
on the probe chain *immediately* after the record stored at its home address.

Note. The position of insertion is *logically*, not necessarily physically, immediately after the record stored at its home address. The link field of the home address
points to the newly inserted record.

If the records X, **D**, and Y are inserted using EISCH in the order given into the
table in Figure 3.3a, the resulting table would appear as shown in Figure 3.6. In
comparing these results of EISCH with the coalescing example with LISCH in Figure

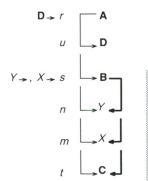

Figure 3.6 Coalescing with EISCH.

3.4a, notice that only a total of 17 probes are needed to retrieve each record once instead of the 19 as before and that the portion of the chain on which coalescing occurs has been reduced. Although more probes are still necessary for retrieval with EISCH than when there is no coalescing, it is an improvement over LISCH.

Let's next reconsider the example with numerical keys using EISCH. Until a chain is greater than two elements, LISCH and EISCH give the same result, since the last position on the chain and the position immediately after the home address record are the *same*. We then begin with the file appearing as

	Record	Link
0		Λ
1		Λ
2	13	Λ
3		Λ
R→ 4	42	Λ
5	27	8
6	28	9
7	18	10
8	16	Λ
9	39	4
10	29	Λ

When we are ready to insert the record with key 17, instead of inserting it at the end of the probe chain, that is, after 42, it is inserted in the probe chain position after the record stored at its home address, that is, after 28. It is simple to make this type of insertion since we are working with linked lists. The resulting file would be

	Record	Link
0		Λ
1		Λ
2	13	Λ
R→ 3	17	9
4	42	Λ
5	27	8
6	28	3
7	18	10
8	16	Λ
9	39	4
10	29	Λ

In this way, only two rather than four probes are required to retrieve 17. The rationale behind early insertion is to reduce the *amount of coalescing* the two synonym chains may have. The earlier insertion postpones the coalescing to a later position in the combined chain so that on the average fewer records need to be probed to locate records on either synonym chain. In this example using EISCH for insertion, the average number of probes needed to retrieve each record in the table once is 1.7 vs. the 1.8 for the records in the table in Figure 3.5, which was constructed using LISCH. Normally, for a 90 percent packing factor, EISCH requires about 4 percent fewer probes than LISCH [6]. The algorithm using early insertion with a cellar is called EICH.

The third way in which we may vary the insertion algorithm is to change the way in which we choose an unoccupied location. As specified in the algorithm, the unoccupied locations are always chosen from the *bottom* of the storage area. (Even if we are using a cellar, when it becomes full, we need to resort to the original algorithm.) By concentrating *all* the overflow items in one area of the table, we *increase* the number of collisions and thereby degrade performance. Hsiao et al. [6] suggest choosing a *random* unoccupied location for the new insertion. In this way the overflow records would be more evenly distributed over the addresses of the table. However, for a 90 percent packing factor, REISCH (R for random) gives only a 1 percent improvement over EISCH (the improvement is greater for lower packing factors as you might expect). Yet another method of choosing the overflow location for a collision insertion is to alternate the selection between the top and bottom of the table [6]. This method is called BLISCH, where the "B" signifies "bidirectional." Experimental results using uniformly distributed key data appear in Table 3.1; it may be used to compare the many variants of coalesced hashing; α is the packing factor, n is the fixed table size, and the address factor for the cellar methods is 0.86.

A means of *eliminating* coalescing is to *move* a record not stored at its home address. In moving an item, we would *not increase the number of probes* for the item moved since we are merely inserting it into a new location but it retains its same position on the probe chain. We refer to this variation of LISCH as direct chaining without coalescing (DCWC). (Calling it coalesced chaining without coalescing just doesn't sound proper.) Later in this chapter we consider a collision resolution method, computed chaining, that applies this notion of moving; so we defer further consideration of it until that time.

TABLE 3.1 MEAN NUMBER OF PROBES FOR SUCCESSFUL LOOKUP ($n = 997$) FOR VARIANTS OF COALESCED HASHING

α method	0.2	0.4	0.6	0.8	0.9	0.95	0.99
EISCH	1.1065	1.2277	1.3684	1.5290	1.6182	1.6653	1.7033
LISCH	1.1063	1.2316	1.3789	1.5657	1.6737	1.7337	1.7827
BEISCH	1.1055	1.2286	1.3721	1.5336	1.6236	1.6728	1.7107
BLISCH	1.1055	1.2341	1.3836	1.5703	1.6818	1.7423	1.7898
REISCH	1.1063	1.2322	1.3693	1.5257	1.6124	1.6614	1.7014
RLISCH	1.1085	1.2384	1.3876	1.5653	1.6723	1.7296	1.7790
EICH	1.1116	1.2256	1.3408	1.4942	1.5867	1.6347	1.6762
LICH	1.1116	1.2256	1.3406	1.4888	1.5801	1.6281	1.6695

Source: Hsiao, Yeong-Shiou, and Alan L. Tharp, "Analysis of Other New Variants of Coalesced Hashing," Technical Report TR-87-2, Computer Science Department, North Carolina State University, 1987.

with links

KEY TERMS

Available space pointer	LISCH
Coalesced hashing	Order of insertion
BLISCH	Probe
Cellar	Probe chain
Coalesced hashing	REISCH
Deleting	Synonym
Early insertion	Synonym chain
Late insertion	Table organization
Direct chaining	Cellar area
Without coalescing	Overflow area
EICH	Primary area
EISCH	Unsuccessful search
LICH	Coalesced hashing

EXERCISES

1. Insert records with the keys

$$73, \ 15, \ 44, \ 37, \ 30, \ 59, \ 49, \text{ and } 99$$

into the smallest table with a packing factor ≤ 85 percent using the key **mod** P hashing function and LISCH for collision resolution.

2. What is the average number of probes needed to retrieve each of the records inserted in Problem 1?

3. **(a)** Repeat Problem 1 using EISCH instead of LISCH. **(b)** What is the average number of probes needed to retrieve each of the records once? **(c)** Are these the results expected relative to those using LISCH in Problem 1?

4. **(a)** Insert the data from Problem 1 into a table of seven primary locations and four cellar locations using the key **mod** 7 hashing function and LICH for collision resolution. **(b)** What is the average number of probes needed to retrieve each record in this table once?

5. **(a)** Repeat Problem 4, but use EICH for collision resolution. **(b)** What is the average number of probes needed to retrieve each record in this table once?

6. In general, would you expect LICH or EICH to give better performance? Explain.

7. All three of the mechanisms to vary coalesced hashing, table organization, manner of linking and choice of empty cells, could be combined to form REICH. How would you expect its performance to compare with REISCH? Explain.

8. Assume that the record with key 39 is deleted from the table in Figure 3.5 as if it were an element of a singly linked list, that is, the link of its predecessor element (at location 6) is changed to point to its successor element (at location 4). What will happen when the record with key 42 is retrieved?

9. Why could a zero *not* be used to represent a null link in a file created using the LISCH collision resolution method?

with links

10. What could happen if simply the last record on the probe chain were moved into the position of a deleted record in LISCH?

11. In the variants of coalesced hashing that used cellars, the cellar space was allocated in advance. What effect would the ability to dynamically allocate additional overflow memory have on the variant LICH?

12. Why is the determination of coalescing a time-consuming process?

PROGRESSIVE OVERFLOW

The primary disadvantage of coalesced hashing is that additional storage is needed for the link fields. When this additional storage is not available, we need to consider a collision resolution technique that uses a convention for where to search next instead of a physical link. One of the simplest conventions, and the one introduced earlier, is that used for *progressive overflow* or linear probing [8]. As the name implies, if a location is occupied, we then look at the *next* location to see if it is empty. We are progressively overflowing into the next location. What happens when we get to the highest address in the table? Consider the table to be a *circular* structure with the first position in the table located immediately after the last one. We continue searching until we find an empty slot or until we encounter the home address of the record a second time; the latter circumstance indicates that the table is full.

For retrieval, we follow the same process. What happens with an unsuccessful search? We must continue searching until we encounter an empty space or the initial search address. Obviously, this algorithm's performance for an unsuccessful search is a poor one. The primary reason for considering it is its simplicity. We later build more efficient algorithms from it.

An Example

We will go through an example inserting the common subset of records from the last example, that is, records with the keys

$$27, 18, 29, 28, 39, 13, \text{ and } 16$$

and again use the hashing function of

$$\text{Hash(key)} = \text{key } \textbf{mod } 11$$

The records with keys 27 and 18 are inserted into locations 5 and 7, respectively, without a collision. When we insert 29, a collision occurs with 18 at location 7. Since location 7 is occupied, we then look at location 8, which is empty. 29 is then inserted into location 8 to give us a file organized as

Key

```
 0 |        |
 1 |        |
 2 |        |
 3 |        |
 4 |        |
 5 |   27   |
 6 |        |
 7 |   18   |
 8 |   29   |     (29 overflows from
 9 |        |          location 7)
10 |        |
```

The record with key 28 hashes to location 6 without a collision. When inserting 39, however, a collision occurs with 28 at location 6. We then progressively overflow to location 7; since it is full, we try location 8. It is also occupied; so we then go to location 9, which is not occupied; so 39 is inserted there giving us the file

Key

```
 0 |        |
 1 |        |
 2 |        |
 3 |        |
 4 |        |
 5 |   27   |
 6 |   28   |
 7 |   18   |
 8 |   29   |
 9 |   39   |
10 |        |
```

The record with key 13 is inserted without difficulty at location 2. The final record with key 16 hashes to location 5, which causes a collision. We locate the first available space at location 10, yielding the final table of

Key

```
 0 |        |
 1 |        |
 2 |   13   |
 3 |        |
 4 |        |
 5 |   27   |
 6 |   28   |
 7 |   18   |
 8 |   29   |
 9 |   39   |
10 |   16   |
```

Discussion

To *retrieve* the record with key 16 we must make *six* probes, which is certainly considerably more than was the case with coalesced hashing. One thing that has happened with the example is that it has been affected by secondary clustering. **Secondary clustering** is the consequence of two or more records following the same sequence of probe addresses. It results in a bunching of records within the table. This phenomenon may be likened to what happens to automobiles on an interstate highway. Often you will find a cluster of cars together. Once you move beyond a bunch, you may have a clear highway for some time again until you encounter another cluster. Secondary clustering of records contrasts with primary clustering which occurs when a large number of records have the same home address. In the example, the secondary clustering was caused by records with *different* home addresses.

The greatest disadvantage of progressive overflow is the large number of retrieval probes associated with it which results to a great extent because of secondary clustering. **Caution:** Because of the high cost of secondary storage accesses, and the relatively large number of retrieval probes associated with it, progressive overflow is *not* a practical method for handling collisions. It may be viewed similarly to the bubble sort, a well-known algorithm for ordering data. Since many people see the bubble sort as an example in elementary textbooks, they assume that it is a reasonable algorithm and use it when they need a sorting procedure. But it has very poor performance. The same is true of progressive overflow. The bubble sort is introduced in elementary textbooks primarily to introduce the concept of nested **loops**. We introduce progressive overflow only as a basis for explaining subsequent collision resolution methods.

To get an impression of its poor performance, the average number of probes to retrieve each record in the file once is 2.3 probes, much higher than the 1.4 average probes for coalesced hashing on the equivalent records. (Later in this chapter, we compare the various collision resolution methods using random data.)

The increment of one is a contributing factor in the secondary clustering. We might be able to reduce the secondary clustering if we *varied* the increment. That is the basic concept behind the next collision resolution procedure that we consider.

In a sense, progressive overflow reduces to a sequential search. The differences are that progressive overflow uses a variable starting point (the home address provided by the hashing function) and it does not need to search the entire file for an unsuccessful search (it terminates on a null record).

To *delete* a record requires that care be taken so that we are still able to retrieve the records that remain in the file. We cannot merely delete a record by reinitializing its storage location to null, for if we did, that might place a gap in a retrieval probe sequence. That gap would then signal termination on a subsequent search which would mean that the sought record would not be found even though it was in the table. What we do to remedy this difficulty is to place an indicator in the location of the deleted record. This indicator is sometimes called a **tombstone,** for it acts as a remembrance of the record that once occupied that location in the table. It tells us that additional records may follow and to keep on searching on a retrieval, *or* that the location may be filled on a subsequent insertion and the indicator removed.

In the previous example, if we desired to delete the record with key 18, we would need to place a tombstone in that position so that we could still retrieve the records with the keys 29, 39, and 16. The contents of the table after deletion would then be

```
        Key
      ┌─────────┐
  0   │         │
  1   │         │
  2   │   13    │
  3   │         │
  4   │         │
  5   │   27    │
  6   │   28    │
  7   │    ◆    │
  8   │   29    │
  9   │   39    │
 10   │   16    │
      └─────────┘
```

where ◆ represents a tombstone.

A technique for speeding the insertion process would be to keep a bit string in main memory with its positions corresponding to the locations in the file. A "one" in the bit string would indicate that the corresponding location was occupied. On insertion, then, instead of making accesses of auxiliary storage, we would only need to check the bit string in primary memory to find the first unoccupied location in the file for inserting the record. This technique would not be applicable if duplicate keys were possible, because we would lose the means of detecting them; it would also not be appropriate if we had to check if an item had been inserted previously.

KEY TERMS

Bubble sort   Secondary clustering
Collision resolution   Table
    Progressive overflow   Circular
Linear probing   Tombstone
Primary clustering   Unsuccessful search
Progressive overflow   Progressive overflow
    Deleting

EXERCISES

1. Insert records with the keys

$$73, 15, 44, 37, 30, 59, 49, \text{ and } 99$$

into the smallest table with a packing factor ≤ 85 percent using the key **mod** P hashing function and progressive overflow for collision resolution.

2. What is the average number of probes needed to retrieve each of the records inserted in Problem 1?

3. (a) Under what conditions is it not necessary to insert a tombstone in the position of a deleted record in a file formed using progressive overflow? (b) Would it be preferable to use a tombstone sometimes or all the time? Explain.

4. Under what condition is progressive overflow the same as a sequential search with a variable starting point?

5. What are the tradeoffs to using tombstones in a deletion procedure for a file formed with coalesced hashing?

USE OF BUCKETS

In the discussions of collision resolution procedures thus far, we have assumed that only a single record may be placed at a storage address. We can reduce the number of accesses of auxiliary storage by storing multiple records at one file address. When a storage location may hold multiple records, it is referred to as a bucket (or block, or page). A **bucket** then is a storage unit intermediate between a record and a file; it is also the unit in which information is accessed and transferred between storage devices. The number of records that may be stored in a bucket is called the **blocking factor**. As the blocking factor increases, the number of auxiliary storage accesses decreases because more colliding records can be stored at one location.

An Example

For comparison purposes, let's insert the records from the progressive overflow example again, but with a blocking factor of two. The keys of the records are

$$27, 18, 29, 28, 39, 13, \text{ and } 16$$

and the hashing function is again

$$\text{Hash(key)} = \text{key } \textbf{mod } 11$$

The records with keys 27 and 18 are inserted without collisions. When we insert 29, a collision occurs with 18 at location 7, but because the bucket size is two, there is still space for storing 29 at that location. The table is then

	Key$_1$	Key$_2$
0		
1		
2		
3		
4		
5	27	
6		
7	18	29
8		
9		
10		

Only when a third record collides at a particular address do we need to move to a new bucket. Because there are never more than two synonyms for a single location in this example, we never have to move from the home address bucket for insertion. The final table would appear as

	Key$_1$	Key$_2$
0		
1		
2	13	
3		
4		
5	27	16
6	28	39
7	18	29
8		
9		
10		

The average number of *probes*, not compares, needed to retrieve each record once is then 1.0, a definite improvement over the previous value of 2.3 when we stored only a single record per address. Comparing only the average number of probes, however, may be misleading because the final table with a bucket size of two has a packing factor of $\frac{7}{22}$, or 32 percent vs. the $\frac{7}{11}$, or 64 percent for the table with a bucket size of one. For more accurate comparisons, the packing factors should be the same. Table 3.2 gives the average number of probes when using progressive overflow for various bucket sizes and packing factors (α).

Discussion

Within a bucket, we need a means of separating the individual records. We can achieve this by knowing the record length for fixed-length records, or by placing a special

TABLE 3.2 MEAN NUMBER OF PROBES FOR SUCCESSFUL LOOKUP, PROGRESSIVE OVERFLOW, VARIOUS BUCKET SIZES

α (in percent)	Blocking factor			
	1	2	5	10
20	1.125	1.024	1.007	1.000
40	1.333	1.103	1.012	1.001
60	1.750	1.293	1.066	1.015
70	2.167	1.494	1.136	1.042
80	3.000	1.903	1.289	1.110
90	5.500	3.147	1.777	1.345
95	10.500	5.600	2.700	1.800

Source: Knuth, Donald. E., *Sorting and Searching*, © 1973, Addison-Wesley Publishing Company, Inc., Reading, MA, p. 536, Table 4. Reprinted with permission.

delimiter between variable-length records. Although some processing time will be required to locate the desired record within a bucket, it will be insignificant when compared with the time required for an access of auxiliary memory.

The use of buckets is especially appropriate in situations in which the record length is a factor of the smallest unit of memory that the operating system will allocate. In such a case, it costs essentially nothing to place multiple records into a storage location.

Buckets can be used to improve *any* of the collision resolution methods considered in this chapter; their use would yield reductions in the average number of retrieval probes similar to those for progressive overflow.

KEY TERMS

Block	Bucket
Blocking factor	Page

EXERCISES

1. **(a)** Reinsert the records in the table in Figure 3.5 using a bucket size of two. **(b)** What is the packing factor for the resulting table?

2. **(a)** What are the advantages of using a blocking factor greater than one for storing records? **(b)** the disadvantages?

3. What modifications would you need to make to the deletion procedure for progressive overflow if a bucket size greater than one were used?

LINEAR QUOTIENT

The primary difference between linear quotient collision resolution [9] and progressive overflow is that with linear quotient we use a *variable increment* instead of a constant increment of one. The purpose of the variable increment is to reduce secondary clustering that occurs with progressive overflow. By reducing the secondary clustering, we also reduce the average number of retrieval probes. Secondary clustering occurs with a hashing scheme in which the incrementing function is a constant or depends only upon the home address of a record. Of the hashing methods that we consider in this text, only progressive overflow has such a characteristic. With linear quotient, the increment is a function of the key being inserted. In fact, that function may be viewed as another hashing function, since we are using it to convert a *key into an increment*. For that reason, linear quotient is a member of a class of collision resolution methods referred to as **double hashing** since we hash once to get the home address (using a function we call H_1) and a second time to get the increment (we call that function H_2). What are some

Record

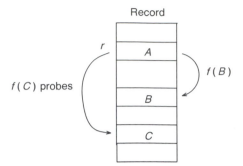

Figure 3.7 Synonym insertion.

possibilities for H_2? Two possibilities among many for determining the increment are

$$H_2 = \text{Quotient(Key} / P)\ \textbf{mod}\ P$$

$$H_2^{'} = (\text{Key}\ \textbf{mod}\ (P - 2)) + 1$$

where P is the prime number table size. Both hashing functions give us a variable increment in the range of the address space. Which one would be preferable in an actual implementation? Consider their execution in terms of machine instructions. H_2 requires the equivalent of two divide operations—one in the quotient and the second in the **mod**. As you may recall, the divide is a relatively slow operation. $H_2^{'}$ requires the equivalent of only a single divide operation—that in the form of the **mod**. So for practical considerations $H_2^{'}$ would be preferable. With that said, we use H_2 for the subsequent example. Why? From past experience, we have found that $H_2^{'}$ is more difficult for *people* to compute. This is an example of a circumstance in which people and computers have different abilities. The home address for a record will then be determined by the *remainder* of the key divided by the table size and the increment for collision resolution by the *quotient* of the same operation.

Unlike coalesced chaining, synonyms are *not* usually on the same probe chain with linear quotient. If A, B, and C are synonyms at r as illustrated in Figure 3.7, B and C will usually have different increments since the increment is a function of the key being inserted. Since different increments will yield different probe chains, secondary clustering will be reduced.

The algorithm for the linear quotient collision resolution method is given in Algorithm 3.2. It *requires* a prime number table size, for otherwise the searching could cycle through a subset of the table several times and then indicate a full table when space was available. For example, if we had a table of size six, with locations 0, 2, and 4 occupied as shown,

if we then attempted to insert a record with a home address of 0 and an increment of two, we would cycle through the three occupied locations twice. We would conclude that we had searched all six locations of the table and that it was full, which it is not. Since the table size of six is not a prime and is evenly divisible by three, it is possible to have a cycle on a subset of the table. Since a prime number is divisible only by itself and one, this prevents a cycle.

An Example

For the example, we will again use the hashing function of

$$H_1(\text{key}) = \text{key } \textbf{mod } 11$$

to locate the home address and the records with the keys

$$27, 18, 29, 28, 39, 13, \text{ and } 16$$

Until a collision occurs, the insertion process will be the same as with progressive overflow. That means that 27 will be inserted into location 5 and 18 into location 7. When we insert the record with key 29, a collision occurs at location 7. But this time instead of using an increment of one, we apply H_2 to obtain an increment of *two*. 29 is then inserted into location 9, giving us the file

	Key
0	
1	
2	
3	
4	
5	27
6	
7	18
8	
9	29
10	

Notice that in this case, there is a gap between the locations of records 18 and 29. What has happened is that the secondary cluster has been broken up. On retrieval then we should ultimately have fewer probes.

28 is inserted without a collision at location 6. With 39, we have a collision at location 6. $H_2(39) = 3$, which gives us the increment. We next attempt to store 39 at location 9, but it is also occupied. We consider the file to be circular, so three locations

Algorithm 3.2
LINEAR QUOTIENT INSERTION

I. Hash the key of the record to be inserted to obtain the home address for storing the record.

II. If the home address is empty, insert the record at that location, else

    A. Determine the increment by obtaining the quotient of the key divided by the table size. If the result is zero, set the increment to one.

    B. Initialize the count of locations searched to one.

    C. While the number of locations searched is less than the table size

        1. Compute the next search address by adding the increment to the current address and then moding the result by the table size.

        2. If that address is unoccupied, then insert the record and terminate with a successful insertion.

        3. If the record occupying the location has the same key as the record being inserted, terminate with a "duplicate record" message.

        4. Add one to the count of the locations searched.

    D. Terminate with a "full table" message.

---

from 9 is location 1, which is empty. Alternatively, you may view the new address location as the

$$\text{New address} = (\text{current address} + \text{increment}) \textbf{ mod } \text{table\_size}$$

or in this case, the new address is equal to

$$(9 + 3) \textbf{ mod } 11 \qquad \text{or} \qquad 1$$

39 is then inserted into that location. The file is then

	Key
0	
1	39
2	
3	
4	
5	27
6	28
7	18
8	
9	29
10	

13 is inserted into location 2 without a collision. The final record with key 16 collides at home address 5. $H_2(16) = 1$. We then look at locations 6, 7, and finally the empty location 8. The final file would then appear as

without links—static positioning

Key

0	
1	39
2	13
3	
4	
5	27
6	28
7	18
8	16
9	29
10	

Discussion

In this example, *retrieving* 16 requires only four probes compared with the six with progressive overflow. Overall we need an average of $\frac{13}{7}$, or 1.9 probes for retrieving the records inserted via linear quotient vs. the 2.3 average probes for progressive overflow and 1.4 for coalesced hashing with LISCH. The variable increment has allowed us to break up the secondary clusters of records which improves the performance of both successful and unsuccessful searches. The mechanism for determining an unsuccessful search with linear quotient is the same as that for progressive overflow. Do you recall what it was? When we encounter an *empty* probe location, we terminate the search unsuccessfully. An unsuccessful search requires fewer probes with linear quotient than with progressive overflow since we are more likely to encounter an empty location to terminate the search as a result of eliminating secondary clusters. As we go through these collision resolution methods, one of the goals is to identify the deficiencies of a procedure and then to discover a means to eliminate the cause of the deficiency. This is problem solving. By noting that secondary clustering was the cause of the high number of probes with progressive overflow, we were able to reduce the secondary clustering and improve the retrieval performance. Linear quotient may not be the only solution or way to reduce the secondary clusters.

We can improve upon linear quotient by observing that the number of retrieval probes is dependent upon the *placement* of the records. If a record is in a location that prevents an incoming record from being placed there, the incoming record may need to be inserted at a position that will require many more retrieval probes. For instance, if we insert a record with a key of 67, it has a home address of 1 which is already filled with 39. We then attempt to insert 67 at locations 7, 2, and 8 until we finally discover an empty slot at location 3. But this means that *five* retrieval probes will be necessary to find 67. What if 39 had not been stored at location 1? We then could have inserted 67 at that location and only *one* probe would have been required to retrieve it instead of the five probes. The basis for the next method we consider is to *move* an item already inserted in the table if the move reduces the average number of probes required to retrieve *all* the records. If we could move 39 to location 4, that would free location 1. 39 would be moved to location 4 since the increment associated with it is three. Without

without links—static positioning

the moving of 39, retrieving 39 and 67 would require a total of *eight* probes, whereas with the move, the retrieval of the two records would require a total of *five* probes. We would then save three probes. The next collision resolution method will illustrate how to check for and move records to reduce the number of retrieval probes.

Deleting a record from a table requires the use of a tombstone, as with progressive overflow, to indicate that additional records may follow on a probe chain.

KEY TERMS

Address space	Probe chain
Collision resolution	Secondary clustering
Linear quotient	Synonym
Cycle	Table size
Double hashing	Prime number
Linear quotient	Unsuccessful search
Deleting	Linear quotient
Performance	Variable increment
Linear quotient	

EXERCISES

1. Insert records with the keys

   73, 15, 44, 37, 30, 59, 49, and 99

   into the smallest table with a packing factor $\leq$ 85 percent using the key **mod** P hashing function and linear quotient for collision resolution.

2. What is the average number of probes needed to retrieve each of the records inserted in Problem 1?

3. **(a)** Reinsert the data from the example using H_2' instead of H_2 as the incrementing function. **(b)** Is the resulting table the same? **(c)** Is the average number of probes the same? **(d)** What conclusion can you draw from your previous answers?

4. Why is a tombstone always required when deleting from a table formed using linear quotient?

5. Instead of using a tombstone for deletion in linear quotient and progressive overflow, what are the tradeoffs to moving the last element on a probe chain into the vacated position as we did with coalesced hashing?

6. Insert records with the keys

   73, 15, 44, 37, 30, 59, 49, and 99

   into a table with 10 locations using the key **mod** N hashing function and linear quotient for collision resolution. Note any interesting occurrences.

7. Reinsert the data from Problem 6 but replace the record with key 99 with one with a key of 29. Note any interesting occurrences.

without links—static positioning

8. What important lesson can be learned from comparing the observations of Problems 6 and 7?

BRENT'S METHOD

All the collision resolution methods that we have considered thus far are *static* methods; that is, an item remains at the address where it is stored initially. When we considered coalesced hashing, one improvement that we mentioned was to *move* an item not stored at its home address. We now want to consider several *dynamic* methods—ones in which an item once stored may be moved. With these methods, any item may be moved; we are not limited to records not stored at their home addresses. These methods require additional processing when inserting a record into the table but reduce the number of probes needed for retrieval. The justification for this additional processing is that we usually insert an item into a table only once but retrieve it many times.

The **primary probe chain** of a record is the sequence of locations visited during the insertion or retrieval of the record. In the previous example with the linear quotient method, the primary probe chain for 39 consisted of the three positions p_1, p_2, and p_3 which corresponded to the addresses (and contents) in the diagram below. p_1, the first position on the primary probe chain, is the home address.

```
p₁  ┐   6 (28)

p₂  ┤   9 (29)

p₃  ┘   1 (empty)
```

Since three positions eventually had to be visited before an empty location was found, three probes were needed for its retrieval. What if its home address were empty? Then 39 could have been inserted at its home address so that only a single probe would have been necessary for its retrieval. We could make the home address available by moving what is stored there. In the example, that would be 28. But where should we move 28? We need to move it to a location such that it could still be retrieved (in this case, we still want to use the linear quotient method for retrieval). We would first try to move 28 to the next location on its (28s) probe chain (which has an increment of two) rather than the probe chain of the item being inserted, 39 (which has an increment of three).

```
p₁  ┐   6 (28) ───────────▶  8 (empty)

p₂  ┤   9 (29)

p₃  ┘   1 (empty)
```

The sequence of positions visited when attempting to *move* a record from the primary probe chain is called the **secondary probe chain**. In the previous diagram, the secondary probe chain, which is represented by the horizontal line, consists of only one entry since the first position visited was empty. Since location 8 is empty, if we move 28 to it, two probes would then be needed to locate 28 but that would free location 6 so that 39 could be stored there and only one probe would be necessary to find 39. 28 will require one more probe for its retrieval but 39 will require two fewer probes for a net reduction of one probe achieved by *moving* 28. We want to minimize the total number of probes for both the item being inserted and the items already in the table. This strategy assumes an equal likelihood of any of the items being retrieved.

Brent's method [10] is the first of several dynamic collision resolution methods that we consider. In each of them, we consider moving a previously stored item to achieve a reduction in the retrieval probes. Figure 3.8 charts how to attempt the moves in Brent's method. The solid vertical line represents the primary probe chain, that is, the addresses that would be considered in storing an item using the linear quotient scheme. The horizontal lines represent the secondary probe chains, that is, the addresses that would be searched in attempting to move an item from a position along the primary probe chain. The q value along the primary probe chain is the increment for the item being inserted whereas the q_i's along the secondary probe chains represent the increments associated with the item being moved. (The q_i's will likely be different for each item on the primary probe chain.) The subscript i gives the number of probes needed to retrieve the item being inserted along its primary probe chain. The subscript j gives the number of additional probes needed to retrieve the item being moved along its secondary probe chain. To minimize the number of retrieval probes, we want to

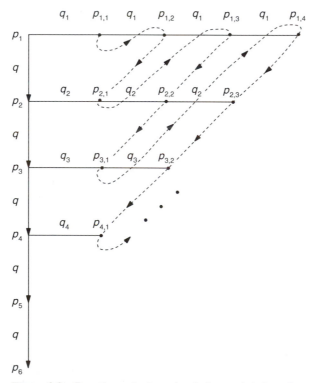

Figure 3.8 Brent's method, probe chains, and their order of processing.

Algorithm 3.3
BRENT'S METHOD INSERTION

I. Hash the key of the record to be inserted to obtain the home address for storing the record.

II. If the home address is empty, insert the record at that location, else

    A. Compute the next potential address for storing the incoming record. Initialize $s \leftarrow 2$.

    B. While the potential storage address is not empty,

        1. Check if it is the home address. If it is, the table is full, terminate with a "full table" message.

        2. If the record stored at the potential storage address is the same as the incoming record, terminate with a "duplicate record" message.

        3. Compute the next potential address for storing the incoming record. Set $s \leftarrow s + 1$.

/* Attempt to move a record previously inserted.*/

    C. Initialize $i \leftarrow 1$ and $j \leftarrow 1$.

    D. While $(i + j < s)$

        1. Determine if the record stored at the ith position on the primary probe chain can be moved j offsets along its secondary probe chain.

        2. If it can be moved, then

            a. Move it and insert the incoming record into the vacated position i along its primary probe chain; terminate with a successful insertion, else

            b. Vary i and/or j to minimize the sum of $(i + j)$; if $i = j$, minimize on i.

/* Moving has failed. */

    E. Insert the incoming record at position s on its primary probe chain; terminate with a successful insertion.

---

minimize $(i + j)$. In the case where $i = j$, we will arbitrarily choose to minimize on i. When should we terminate this process of attempting to move an item? When we can no longer achieve a reduction in the number of retrieval probes. Let s be the number of probes required to retrieve an item if nothing is moved. We then try all combinations of $(i + j) < s$ such that we minimize $(i + j)$. On equality, since there would be no reduction in the number of probes, no movement would occur. Figure 3.8 shows the order in which the various movements would be attempted (follow the dotted line). The algorithm for insertion into a file using Brent's method appears in Algorithm 3.3.

An Example

Let's insert the records used in the previous examples, that is, those with the keys

$$27, 18, 29, 28, 39, 13, \text{ and } 16$$

and again use the hashing function of

$$\text{Hash(key)} = \text{key } \textbf{mod } 11$$

and the incrementing function of

$$i(\text{key}) = \text{Quotient(Key / 11) } \textbf{mod } 11$$

Since we know that the first record will not cause a collision, 27 may be inserted immediately at its home address of 5. 18 hashes to location 7, which is empty so it too does not require collision resolution.

Since 29 has the same home address as 18, we have a collision at location 7. Should we try moving 18 to reduce the overall number of probes? The s value for inserting 29 is two; that is, since we could insert 29 into the empty location 9, two probes would be necessary to retrieve it. From Figure 3.8, is there any combination of $i + j < 2$? No, so we do not even need to attempt moving an item. Only if s were three or greater would we be able to reduce the number of probes by moving an item. The table at this point is the same as in the linear quotient example, that is,

	Key
0	
1	
2	
3	
4	
5	27
6	
7	18
8	
9	29
10	

The home address for 28 of 6 is empty, so the record may be stored there. Since 39 also has a home address of 6, a collision occurs. After computing its s value of three, we first try a move with $i = 1$ and $j = 1$, which means to try to move what is at the home address of 6 one offset along its probe chain, or in this case, we try moving 28 to location 8.

Note. We use the increment associated with the item being *moved* and not with the record being inserted.

Because location 8 is empty, we complete that move so that 39 can be inserted into location 6, thus reducing the total number of probes by one. The table then appears as

```
      Key
    ┌──────┐
 0  │      │
 1  │      │
 2  │      │
 3  │      │
 4  │      │
 5  │  27  │
 6  │  39  │
 7  │  18  │
 8  │  28  │
 9  │  29  │
10  │      │
    └──────┘
```

where **boldface** indicates that a record has been moved from its original location.

13 hashes to the unoccupied home address of 2 so it can be stored there. To resolve the collision that occurs when attempting to insert 16 at its home address of 5, we first compute its s value of six. How many combinations of $i + j$ are less than six? 10, so there are a lot to try, but remember that we always minimize the sum of $i + j$. With $i = 1$ and $j = 1$, we try to move what is at the first or home address on the probe chain of 16 to the next position on *its* probe chain. Specifically, we try to move 27 to location 7 from location 5. 7 is filled. For $i = 1, j = 2$, we try to move 27 to location 9. Again the position is filled. For $i = 2, j = 1$, we try to move what is in the second position on the probe chain for 16, in this case, 39, to the next location on *its* (39's) probe chain, that is, 9. Again there is a failure. Next for $i = 1$ and $j = 3$, we try to move 27 to location 0. Because that location is empty, we can move 27 there, which in turn frees location 5 so that 16 may be inserted into it. The table then appears as

```
      Key
    ┌──────┐
 0  │  27  │
 1  │      │
 2  │  13  │
 3  │      │
 4  │      │
 5  │  16  │
 6  │  39  │
 7  │  18  │
 8  │  28  │
 9  │  29  │
10  │      │
    └──────┘
```

How many probes overall are saved by making this move? Without moving 27, the total number of probes needed to retrieve both 27 and 16 would have been seven. By moving 27 we need only five probes to retrieve it and 16. The average number of

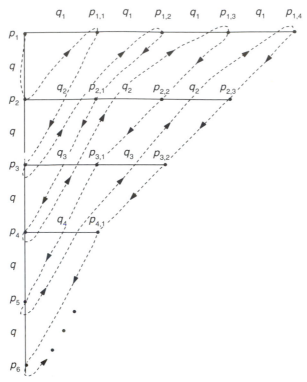

Figure 3.9 Revised order of processing.

probes needed to retrieve each record in the file once is 1.7 probes vs. the 1.9 of linear quotient or the 2.3 of progressive overflow or the 1.4 for coalesced hashing with LISCH.

Discussion

The improvement in retrieval performance resulting from applying Brent's method for collision resolution is predicated on the assumption that (1) all records are equally likely to be retrieved and (2) the retrieval frequency for a record will be significantly greater than one to offset the additional insertion costs. As a consequence of a record being moved to a different location during the insertion process for another record, the moved record will require *more* retrieval probes. The advantage gained, however, is that the increase in the number of retrieval probes will be more than offset by a *reduction* in retrieval probes for the inserted record. Brent's method only pertains to the insertion process; *the linear quotient method is used for retrieval*.

The insertion process can be modified [11] to eliminate the need to compute s prior to trying the various move possibilities. In doing so we also eliminate the need to traverse the primary probe chain when computing s. Eliminating this traversal improves the insertion performance. If we follow the processing order represented by the dashed lines in Figure 3.9, a record will be inserted into a location that minimizes the number of retrieval probes for both the item being inserted and the items along its primary probe chain. In other words, we will move a record when a move is advantageous and we will not when it isn't. We are interleaving the process of computing s and the process of testing for improvement and moving previously stored records where appropriate. The processing order of Figure 3.9 attempts insertions along the primary probe chain *unless* a movement would reduce the total number of probes. Then those movements are investigated for achievability. If not attainable, the insertion process then continues at the next position on the primary probe chain. Then movements are considered, etc. This alternating between *inserting on the primary probe chain* and *moving an item along its secondary probe chain* continues until the insertion is made or until it is discovered that the table is full. This modification only changes the control mechanism for insertions; it does *not* affect the locations of the records or the number of retrieval probes. Let's reconsider the insertion of the record with key 16 into the file

	Key
0	
1	
2	13
3	
4	
5	27
6	39
7	18
8	**28**
9	29
10	

using the modified insertion process following the path of Figure 3.9. We first attempt the insertion at p_1, the home address, which is location 5. That location is occupied so we follow the dashed lines to the next alternative which is p_2, or location 6. It is also filled so we continue along the black dashed line (not as colorful as a yellow brick road but it serves the same function of directing us to our goal). The next possibility is $p_{1,1}$ which indicates a move of one offset of the record stored at p_1; in this example, that is an attempt to move 27 to location 7. That location is also filled, so we continue. What is the next alternative on the chart? We go back to the primary probe chain at p_3. Can we insert 16 at p_3? p_3 corresponds to location 7 which is filled. We keep on going to $p_{1,2}$ which symbolizes an attempt to move the record stored at p_1 two offsets, or to move 27 to location 9. Again we can't do it. We then follow the dashed line to $p_{2,1}$ which signifies an attempt to move what is stored at p_2 one offset, or to move 39 to location 9. Again, we have a failure, so we proceed to p_4, which corresponds to location 8. We have another failure. The next possibility along the dashed line is $p_{1,3}$ which signifies an attempt to move the record stored at p_1 three offsets or to move 27 to location 0. We can do that, so we do, which in turn frees location 5 for inserting 16. Notice that we did not need to compute s in advance and that the result is the same as we obtained before.

```
          Key
      ┌───────────┐
  0   │    27     │
  1   │           │
  2   │    13     │
  3   │           │
  4   │           │
  5   │    16     │
  6   │    39     │
  7   │    18     │
  8   │    28     │
  9   │    29     │
 10   │           │
      └───────────┘
```

Because linear quotient is used for retrieval with Brent's method, *deleting* a record would require the placement of a tombstone at the position of the deleted record.

KEY TERMS

Brent's method   Brent's method
   Deleting   Dynamic methods
   Insertion modification   Primary probe chain
Collision resolution   Secondary probe chain

EXERCISES

1. Insert records with the keys

73, 15, 44, 37, 30, 59, 49, and 99

into the smallest table with a packing factor $\leq$ 85 percent using the key **mod** P hashing function and Brent's method for collision resolution.

2. What is the average number of probes needed to retrieve each of the records inserted in Problem 1?

3. Is the concept of a secondary probe chain relevant to retrieval when using Brent's method? Explain.

4. Under what circumstances will the average number of probes for a given set of data be the same for Brent's insertion vs. linear quotient insertion?

5. How would you implement that part of Brent's insertion algorithm which varies the values of i and j?

6. What is a complication with using the modified insertion process for Brent's method when deletions and duplicate records may occur?

BINARY TREE

A question that is often asked when considering Brent's collision resolution method is, "If it is a good idea to move an item on a primary probe chain, why not carry this concept one step further and move items from secondary and subsequent probe chains?" For many years, the answer seemed to be that this would make the algorithm much too complex and/or it wouldn't improve performance that much. Excuses. Two features of the binary tree collision resolution method make it worth considering:

- It needs fewer retrieval probes than Brent's method.
- Perhaps more importantly, it illustrates the importance of choosing an appropriate data structure in order to be able to solve a problem effectively.

The method was developed by Gonnet and Munro [12]. The key component of their method is the use of a binary tree structure to determine when to move an item and where to move it. A binary tree is appropriate since there are essentially two choices at each probable storage address—*continue* to the next address along the probe chain of the item being inserted or *move* the item stored at that address to the next position on *its* probe chain. A left branch in the binary tree signifies the *continue* option and a right branch the *move* option.

The binary decision tree is generated in a breadth first fashion from the top down left to right as shown:

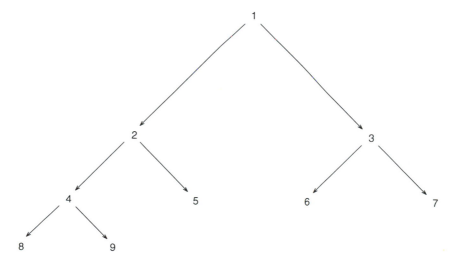

without links—dynamic positioning

Note. The binary tree is used *only* as a control mechanism in deciding where to store an item and is *not* used for storing records.

A different binary tree is constructed for each insertion of a record. Encountering either an empty leaf node in the binary tree or a full table terminates the process.

By moving items from secondary and subsequent probe chains, we are achieving a placement of records that will further reduce the average number of retrieval probes when compared with Brent's method. We are performing more preprocessing on the file with the binary tree method than with Brent's method, but remember that we insert once but retrieve often. The extra effort at insertion time will pay dividends if the file will exist for some time with many retrievals on it. The algorithm for binary tree

Algorithm 3.4
BINARY TREE INSERTION

I. Hash the key of the record to be inserted to obtain the home address for storing the record.

II. If the home address is empty, insert the record at that location, else

    A. Until an empty location or a "full table" is encountered,

        1. Generate a binary tree control structure in a breadth first left to right fashion. The address of the lchild of a node is determined by adding (i) the increment associated with the key of the record *coming in* to the node to (ii) the current address. The address of the rchild of a node is determined by adding (i) the increment associated with the key of the record *stored* in the node to (ii) the current address.

        2. At the leftmost node on each level, check against the record associated with it for a duplicate record. If found, terminate with a "duplicate record" message.

    B. If a "full table," terminate with a "full table" message.

    C. If an empty node is found, the path from the empty node back to the root determines which records, if any, need to be moved. Each right link signifies that a relocation is necessary. First, set the current node pointer to the last node generated in the binary tree and set the empty location pointer to the table address associated with the last node generated.

    D. Until the current pointer equals the root node of the binary tree (bottom up), note the type of branch from the parent of the current node to the current node.

        1. On a right branch, move the record stored at the location contained in the parent node into the location indicated by the empty location pointer. Set the empty location pointer to the newly vacated position and make the parent node the current node.

        2. On a left branch, make the parent node the current node.

    E. Insert the record coming into the root position into the empty location. Terminate with a successful insertion.

---

without links—dynamic positioning

collision resolution appears in Algorithm 3.4; it may be helpful to consider it with the following example.

As with Brent's method, the special processing is only for the insertion of items; *retrieval* is handled through the linear quotient hashing scheme.

An Example

The easiest way to understand this technique may be to go through an example. Let's again insert those records that we have used in the previous examples, those with the keys of

$$27, 18, 29, 28, 39, 13, 16$$

plus three additional records with keys of

$$41, 17, \text{ and } 19$$

We also use the same table size of 11 locations and the hashing function of

$$\text{Hash(key)} = \text{key } \textbf{mod } 11$$

The incrementing function will again be

$$i(\text{key}) = \text{Quotient(Key / 11) } \textbf{mod } 11$$

Since a collision cannot occur on the first insertion into a table, the record with key 27 may be inserted directly into its home address of 5. The next record with key 18 hashes to location 7 [since Hash(18) = 7] which is also empty so that record may be inserted without difficulty. When we hash 29 a collision occurs at location 7 which is already filled with 18. We now generate a binary tree to determine which, if any, records to move. The root of the binary tree is the home address for 29, which is 7. The left branch from that node points to the next location we would encounter if we *continue* along the primary probe chain for inserting 29; that would be location 9 [since the increment or $i(29)$ is two]. Location 9 is empty; so the insertion process terminates. The binary tree then appears as

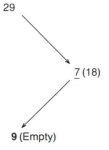

29

7 (18)

9 (Empty)

The **boldface** numbers represent locations in the table and the values within parentheses are the keys stored at those locations. The *underlined* node is the root node. The tree tells us to attempt to *move* 29 into location 7. That location is occupied so we *continue* along the primary probe chain until we reach location 9, which is empty, which allows 29 to be inserted there. The table then appears as

	Key
0	
1	
2	
3	
4	
5	27
6	
7	18
8	
9	29
10	

The record with key 28 hashes to the empty location 6; so it can be inserted without collision resolution. The next record, with key 39, also hashes to location 6. This time then we need to generate a binary tree to determine the placement of the records. The binary tree would be

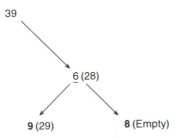

Encountering the empty location terminates the process. This binary tree requires a *move* because a right branch is needed to reach the node corresponding to the empty location. In general we need to make a move for each right branch on the path from the first empty location encountered in generating the binary tree back to the root node. The binary tree says to attempt to *move* 39 into location 6. It is filled; so we then *continue* and attempt to place it at the next location along its primary probe chain which is location 9. That location is also occupied. The next branch (a right one) says to *move* what is stored at location 6 to the next location along its probe chain, which sends us to an empty location 8. Only the path from the leaf node corresponding to the empty location back to the root node, which specifies the necessary actions for insertion, needs to be considered. When 28 is moved to location 8, that frees location 6 so that 39 may be inserted there directly. The table then becomes

without links—dynamic positioning

Key

0	
1	
2	
3	
4	
5	27
6	39
7	18
8	**28**
9	29
10	

The **boldface** in the table represents a record that has been moved.

The record with key 13 is inserted into location 2 without requiring collision resolution. 16 hashes to location 5 which contains 27. We then generate the binary tree

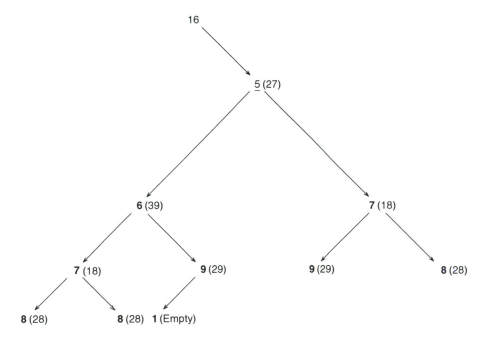

Note. The path formed by the leftmost branch at each level of the binary tree is equivalent to the primary probe chain of the linear quotient method and of Brent's method.

Working back from the first empty leaf node found in the generation process to the root node, we notice that we attempt to *move* 39 from location 6 to location 9

without links—dynamic positioning

[because $i(39) = 3$] but that position is filled so we then *continue* along the probe chain for 39 which leads us to the empty location 1. 39 is *moved* to location 1 which then frees location 6 so that 16 may be inserted there. The table then looks like

	Key
0	
1	**39**
2	13
3	
4	
5	27
6	16
7	18
8	**28**
9	29
10	

We have now inserted the common set of records that we have been using with all the collision resolution methods. The average number of probes needed to retrieve each record once is 1.7 probes. This compares with 1.4 probes for coalesced chaining with LISCH and 1.7 for Brent's method. Why have we not experienced an improvement over Brent's method at this stage? It could be the placements of this set of data items or it could be that we have made at most only a single move of a previously stored data item per insertion. The next insertion record with key 41 collides at location 8 and eventually is placed at empty location 0. The next record with key 17 collides at location 6, and we generate the binary tree

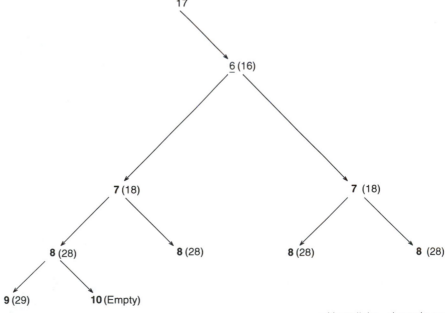

without links—dynamic positioning

28 is then *moved* to empty location 10 and 17 is inserted into location 8. The table is then

	Key
0	41
1	**39**
2	13
3	
4	
5	27
6	16
7	18
8	17
9	29
10	**28**

Notice that 28 has now been moved twice. We are making the assumption that all items will be retrieved equally often. So even though we now require more probes to retrieve 28 we need fewer probes to retrieve 17 and 16.

The final insertion record, which has a key of 19, collides at location 8. The collision resolution process then generates the binary tree

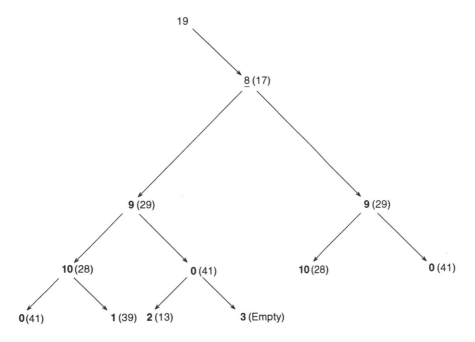

What is different about this binary tree compared with the previous ones is that we need to make two *moves*. Each right branch on the path from the terminating node back to the root signifies a *move*. We move 41 from location 0 to the empty location 3. Freeing location 0 allows us to move 29 there. That in turn allows us to insert 19 at

without links—dynamic positioning

the now empty location 9. This is the first situation where we go beyond Brent's method. In Brent's method, we only made *one* move per insertion. With the binary tree collision method, we may make any number of moves. The movement of 41 from location 0 to location 3 corresponds to a move from a secondary probe chain in the terminology of Brent's method. The movement of 29 from location 9 to location 0 is a move from a primary probe chain. The final table is then

	Key
0	**29**
1	**39**
2	13
3	**41**
4	
5	27
6	16
7	18
8	17
9	19
10	**28**

Implementation

One significant aspect of the binary tree used in inserting items into the table is that it is a **complete binary tree**.

> The *n* nodes of a **complete binary tree** correspond to the first *n* nodes of a full binary tree numbered in a breadth first fashion (top down, left to right).

> A **full binary tree** has the maximum number of nodes for a tree of its depth.

An advantage of a complete binary tree is that it can be readily implemented as a *sequential* structure and in using the following relationships, we can easily move about the tree.

- lchild(i) = 2 * i
- rchild(i) = 2 * i + 1
- parent(i) = $\lfloor i / 2 \rfloor$

where i is the address of a node in the binary tree.

How large can the binary tree grow? The worst case can be somewhat large.

Gonnet and Munro [12] have shown that the expected *depth* of the binary tree is bounded by $O(\log_2 n + c)$ where n is the number of entries in the table and c is usually one and sometimes two. In actual practice, you would not have a full table nor would you encounter the worst case (unless your name is Murphy), but you need to be prepared for it should it arise. One thing that could be done to provide sufficient storage is that after the tree gets to be a certain size, secondary storage could be used to store the tree from that point on. Because the size of the binary tree can become large, the method is not practical for a large number of records at a packing factor approaching 100 percent. An implementation technique to conserve storage is to store only the bottom two levels of the tree. **Remember** that when space is at a premium, store only that information which still needs to be accessed and cannot be computed.

How would you implement the algorithm to check for a full table? One method is to keep a counter of the depth of the binary tree and when it exceeds $\log_2 n + 2$, you know that the table is full. But that requires the generation of a possibly enormous binary tree. A better solution is to merely keep a counter of how many records have already been inserted into the table and then to check that number *before* generating the binary tree.

Discussion

The binary tree data structure is used *not* for storing information but for controlling the algorithm for storing information. It enables the algorithm to provide improved performance *without* increased complexity as compared with Brent's method. Many students say that the binary tree collision resolution method is actually easier to comprehend than Brent's method. That contradicts the myth that greater power requires greater complexity. *Power and simplicity are not mutually exclusive goals.*

As with Brent's method, the linear quotient method is used for retrieval. That is why it is necessary to use the Q value or increment associated with the item being moved in the table. Otherwise it would not be possible to subsequently retrieve the record. Even though the record with key 28 was moved twice during the insertion process, we are still able to locate it by applying linear quotient. The home address of 28 is 6 and its increment is two. We then search locations 6, 8, and finally 10.

Also, because of using linear quotient for retrieval, a tombstone would be inserted in the position of a *deleted* record to ensure that records appearing later in the probe chain could still be retrieved.

KEY TERMS

Collision resolution
    Binary tree
Complete binary tree
Full binary tree

Secondary probe chain
Sequential structure

EXERCISES

1. Insert records with the keys

$$73, 15, 44, 37, 30, 59, 49, \text{ and } 99$$

   into the smallest table with a packing factor ≤ 85 percent using the key **mod** P hashing function and the binary tree collision resolution method.

2. What is the average number of probes needed to retrieve each of the records inserted in Problem 1?

3. What percentage of the overall storage is saved by storing only the bottom two levels of the binary tree?

4. What is the order of magnitude of the maximum depth of the binary trees used to insert n nodes into a table with a packing factor < 1?

5. If at most only one right branch occurs in the binary trees used to control record movement in a binary tree insertion, will the resulting table be identical to that obtained from inserting the same data with Brent's method? Explain.

6. Explain why only the leftmost node at each level of a binary tree needs to be checked for a duplicate.

7. What efficiencies in generating a binary tree are possible when the increments associated with *both* the left and right branches are the same? Examine the binary trees in the example for a clue.

COMPUTED CHAINING

We have now considered two types of collision resolution procedures—those which use a link field (for example, coalesced hashing) and those which do not (for example, progressive overflow, linear quotient, Brent's method, and binary tree). The methods that use a link field provide better performance but require more storage because of the additional field for keeping the link. The methods that do not use a link field require less space but produce poorer performance. We then have a dilemma: there is no one best method for all circumstances. We therefore want to consider a third class of techniques that are between the extremes of the other two classes in both performance and storage. The unifying characteristic of this third class is that instead of storing an *actual* address in the link field of a node, they store a *pseudolink* field to locate the next search address. A collision resolution technique called pseudochaining introduced the notion of a pseudolink [13]. The link name is prefaced with "pseudo" to indicate that the value contained therein requires additional processing before it yields an actual address. What we are doing is *computing* information rather than *storing* it. This is a principle that we will see again later. If we are limited in the storage available, in many cases the information that we need can be computed rather than stored.

The performance of the pseudolink methods are better than the nonlink collision resolution methods since they eliminate the probes of locations that occur between a stored record and its successor item on its probe chain. The number of offsets or locations which would need to be searched to locate the successor item are stored in the pseudolink field so that we can then compute the successor's actual address. In that way, the intermediate locations do not have to be searched and eliminating those probes improves performance. The storage requirements of the pseudolink methods are less than those for coalesced hashing because it takes fewer bits (less storage) to store the number of offsets rather than a full address. It is a few bits compared with typically three or more bytes. On the opposite side of the coin, the pseudolink methods do not perform as well as the coalesced hashing methods and they require more storage than the nonlink methods, since it is necessary to store the pseudolink value in addition to that of the record. But these intermediate, pseudolink methods are useful in situations in which there is not sufficient space to store an entire link field and efficient performance is important.

We will focus on a pseudolink method called computed chaining [14]. Computed chaining follows the process of coalesced hashing (see Figure 3.3a) in that the first item on a probe chain points to its immediate successor and that successor in turn points to its immediate successor. To eliminate the coalescing of chains, computed chaining also *moves* a record stored at another record's home address. In this way all the records on a probe chain have the same home address; there is no coalescing. That in itself improves performance. The number of probes needed to retrieve a record is equivalent to the position of the record in its probe chain. Figure 3.10 illustrates the collision resolution process.

Note.   Unlike the nonlink collision resolution methods, computed chaining uses the key of the record *stored* at a probe address to locate the next probe address and not the key of the record being *inserted* or *retrieved*.

In Figure 3.10, the $i(B)$ means that the increment function is applied to what is stored at location s to locate the successor probe address of t. Using a function of the item stored at a location ensures that only one actual probe will be needed to locate the successor record. The algorithm for a computed chaining insertion appears in outline form in Algorithm 3.5 and in pseudocode in Algorithm 3.6. You may want to attempt to write a pseudocode version independently before you study Algorithm 3.6. The **probe** function *computes* the address of the successor element given the key of the

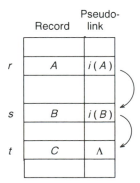

Figure 3.10 Synonym chain.

Algorithm 3.5
COMPUTED CHAINING INSERTION

I. Hash the key of the record to be inserted to obtain the home address for storing the record.

II. If the home address is empty, insert the record at that location.

III. If the record is a duplicate, terminate with a "duplicate record" message.

IV. If the item stored at the hashed location is not at its home address, move it to the next empty location found by stepping through the table using the increment associated with its predecessor element, and then insert the incoming record into the hashed location, else

    A. Locate the end of the probe chain and in the process, check for a duplicate record.

    B. Use the increment associated with the last item in the probe chain to find an empty location for the incoming record. In the process, check for a full table.

    C. Set the pseudolink at the position of the predecessor record to connect to the empty location.

    D. Insert the record into the empty location.

---

record stored at the current location and the pseudolink of the current location. The incrementing scheme of linear quotient is used to calculate the successor position.

An Example

We will again use the records from the previous examples with the keys of

$$27, 18, 29, 28, 39, 13, 16$$

plus two additional records with the keys of

$$38 \text{ and } 53$$

The table size will again be eleven and the hashing function will be

$$\text{Hash(key)} = \text{key } \mathbf{mod} \text{ } 11$$

The incrementing function will be

$$i(\text{key}) = \text{Quotient(Key / 11) } \mathbf{mod} \text{ } 11$$

The first two records with keys of 27 and 18 are inserted without a collision into locations 5 and 7, respectively, as with the previous examples. 29 collides at location 7. We first check to see if the record stored at that location is stored at its home address. How do we do that? We process the key of that record, that is, 18, through the hashing

with pseudolinks

Algorithm 3.6
PSEUDOCODE COMPUTED CHAINING INSERTION

proc   computed—chaining—insert

/*   Inserts a record with a key x according to the computed chaining hashing method. Table[r] refers to the contents at location r in the file and pseudolink[r] contains the offset value associated with that location.   */

```
1    h ← Hash(x)   /* Locate the home address. */
2    if Table[h] = Λ then [Table[h] ← x; return] /* The home address is empty, so insert
                      the item.*/
3    if Table[h] = x then return /* The item is a duplicate.*/
4    if Hash(Table[h]) ≠ h then move the item /* The item stored at h is not stored at
                      its home address, so move it and store x at h. */
5    while pseudolink[h] ≠ Λ
6        do
7            h ← probe(Table[h],pseudolink[h],h) /* probe is a function to locate the
                      address of the next item in the probe chain */
8            if Table[h] = x then return /* The item is a duplicate*/
9        end
10   i ← 1    /* Initialize a loop variable to locate the first empty cell for storing x. */
11   j ← probe(Table[h],i, h) /* Locate the next probe address.*/
12   while Table[j] ≠ Λ
13       do
14           i ← i + 1
15           if i > table—size then [print "table full"; stop]
16           j ← probe(Table[h],i,h) /*Locate the next probe address. */
17       end
18   pseudolink[h] ← i /* Insert the probe number */
19   Table[j] ← x /* Store the item */
```

end   computed—chaining—insert

function to see if it yields the current address. If it does, the item is stored at its home address. Since 18 is stored at its home address, we then need to determine if it has an immediate successor, and if not, we then need to find a place for inserting the incoming record. A procedure for determining if a record has an immediate successor is to look at its pseudolink field. If it is null or empty, we know that the current record does not have one; if it is not null, we can use the pseudolink value to compute the address of the successor element so that we can locate it in one probe. Since the pseudolink field at location 7 is null, we know that 18 does not have a successor. We use the increment of one that is associated with the item stored at location 7, that is, 18, and not that associated with the item being inserted, that is, 29, to locate a position for the successor. This increment of one sends us to location 8, which is empty; thus, the record with key 29 is inserted there. One last step that is necessary is to set the pseudolink of 29's predecessor location so that we can retrieve the record with only a single additional actual probe. Since we had to move one position on the probe chain, the pseudolink value at location 7 is set to one. The table is then

	Key	nof
0		
1		
2		
3		
4		
5	27	
6		
7	18	1
8	29	
9		
10		

where nof represents the number of offsets, or the pseudolink value.

The next record with key 28 is inserted without a collision at location 6. The record with key 39 collides at location 6. The record currently stored at location 6, that is, 28, is at its home address. Since the pseudolink field indicates that 28 does not have a successor, we need to locate an empty space to store 39 as 28's successor. We use the increment of two associated with 28 to determine the next probe address. Since location 8 is occupied, we then look at location 10, which is empty, so 39 can be inserted there. The increment is changed only when we reach the successor element on a probe chain. The current increment is always the one associated with the most recently visited record *on the current probe chain*. One last step is to set the pseudolink value for location 6. In this case, *two* additional locations had to be searched, so we place a two in the pseudolink field at address 6. The table then appears as

	Key	nof
0		
1		
2		
3		
4		
5	27	
6	28	2
7	18	1
8	29	
9		
10	39	

Notice what we have gained over the nonlink collision resolution methods. On *retrieval*, only *two* probes are necessary to locate 39 and not three. That is because once we arrive at location 6, we can *compute* its successor address of 10 directly since we know the current address is 6, the increment is two, and the number of offsets is two. That information then sends us directly to location 10 from location 6. We therefore need a total of only two probes to find 39. We have eliminated the need to probe address 8.

The record with key 13 is inserted without a collision into location 2. 16 collides at location 5. The record stored there, 27, is at its home address. Since 27 does not have a successor, we use its increment of two to find a place for inserting 16. We then look at location 7; since it is occupied, we then consider location 9.

Note. We don't change increments when we encounter an occupied location; we continue with an increment of two until we find an empty location or until we find that the table is full.

Location 9 is empty; so 16 can be stored there. Since two additional locations needed to be searched, the pseudolink value for location 5 is *two* and the table becomes

	Key	nof
0		
1		
2	13	
3		
4		
5	27	2
6	28	2
7	18	1
8	29	
9	16	
10	39	

At this stage, we have inserted the common set of records that we have been using to compare the various collision resolution methods. The average number of probes needed to retrieve each record in the table at this stage is 1.4. This compares with the 1.7 for the binary tree method and the 1.4 for LISCH. Since LISCH and computed chaining give the same average retrieval performance for this set of data, can you think of an advantage other than performance of computed chaining over LISCH? Yes, the pseudolink field in computed chaining will take less storage than the complete address field in coalesced hashing with LISCH. We only need 2 bits to store the largest pseudolink value (that of two), whereas we would typically use 3 or 4 bytes to store a complete address using the pointer structure of a higher-level language. Even in its most compact form, we would need 4 bits to store the largest address (10) in the example table.

The next record, 38, collides with 27 at location 5. Since the link field at location 5 contains a two, we know that 27 has an immediate successor. We can locate the successor at address 9 in *only one additional probe* since we can compute its address. The record stored there, 16, does not have a successor since its pseudolink field is null. We use the increment of one associated with 16 to find an empty location for inserting 38; we eventually find one at address 0. Which pseudolink field do we need to set? Yes, the one associated with 16. Since we had to look at two additional locations, the pseudolink value is two. The table is now

	Key	nof
0	38	
1		
2	13	
3		
4		
5	27	2
6	28	2
7	18	1
8	29	
9	16	2
10	39	

Only three probes are necessary to *retrieve* the record with key 38.

We finally add the record with key 53. It hashes to location 9. Notice that we have a different circumstance with this insertion. The record stored at location 9, that is, 16, is *not* stored at its home address. We know that since Hash(16) = 5 and not 9. So we observe a *move* process. We move 16 *plus* all its successors. Otherwise we would no longer be able to retrieve them. Are there any successors to 16? Yes, but how do we know? The nonnull pseudolink field tells us that there is an immediate successor for 16, and using the value stored in that field, we can compute its address and go there directly, that is, to location 0. The record stored at location 0, 38, does not have any immediate successors since its pseudolink field is null. We then need to temporarily store the records with keys 16 and 38 so that we can reinsert them later. By moving 16, the record with key 53 may be inserted directly at location 9 so that only one retrieval probe will be necessary to locate it. We now need to reinsert the two records that were displaced by the insertion. We need to keep a pointer or remember location 9 so that we can find the end of the chain that those two records were on previously; we need to identify the immediate predecessor to the item that was removed from location 9 so that we can relink it to the new location for the removed record. We know where to begin the search by hashing 16 which yields location 5. We search from that location until we find a pseudolink to location 9. We find that immediately at location 5. In general, the search for the immediate predecessor of a removed item may be represented as

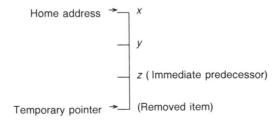

We do not know how many items in the probe chain precede the item that was removed. We begin the search at the home address which contains a record, for example, x, and continue the search until we find a record, in this case z, that points to the location of the removed item which is indicated with the temporary pointer. We now know the location of the pseudolink field that needs to be modified in the reinsertion process.

In returning to the insertion example, we then need to use that increment associated with 27 until we can find another empty space to insert the record with key 16. We eventually find that at location 0. How many positions had to be searched in going from address 5 to address 0 in increments of two? That would be three, or we could apply the fact that the pseudolink was previously two to locate position 9, and we need to go one more offset to locate position 0; so that would yield a pseudolink value of three. We now need to reinsert 38 from location 0 using the increment of one associated with the record stored there, that is, 16. We find an empty slot at address 1. 38 is inserted there and the pseudolink field at location 0 is set to one. The final table becomes

	Key	nof
0	16	1
1	38	
2	13	
3		
4		
5	27	3
6	28	2
7	18	1
8	29	
9	53	
10	39	

Did we degrade the performance for retrieving 16 by moving it? **No**, before inserting 53, two retrieval probes were required to locate 16. That same number is required after inserting 53. We essentially have a linked list such that the actual physical location of a record is immaterial. Only one actual probe is needed to go from a record to its immediate successor.

Discussion

The advantage of computed chaining is that we obtain better performance than with the nonlink collision resolution methods without the space requirements of coalesced hashing. With coalesced hashing, the link field must contain enough bits to designate any location in the table. With computed chaining, the pseudolink field is normally only of a size sufficient to store the largest possible pseudolink value. But what if we do not know what that largest value is? Or what if only a limited, fixed number of bits is available for the pseudolink field? Computed chaining allows the size of the pseudolink field to be any number of bits, but that number is the same for all entries in the table; if insufficient bits exist to store the actual pseudolink value, its greatest factor that can fit into the given number of bits is stored instead. Storing a factor instead of the full value will, however, increase the number of retrieval probes. For example, if the pseudolink value for record A should be six but only two bits are available for storing the pseudolink, then its greatest factor, that is, three, would be stored. Instead of needing only one probe to find the successor to A, we would need two, that is,

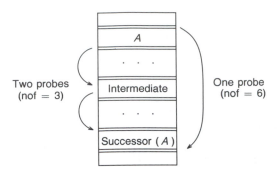

The effects of varying the size of the pseudolink field are considered in the **Comparison** section of this chapter. See [14] for the implementation details when insufficient bits exist for storing the full pseudolink value.

With computed chaining, the number of probes required to *retrieve* a record is equal to the position of that record in its probe chain. The worst case retrieval performance is *bounded* by the number of records that collide at any one home address. The other collision resolution methods that we considered have worst case performances based upon the placement of the records inserted previously into the table. The worst case performance could be quite poor depending upon the order of insertion and key values of the records. In some applications, such as those performed in real time, a bounded worst case performance is essential. With computed chaining, the worst case retrieval performance can be improved by choosing a hashing function that evenly distributes the probable addresses for the given data. A hashing function that provides an even distribution for one type of data may not for another.

An unsuccessful search terminates when a Λ or null value, signifying the end of the probe chain, is encountered in the pseudolink field of a node in the probe chain originating with the home address of the unsuccessfully sought item. As with the other collision resolution methods, in computed chaining also, an unsuccessful search takes more time than a successful one because the entire probe chain needs to be searched before termination.

To *delete* a record from a table formed using computed chaining requires the reinsertion of *all* the records that follow the deleted record on that probe chain.

Computed chaining, as with several of the techniques that we have considered or will consider, requires more time for inserting and deleting so that less time may be needed for retrieval. The rationale is that we insert or delete once but retrieve often.

Variant

One theme of this text is that it is important to continually seek alternative and potentially better solutions. Just because we have a solution doesn't mean that it cannot be improved upon. Can computed chaining be improved upon? We will consider a modification of computed chaining that illustrates a general principle that we will see many times again.

When we search for a record, we want to complete the task as quickly and efficiently as possible. You want to do the same thing in your own day-to-day activities. If you are searching for this textbook among all your possessions, you want the search

with pseudolinks

to take as little time as possible. If you only knew in which room of your home that you left the book, you could then confine your search to that one room. Searching only a single room is certainly faster than searching the entire house. We want to apply such a *divide and conquer* strategy to retrieving data also. Instead of searching an entire file of records or an entire probe chain of records, if we could *divide* the records into smaller groups, we could narrow the search considerably if we knew the *one* group that the record must be in if it were in the file. We could *conquer* the retrieval task by searching a more manageable group of records.

Nishihara and Ikeda [15] introduced the concept of having *multiple* probe chains instead of a single probe chain for organizing records. Their reasoning is that if several probe chains exist for the records that have the same home address then each chain will be smaller and the searching will be faster. For example, instead of searching a chain of eight records, it would be easier to search *one* subset of the chain that contained only two or three records.

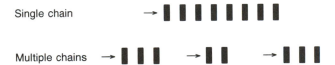

We observed this same concept when we considered coalesced hashing. In that situation, it was preferable to eliminate the coalescing so that we would have several shorter chains to search rather than one long coalesced chain. We now apply this divide and conquer concept to computed chaining.

In addition to the hashing function, Hash(key), to obtain the home address for a record and the incrementing function, $i(\text{key})$, to obtain an increment to locate another probable address, we introduce a *third* hashing function, $g(\text{key})$, to tell us which probe chain to insert into or to search.

$$g(\text{key}) \rightarrow 0, 1, \ldots, R - 1$$

where R is the number of subgroupings. A simple function for $g(\text{key})$ is

$$g(\text{key}) = \text{key } \textbf{mod } R$$

An Example with Multiple Chains

Let us go through the previous example but use two chains this time instead of one. The keys again are

$$27, 18, 29, 28, 39, 13, 16, 38, \text{ and } 53$$

with a table size of eleven. The hashing function,

$$\text{Hash(key)} = \text{key } \textbf{mod } 11$$

and the incrementing function,

$$i(\text{key}) = \text{Quotient(Key / 11) } \textbf{mod } 11$$

remain the same. The chain selection function will be

$$g(\text{key}) = \text{key } \textbf{mod } 2$$

If a key is even, it will go on the zeroth chain, if it is odd, it will go on the first chain. The records are inserted as before with the only difference being the placement of the pseudolink values: in chain zero for even key values and in chain one for odd key values. The table then appears as it did previously except for the placement of the pseudolinks.

	Key	nof 0	1
0	16	1	
1	38		
2	13		
3			
4			
5	27	3	
6	28		2
7	18		1
8	29		
9	53		
10	39		

with pseudolinks

At this point in the example, we have gained nothing in retrieval performance since all the records that collided at a particular address mapped into one or the other of the chains, so having two chains was not necessary.

To illustrate the advantage of multiple chains, let's add a record with **key 49**. This record collides at location 5, but since it is odd it goes to chain one instead of chain zero. It is inserted into location 4 and the nof field at location 5 for chain *one* is set to a value of five. The final table is then

	Key	nof 0	nof 1
0	16	1	
1	38		
2	13		
3			
4	49		
5	27	3	5
6	28		2
7	18		1
8	29		
9	53		
10	39		

The advantage of having multiple chains is that only *two* probes are necessary to retrieve 49, whereas *four* probes would have been required with the original example containing only one chain. 3 bits would be necessary in chain 1 to store the pseudolink value of 5.

The advantage of the multiple probe chains is that they reduce the number of retrieval probes needed to locate a record by dividing the one long chain into several smaller ones. The disadvantage, of course, is that the extra pseudolink fields require more space. On the plus side though, a pseudolink requires less space than an actual link. The use of multiple chains is an example of triple hashing: first to obtain the home address, second to get the increment, and third to obtain the chain number.

KEY TERMS

Actual probe
Applications
    Real time
Collision resolution
    Computed chaining
Computed chaining
    Deleting
Computing information
Divide and conquer
Immediate successor
Move process
Multiple chains
    Computed chaining

Number of offsets
Performance
    Computed chaining
    Worst case
Probe chains
    Multiple
Pseudochaining
Pseudolink
Triple hashing
    Computed chaining
Unsuccessful search
    Computed chaining

with pseudolinks

EXERCISES

1. Insert records with the keys

    73, 15, 44, 37, 30, 59, 49, and 99

   into the smallest table with a packing factor $\leq$ 85 percent using the key **mod** P hashing function and the computed chaining collision resolution method.

2. What is the average number of probes needed to retrieve each of the records inserted in Problem 1?

3. Repeat Problem 1 using two chains instead of one.

4. What is the average number of probes needed to retrieve each of the records inserted in Problem 3?

5. Repeat Problem 3 but insert a record with key 48 instead of 59.

6. What is the average number of probes needed to retrieve each of the records inserted in Problem 5?

7. How many bits are necessary to contain the nof field for the file in Problem 5?

8. Modify Algorithm 3.6 to handle multiple chains.

9. Write an algorithm for retrieving a record from a file that was created using computed chaining.

10. What are the tradeoffs of a modification to the computed chaining insertion procedure in which instead of moving a record *plus* all its successors, only the record is moved and the pseudolink value of its predecessor is recalculated to point to the relocated record's successor element?

11. What is the greatest factor of 27 that will fit into two bits?

COMPARISON OF COLLISION RESOLUTION METHODS

By now you may know more about collision resolution methods than you ever cared to. But the purpose of studying such a variety of methods was not merely to explain the concept of collision resolution or to provide you with a procedure to use in the future; that could have been accomplished in a method or two. Instead, the purpose was to illustrate, in depth, with one topic, how changing data structures can have significant effects on software performance. In addition, you were exposed to a wide variety of techniques, or "tools," that may help you to become a better software developer in the future. The individual collision resolution methods that were presented are not as important as the problem solving exercise and the structuring techniques used to improve their performance. Many of the ideas introduced in this section are general and can be applied to many different situations. In fact we will see several of the techniques applied repeatedly in the remainder of the text but in quite different contexts.

   Having considered several collision resolution methods then, we now want to

TABLE 3.3 COMPARISON OF MEAN NUMBER OF PROBES FOR SUCCESSFUL
LOOKUP (n = 997; α = packing factor)

α (percent)	DCWC (10-bit link)*	LISCH (10-bit link)	Progressive overflow†	Linear quotient	Brent's method	Binary tree	Computed chaining (6-bit link)	(2-bit link)
20	1.100	1.106	1.125	1.15	1.102	1.102	1.070	1.070
40	1.200	1.232	1.333	1.277	1.217	1.217	1.168	1.214
60	1.300	1.379	1.750	1.527	1.367	1.364	1.264	1.381
70	1.350	—	2.167	1.720	1.444	—	1.323	1.528
80	1.400	1.566	3.000	2.011	1.599	1.579	1.356	1.715
90	1.450	1.674	5.500	2.558	1.802	1.751	1.408	2.062
95	1.475	1.734	10.500	3.153	1.972	1.880	1.433	2.414
99	1.495	1.783	50.500	4.651	2.242	2.049	1.601	3.330
100	1.500	—	—	6.522	2.494	2.134	—	—

*Theoretical results; mean probes = $1 + \alpha/2$.
†Theoretical results; mean probes = $(1 - \alpha/2) / (1 - \alpha)$

compare them so that you can readily decide which method to use when. You may consider this section the *Consumer Reports*[1] of collision resolution methods. When you are shopping for a new automobile or refrigerator, you want considerable amounts of information in a condensed form so that you can make a wise and quick determination. This section condenses much of the information that you have already seen in this chapter and introduces a few new comparison metrics.

Table 3.3 provides the average number of retrieval probes for *successful* searches on a table of 997 records with a uniform distribution of keys. Figure 3.11 graphically displays this performance data for all methods except for computed chaining with a 2-bit link field. The average number of probes is a useful metric since each probe is a relatively costly access of secondary storage in many applications. The fewer the probes the better performance that the method will have. Especially notice the wide variance in performance at packing factors $\geq$ 90 percent. Although the differences may not seem that great for a single retrieval, multiply those numbers by 10,000 or 1,000,000. For the methods that require links, the number of bits used for the link field is given. 10 bits are needed to address 997 locations. Experimental results are used as the primary comparison standard for most of the methods since theoretical analyses do not always produce a function that both models the method and readily yields an answer. The result of computed chaining with a 20 percent packing factor is less than that for DCWC (direct chaining without coalescing). An explanation is that the computed chaining result is experimental whereas DCWC is theoretical; Lum, Yuen and Dodd [16] indicate a similar performance with the linear quotient method.

Table 3.4 offers additional useful comparison criteria. The successful search criteria give the minimum and maximum number of probes necessary to retrieve an item. As you would expect, the best that we can do is to not have a collision that would

[1]Copyright, Consumers Union of United States Inc.

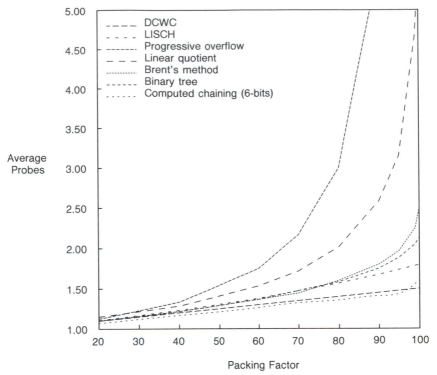

Figure 3.11   Performance of collision resolution methods.

mean only one probe is needed for retrieval. The information on worst case search performance may be useful for real time applications. It may be that an application may have a limit on the amount of time that can be expended in retrieving a record, for example, in a manufacturing application in which a product must be processed in a fixed amount of time. The range for worst case performance varies from ln n to n. Although the worst case performance for locating a record with both LISCH and computed chaining is n, their typical performances would be better, because only records on one chain need to be searched. With computed chaining, all the records on a chain must have the same home address, which further limits the size of the chain. Whether or not a method moves an item once it has been stored can be important if other pieces of information reference it. The possibility of moving a record may dictate what type of referencing mechanism is used, for example, actual addresses or key values. And finally, the need for extra storage may eliminate from consideration those methods which require it in applications where space is at a premium.

What is the *best* method? Unfortunately, there is no *one* best method for *all* purposes. Instead, we need to consider the constraints and goals. From Table 3.3, the method that provides the lowest average number of probes, and thus the best performance, in general, is DCWC. The method with the second lowest average number of retrieval probes is computed chaining. What is the primary reason for computed chaining giving better results over LISCH? Computed chaining eliminated coalescing

TABLE 3.4 SEARCH, RELOCATION, AND STORAGE COMPARISONS

Criteria	LISCH	Progressive overflow	Linear quotient	Brent's method	Binary tree	Computed chaining
Successful search						
Best case	1	1	1	1	1	1
Worst case	n	n	n	n	n	n
Move an item	No	No	No	Yes	Yes	Yes
Extra storage	Yes	No	No	No	No	Yes

whereas LISCH did not. Without coalescing, LISCH is DCWC and does perform better than computed chaining. When the packing factor becomes very large, that is, $\geq$ 99 percent, DCWC outperforms computed chaining with 6 bits for the pseudolink field. The explanation for this is that with a table of 997 locations, 6 bits are insufficient for storing all possible pseudolink values. For a record that would normally have a very large number of locations visited in locating its successor, it may be necessary to store a factor of the number of offsets rather than the actual number because of insufficient bits. Such a situation then requires more probes on retrieval. That is the reason that computed chaining with 6 bits outperforms computed chaining with 2 bits. The advantage of computed chaining with 6 bits over DCWC is that it needs less space for the link field (6 bits versus at least 10). If storage is somewhat scarce, computed chaining will then have an advantage over DCWC.

If space is definitely limited, the method that gives the best performance without requiring additional storage for the records is the binary tree method. It however requires more processing time and temporary storage for insertions than do the other methods that do not use a link or pseudolink field.

Remember the reason that a method that requires more processing time at insertion is viable and often the preferred method is that retrieval of a record from auxiliary storage is an order of magnitude greater than retrieval of a record from primary memory, *and* we insert once but retrieve often.

Before examining the final comparison table (Table 3.5), see if you can construct one on your own. It contains the advantages, disadvantages and "when to use" data for the collision resolution methods that we have considered. This information is quite important when one is confronted with a situation where a decision must be made on which method to use. Knowing how a method works is not enough, we also need to know when to apply it. After finishing your table, compare it with the one following. Note any difference between your table and the subsequent one and try to understand why they are different. Is it because of a different interpretation, or did you make different assumptions, or was it the result of a misunderstanding?

Not only has it been useful to be introduced to several collision resolution methods so that an optimum one may be chosen for a subsequent application, but it has also given us an opportunity to sharpen our problem solving skills and to observe a variety of data structuring techniques. It is difficult to say when one of these techniques may be just the right "tool."

TABLE 3.5 ADVANTAGES, DISADVANTAGES, AND WHEN TO USE VARIOUS COLLISION RESOLUTION METHODS

Method	Advantages	Disadvantages	When to use
Coalesced hashing	Excellent performance	Requires space for link field	When ample space is available
[DCWC]	Superior performance	Requires space for link field	When ample space is available and insertion time is not critical
Progressive overflow	No additional space; simple algorithm	Very poor performance	Almost NEVER
Linear quotient	No additional space; reasonable performance	Performance below that of other methods	When space is at a premium; retrieval performance is not critical
Brent's method	No additional space; good performance	Performance below that of linked methods	When space is at a premium, even space for a binary tree
Binary tree	No additional space; very good performance	Performance below that of linked methods	When file space is limited but temporary space is available for tree; performance is important
Computed chaining	Small amount of additional space; excellent performance	Requires some space for pseudolink field; performance below that of DCWC	When some extra file space is available and performance is important; insertion time is not critical

KEY TERMS

Binary tree
Brent's method
Coalesced hashing
Collision resolution
    Comparison
Computed chaining
DCWC
Direct chaining
    Without coalescing

Hashing
    Moving an item
    Storage
Linear quotient
LISCH
Progressive overflow

EXERCISES

1. Which of the collision resolution techniques use double hashing?

2. Can a direct-access file be accessed sequentially? If yes, would it take more *or* less time than a sequentially organized file?

3. Can a sequentially organized file be accessed directly? If yes, would it take more *or* less time than with a direct-access file?

4. Write an algorithm for direct chaining without coalescing.

5. Rank the collision resolution algorithms considered in this chapter in terms of algorithm simplicity. List the methods from simplest to least simple based upon (a) the ease of insertion, (b) the ease of deletion, and (c) the ease of locating a record.

6. Classify the collision resolution algorithms of this chapter in terms of **(a)** insertion costs (from most costly to least costly), and **(b)** retrieval costs (from least costly to most costly). Assume that the same amount of storage is available for each method and that the keys of the records stored have an even distribution.

7. All the collision resolution methods that were discussed assumed an equal likelihood of each record's being retrieved. How might you modify your use of each of these collision resolution methods in a case in which the chances for retrieval of the records are not equally likely *and* in which you have knowledge of the relative frequencies of retrieval in advance of storing the records? Note those methods which can be readily used in such a way to take advantage of this a priori information and those which cannot.

PERFECT HASHING

With the adjective "perfect" describing this type of hashing, one would assume that there is something special about it. What makes perfect hashing distinctive is that it is a process for mapping a key from a key space to a *unique* address in a smaller address space, that is,

$$\text{hash(key)} \rightarrow \textbf{unique address}$$

Previous hashing functions yielded a *probable* address. Since perfect hashing yields a unique address for each key, both primary and secondary clustering are eliminated. In considering the various collision resolution techniques, we have sought to improve retrieval performance by reducing the average number of probes necessary to retrieve a record. With perfect hashing, we need only a single probe to retrieve a record that in turn eliminates the need for a collision resolution technique. Does that mean that all the techniques for collision resolution that we have considered previously are for naught? Has the ultimate in hashing functions been discovered? Unfortunately, that is not the case, for perfect hashing is currently applicable to files with only a relatively small number of records. Thus with most of your applications, you will not be able to use perfect hashing. Where it is applicable is in storing a set of items that are static and frequently accessed, for example, reserved words in a programming system, or the most common words in a language for use in a spelling checker. (In a later chapter, we examine a technique called signature hashing that organizes a file of any number of records such that a record may be retrieved in only a single access of auxiliary storage.)

> A **perfect hashing function** maps a key into a unique address. If the range of potential addresses is the same as the *number* of keys, the function is a minimal (in space) perfect hashing function.

Not only does a perfect hashing function improve retrieval performance, but a minimal perfect hashing function would provide 100 percent storage utilization. *The challenging aspect of perfect hashing is discovering such functions.* Several procedures have been proposed for constructing perfect and minimal perfect hashing functions [17, 18, 19, 20, and 21].

Note. A separate hashing function needs to be devised for *each* set of keys. If one or more of the keys change, a new hashing function must be constructed.

Let's examine the process for creating a perfect hash function. For a table of size N, Sager [17] characterizes the general form of a perfect hashing function as

$$\text{p.hash(key)} = (h_0(\text{key}) + g[h_1(\text{key})] + g[h_2(\text{key})]) \textbf{ mod } N$$

What needs to be decided are the functions h_0, h_1, h_2, and g. These component functions should be efficient, because they will be executed frequently. The reason that perfect hashing is only applicable to sets with a small number of records is that the computing time needed to determine these component functions is considerable. The worst case computational complexity of Sager's minicycle algorithm, which is the most efficient method to date, is $O(r^6)$ where r is the number of records in the set to be stored. With the computational limitations of current computers, an upper bound for r seems to be ≈ 512. Rather than consider Sager's minicycle algorithm, which is based upon graph theory, we look at a special case of it that was developed earlier by Cichelli[18]. Cichelli's algorithm is not as efficient as that of Sager; it is only appropriate for an r value up to 60, plus it has other disadvantages that we will mention later. But its strong suit is that it is straightforward to understand. *Simplicity* is certainly an important criterion when choosing a data structure or processing technique. For larger amounts of data or in situations in which Cichelli's algorithm is not applicable, the minicycle algorithm would be appropriate.

From a practical standpoint, perfect hashing is not yet very useful because the file size must be small. In those cases, the entire file can probably be stored in primary memory and the savings in probes would be main memory probes rather than auxiliary storage probes. A reduction in probes of primary memory would not affect performance as much as a reduction in the number of probes of auxiliary storage. Nevertheless, the size of the sets for which perfect hashing is applicable has steadily increased over time.

Cichelli's Algorithm

In Cichelli's algorithm, the component functions are

$$h_0 = \text{length(key)}$$

$$h_1 = \text{first\_character(key)}$$

$$h_2 = \text{last\_character(key)}$$

and
$$g = T(x)$$

where T is a table of values associated with individual characters x which may appear in a key. The time-consuming part of Cichelli's algorithm is determining T. Constructing Table 3.6 for the values associated with the characters that may appear in the 36 reserved words in Pascal required 40 minutes on a PDP-11/45. In the table, a value may be assigned to more than one character.

TABLE 3.6 VALUES ASSOCIATED WITH THE CHARACTERS
OF THE PASCAL RESERVED WORDS

A = 11	B = 15	C = 1	D = 0	E = 0	F = 15
G = 3	H = 15	I = 13	J = 0	K = 0	L = 15
M = 15	N = 13	O = 0	P = 15	Q = 0	R = 14
S = 6	T = 6	U = 14	V = 10	W = 6	X = 0
Y = 13	Z = 0				

To demonstrate how Cichelli's perfect hashing function works, let's apply it to the keyword **begin** using this table,

$$
\begin{aligned}
\text{p.hash}(\textbf{begin}) &= 5 + T(h_1(\text{key})) + T(h_2(\text{key})) \\
&= 5 + T(b) + T(n) \\
&= 5 + 15 + 13 \\
&= 33
\end{aligned}
$$

The keyword **begin** would be stored in location 33. Since the hash values run from 2 through 37 for this set of data, the hash function is a *minimal* perfect hashing function.

The algorithm for constructing T appears in Algorithm 3.7. It involves ordering the keys based upon the frequencies of occurrence of the first and last characters in the keys. Assignments of values are made to the characters in the first and last positions of the keys from the top of the ordering to the bottom. The algorithm uses an exhaustive search with backtracking. If one assignment of a value yields a conflict or collision with a previously processed key, another value is tried. When all possibilities for the current key have been tried, the processing backs up to the previously tried key. Processing continues with the next alternative value for a character in the previous key. The algorithm terminates when all the characters have been assigned values or when all possible assignments have been tried, in which case there is no solution using the algorithm.

An Example

To help you to better understand how values are assigned to characters using Cichelli's algorithm and to elaborate on it, let's work through an example using the following names of creatures as key data:

cat, ant, dog, gnat, chimp, rat, toad

We assume that the maximum value that may be assigned to a character is four. If we cannot find a solution using four, we would need to try a larger value. If this maximum value is too small, we will either have no solution or a great amount of backtracking. If the value is too large we will not obtain a minimal solution.

The frequencies of occurrence of the first and last characters are

$$
a = 1 \quad c = 2 \quad d = 2 \quad g = 2 \quad p = 1 \quad r = 1 \quad t = 5
$$

Algorithm 3.7
CICHELLI'S ALGORITHM FOR DETERMINING A PERFECT HASH FUNCTION

I. Order the set of keys based upon the sum of the frequencies of occurrence of the first and last characters of the keys in the entire set. Place the keys with the most frequently occurring first and last characters at the top of the ordering.

II. Modify the ordering. Place a key whose first and last characters both appear in prior keys as high in the ordering such that this condition holds.

III. Assign the value 0 to the first and last characters of the first key in the set.

IV. For each remaining key in the set, in order,

    A. If the first and last characters have already been assigned values, hash the key to determine if a conflict arises. If yes, discontinue processing on this key and recursively apply this step with the previous key (backtrack).

    B. If either the first *or* last character has already been assigned a value, assign a value to the other character by trying all possibilities from 0 to the maximum allowable value. If a conflict still exists after trying all possible values for the other character, then discontinue processing on this key and recursively apply this step with the previous key (backtrack).

    C. If both the first and last characters are as yet unassigned, vary the first and then the last character, trying each combination. If all combinations cause conflicts, then discontinue processing on this key and recursively apply this step with the previous key (backtrack).

V. If all keys have been processed, terminate successfully, else terminate unsuccessfully.

Using these values, we can compute the sums of the frequencies of occurrence of the first and last characters of each key. We obtain

cat	7
ant	6
dog	4
gnat	7
chimp	3
rat	6
toad	7

We next order the keys in descending order based upon the sum of the frequencies to obtain

toad	7
gnat	7

$$\begin{array}{ll}
\underline{\text{cat}} & 7 \\
\underline{\text{rat}} & 6 \\
\underline{\text{ant}} & 6 \\
\underline{\text{dog}} & 4 \\
\underline{\text{chimp}} & 3
\end{array}$$

The ordering of the keys within a frequency is immaterial but for this example we arbitrarily chose the descending order of the keys; a random ordering would have been equally acceptable. The next step is to check the ordering to see if any keys exist that have both their first and last characters appearing in previous keys. Notice that the d and g of "dog" have appeared previously so that "dog" would then be moved to the position following "gnat." The ordering then becomes

$$\begin{array}{ll}
\underline{\text{toad}} & 7 \\
\underline{\text{gnat}} & 7 \\
\underline{\text{dog}} & 4 \\
\underline{\text{cat}} & 7 \\
\underline{\text{rat}} & 6 \\
\underline{\text{ant}} & 6 \\
\underline{\text{chimp}} & 3
\end{array}$$

The ordering of the keys prior to the assignment of the values causes the inescapable conflicts to occur as early as possible in the assignment process. We are now ready to begin assigning values to the characters. T and d are initially assigned the value of zero. With those values assigned, "toad" would hash to location 4.

$$t = 0 \quad d = 0$$

$$\text{p.hash(toad)} = 4$$

In continuing the assignment process, if we assign the g in "gnat" a zero, we then have a conflict, for it would also hash to location 4. We next try a value of one for g, which would cause "gnat" to hash to location 5.

$$t = 0 \quad d = 0 \quad g = 1$$

$$\text{p.hash(toad)} = 4$$
$$\text{p.hash(gnat)} = 5$$

Since the first and last letters of "dog" have previously been assigned values, "dog" hashes to location 4. This result causes a conflict. We then backtrack to the previous key, that is, "gnat," and try the next value for g, which would be a two. We would then have

$$t = 0 \quad d = 0 \quad g = 2$$

$$p.hash(toad) = 4$$
$$p.hash(gnat) = 6$$

Continuing in a forward direction, "dog" now hashes to location 5, so there is no conflict. The assignments become

$$t = 0 \qquad d = 0 \qquad g = 2$$

$$p.hash(toad) = 4$$
$$p.hash(gnat) = 6$$
$$p.hash(dog) = 5$$

We continue and obtain

$$t = 0 \qquad d = 0 \qquad g = 2 \qquad c = 0$$

$$p.hash(toad) = 4$$
$$p.hash(gnat) = 6$$
$$p.hash(dog) = 5$$
$$p.hash(cat) = 3$$

We need to assign r a value of four to avoid a conflict. We now have

$$t = 0 \qquad d = 0 \qquad g = 2 \qquad c = 0 \qquad r = 4$$

$$p.hash(toad) = 4$$
$$p.hash(gnat) = 6$$
$$p.hash(dog) = 5$$
$$p.hash(cat) = 3$$
$$p.hash(rat) = 7$$

However, we are not able to assign a value to a without causing a collision, so we must backtrack. Eventually after going forward and then backtracking again we get back to the point at which g is assigned a value. We then try $g = 3$. Continuing in a forward direction we would obtain

$$t = 0 \qquad d = 0 \qquad g = 3 \qquad c = 0 \qquad r = 2$$

$$p.hash(toad) = 4$$
$$p.hash(gnat) = 7$$
$$p.hash(dog) = 6$$
$$p.hash(cat) = 3$$
$$p.hash(rat) = 5$$

Again we have a problem when attempting to assign a value to a without a conflict; we have to backtrack and eventually reach the point at which we try a new value for g. This time we assign a value of four to g and then after several steps obtain

$$t = 0 \qquad d = 0 \qquad g = 4 \qquad c = 0 \qquad r = 2 \qquad a = 3$$

p.hash(toad) = 4
p.hash(gnat) = 8
p.hash(dog) = 7
p.hash(cat) = 3
p.hash(rat) = 5
p.hash(ant) = 6

The final solution is

$$t = 0 \qquad d = 0 \qquad g = 4 \qquad c = 0 \qquad r = 2 \qquad a = 3 \qquad p = 4$$

p.hash(toad) = 4
p.hash(gnat) = 8
p.hash(dog) = 7
p.hash(cat) = 3
p.hash(rat) = 5
p.hash(ant) = 6
p.hash(chimp) = 9

Since this solution maps the seven keys into seven consecutive locations, it is a minimal perfect hashing function.

Discussion

As you can observe from the previous simple example, the amount of backtracking in Cichelli's algorithm can be considerable, which in turn will make the computation time substantial. On the positive side, though, the parameter values of a minimal perfect hashing function need to be computed only once for a given set of keys. Therefore, if the keys are to be accessed frequently, especially in real time situations, it might be worthwhile to spend the extra time initially to build a minimal perfect hashing function.

The disadvantages of Cichelli's algorithm are that:

- It requires that no two keys of the same length share the first and last characters. (Is it clear why that is necessary?)
- For a list with more than 45 elements, it may be necessary to segment it into sublists.
- For certain lists, it may be impossible to tell in advance if the method will yield a minimal perfect hashing function.

As previously mentioned, when Cichelli's algorithm is not applicable, the minicycle algorithm described in [17] could be used.

An observation on Cichelli's algorithm is that it is perhaps the first one that we have encountered thus far in the text in which for simple problems, it is probably easier for a person to write a program to let a computer execute the algorithm rather than for a person to do it directly himself.

As the maximum applicable file size for minimal perfect hashing increases, this technique will become more useful. Note that this maximum has increased from 60 to 512 in 5 years. But even as the applicable file size increases, the other hashing techniques that we have considered will still have their place, for minimal perfect hashing requires a *static* set of keys, and in many circumstances that requirement cannot be met. We continue to develop new tools and techniques, but they do not always replace existing ones completely. We are continually expanding our set of tools so that we can use just the right one in the proper place.

KEY TERMS

Backtracking	Minicycle algorithm
Cichelli's algorithm	Minimal perfect hashing function
Exhaustive search	Perfect hashing function

EXERCISES

1. (a) Construct a perfect hashing function using Cichelli's algorithm with the data in the example, but use a maximum value of five for assigning to characters. **(b)** Is the result minimal?

2. (a) Construct a perfect hashing function using Cichelli's algorithm with the data from [17] of AA, AB, BAA, BB, AAD and FA using a maximum value of five for assigning to characters. **(b)** Is the result minimal?

3. (a) Construct a perfect hashing function with Cichelli's algorithm using as keys the first three letters of the months in the first half of a year. Use a maximum value of four for assigning to characters. Change to $h_1 = $ second—character(key). **(b)** Why is this modification of h_1 necessary?

4. (a) Why is perfect hashing currently of limited practical use? **(b)** What is its promise?

REFERENCES

1. Horowitz, Ellis, and Sartaj Sahni, *Fundamentals of Data Structures*, Computer Science Press, Rockville, MD, 1982.
2. Reingold, Edward M., and Wilfred J. Hansen, *Data Structures in Pascal*, Little, Brown and Company, Boston, MA, 1986.
3. Knuth, D. E., *Sorting and Searching*, Addison-Wesley, Reading, MA, 1973.
4. Williams, F. A., "Handling Identifiers as Internal Symbols in Language Processors," *CACM*, Vol. 6, No. 6 (June 1959), pp. 21–24
5. Vitter, J. S., "Analysis of the Search Performance of Coalesced Hashing," *JACM*, Vol. 30, No. 2 (April 1983), pp. 231–258.
6. Hsiao, Yeong-Shiou, and Alan L. Tharp, "Analysis of Other New Variants of Coalesced Hashing," Technical Report TR-87-2, Computer Science Department, North Carolina State University, 1987.
7. Chen, W., and Jeffrey Vitter, "Analysis of New Variants of Coalesced Hashing," *ACM TODS*, Vol. 9, No. 4 (December 1984), pp. 616–645.

8. Peterson, W. W., "Addressing for Random-Access Storage," *IBM J. Res. Dev.*, Vol. 1 (1957), pp. 130–146.

9. Bell, J. R., and C. H. Kaman, "The Linear Quotient Hash Code," *CACM*, Vol. 13, No. 11 (November 1970), pp. 675–677.

10. Brent, R. P., "Reducing the Retrieval Time of Scatter Storage Techniques," *CACM*, Vol. 16, No. 2 (February 1973), pp. 105–109.

11. Tharp, Alan L., "Further Refinement of the Linear Quotient Hashing Method," *Information Systems*, Vol. 4, No. 1 (1979), pp. 55–56.

12. Gonnet, Gaston, and J. I. Munro, "Efficient Ordering of Hash Tables," *SIAM J. Computing*, Vol. 8, No. 3 (August 1979), pp. 463–478.

13. Halatsis, C., and G. Philokyprou, "Pseudochaining in Hash Tables," *CACM*, Vol. 21, No. 7 (July 1978), pp. 554–557.

14. Tai, K. C., and A. L. Tharp, "Computed Chaining—A Hybrid of Direct Chaining and Open Addressing," *Information Systems*, Vol. 6, No. 2 (1981), pp. 111–116.

15. Nishihara, S., and K. Ikeda, "Reducing the Retrieval Time of Hashing Method Using Predictors," *CACM*, Vol. 26, No. 12 (December 1983), pp. 1082–1088.

16. Lum, V. Y., P. S. Yuen, and M. Dodd, "Key-to-Address Transform Techniques: A Fundamental Performance Study on Large Existing Formatted Files," *CACM*, Vol. 14, No. 4 (April 1971), pp. 228–239.

17. Sager, Thomas J., "A Polynomial Time Generator for Minimal Perfect Hash Functions," *CACM*, Vol. 28, No. 5 (May 1985), pp. 523–532.

18. Cichelli, Richard J., "Minimal Perfect Hash Functions Made Simple," *CACM*, Vol. 23, No. 1 (January 1980), pp. 17–19.

19. Sprugnoli, Renzo, "Perfect Hashing Functions: A Single Probe Retrieving Method for Static Sets," *CACM*, Vol. 20, No. 11 (November 1977), pp. 841–850.

20. Jaeschke, G., "Reciprocal Hashing: A Method for Generating Minimal Perfect Hashing Functions," *CACM*, Vol. 24, No. 12 (December 1981), pp. 829–833.

21. Chang, C. C., "The Study of an Ordered Minimal Perfect Hashing Scheme," *CACM*, Vol. 27, No. 4 (April 1984), pp. 384–387.

Chapter 4

Indexed Sequential File Organization

BACKGROUND

We have seen that a sequential file organization functions well for sequential access and that direct file organization works well for direct access, but the reverse is just the opposite. Direct access on a sequentially organized file gives poor performance and sequential access on a directly organized file is slow and/or requires additional storage to provide an ordering of the records. We now consider indexed sequential file organization. It is a hybrid of sequential and direct file organization and is suitable for those applications in which we must access data both directly and sequentially. It is useful when we need to search a single record at some times and when we need to look at many records at other times. Indexed sequential file organization provides something for everyone. If you are an optimist, you will observe that it facilitates better performance for sequential access than a directly organized file and better performance for direct access than a sequentially organized file. If you are a pessimist, you will note that it takes longer for direct access than a directly organized file and longer for sequential access than a sequentially organized file. What these observations underscore is that indexed sequential is *only* for those applications in which we need *both* types of access.

What is the problem in performing direct access on a sequentially organized file? On the average, we have to search half of the records to locate the one that we want. What do we do when we have a great deal of information to access? A sequential search of the information would be too slow. Inherently, the information may appear to have a sequential form to it as

To accelerate the search, we *order* the information and put tabs or an *index* to groups of like information, for example, file folders, so that the information appears as

where the ǀ represents a tab into the information. Then instead of searching the records *individually,* we search the *tabs* until we find the group that the desired record should be in. We then need to search exhaustively *only* those records in that group. Since there are fewer tabs than records, the search is faster. We have all experienced this phenomenon. What do we do when we have too many groups of information to sequentially search the tabs efficiently? We apply this grouping process a second time and organize the current groups into larger ones, for example, file drawers. The information may then be represented as

where || represents the tabs at the next higher level. Now instead of searching a large number of tabs, we only have to search a much smaller number of higher-level tabs (five in the diagram). When we locate the major grouping that the sought record should be in, we then search the tabs of the information in that group. We next proceed as before to locate the specific group of information and then to locate the record. If we wish to locate the record ▓ in the previous diagram, we need only to search tabs *A, B,* and *C* at the higher level, and then tabs 6 and 7 at the next level. We continue by searching the records within group 7 until we locate ▓. We had to make a total of nine comparisons. If we did not have the higher level of the index, we would have had to make 11 comparisons. And if we didn't have an index at all, we would have had to make 38 comparisons. We have reduced the search effort by *organizing* the information and *eliminating* the need to search many of the records. We do the job faster by *doing less.* As the information expands, we can continually add levels to the index to maintain an efficient searching process. The tab or index structure is, as you have observed, a *tree* structure. In Part Three, we consider tree structures in detail. One of the advantages of the tree index is that we can narrow the search effectively. When we encounter a tab value at the higher level of the index that does *not* match what we are searching for, we do *not* need to search *any* of the information in the file below that tab value. What we are able to do is to *prune* the branches of the tree, and that leads to efficient searching.

How should we proceed if we wish to insert a record S whose ordered position would be just prior to ▓ in the previous diagram? Do we need to move all the records between the position of insertion and the end of the file? Obviously that would be inefficient. Instead, what we do is to associate an *overflow* area with each original or *primary* storage area for a group of records; so at most we only have to move the number of records in such a group. The previous diagram becomes

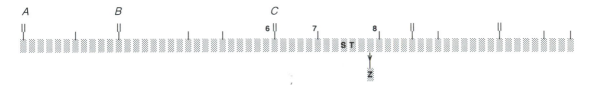

which illustrates that the records beyond the point of insertion in group 7 are moved one position to the right and the rightmost record, Z, is moved to an overflow location. By having the *overflow* areas associated with each grouping of records, the insertion performance is not degraded much. You can imagine the time it would take otherwise if we had a file with hundreds of thousands of records in it.

THE BASIC STRUCTURE

Within a computer, instead of talking about file folders and file drawers, we use the physical structures internal to a computer. Usually, the information that we need to store is too large to fit into primary memory; so we need to use an auxiliary storage device such as a disk to store the records. If we are using a computer with block addressable disks, we then use *tracks* as the lowest level of grouping information, the next higher level is a grouping by *cylinders,* and then we could have additional levels of *master* indexes. We limit this discussion to cylinder and track indexes. In the previous diagram, ‖ represents the cylinder index and a | represents the track index. A file organized with an indexed sequential structure is commonly referred to as an ISAM (for indexed sequential access method and pronounced i-sam) file. An access method is a system program that manages the transferring of data between an application program and the computer system.

An Example

Let's consider an example using an indexed sequential file organization with block addressable formatting of auxiliary storage. We concentrate on records within a single cylinder, but the cylinder index contains pointers to several cylinders. A pair of entries contains the information for each cylinder in the cylinder index, that is,

where the *key* is the highest key of any record on that cylinder and *ptr.* is a pointer to the track index for that cylinder. Each track in a cylinder has two pairs of entries associated with it in the track index. One pair contains information on the primary storage area and the other pair has information on the overflow records associated with the track. For each track, the associated entries will appear as

Primary   Overflow

The key in the first pair provides the highest key on that track in the *primary* area, and the key in the second pair of entries provides the highest key in the *overflow* area associated with that track. The primary pointer indicates the track containing the primary records, and the overflow pointer indicates the first overflow record, if one exists for that primary track.

In Figure 4.1, the cylinder index gives the explicit entries for three cylinders (1,

2, and 3) and implicitly contains more. The highest key on cylinder 1 is 350. In the pointer entry, the notation is *x-y,* where *x* gives the cylinder number and *y* gives the track number where the track index for that cylinder is stored. So the pointer 1-0 means track 0 on cylinder 1. The track index would not normally require an entire track of storage; so the remaining area could be used for additional primary or overflow storage. The value of a pointer in the track index points to a specific track number. At this point, the highest key values in both the primary and overflow entries in the track index are the same. That is because we have yet to insert any records into the overflow area since initially we place all records into a primary storage area. If we know how many records we will insert, we can determine how many tracks of primary storage are required and then specify the file size accordingly. For simplicity, we assume that the records in primary storage have a blocking factor, or bucket size, of one and that all the primary storage is filled. In practice, the blocking factor would be greater than one. As we subsequently add records, we need to use the overflow storage. The Λ (NULL) in the overflow pointer indicates that no overflow records currently exist. We must maintain the order of the records to preserve the retrieval mechanism; so when we insert a new record we place it into its proper lexicographic position. Since we can have overflow records associated with each track, we never have to *move* more than the number of records on a single track. When we insert new records into the primary area of a track, that may bump a current primary record into an overflow area. That means that the highest key value for the *primary* entries will change but not that for the overflow entry. In fact, the highest key value for an overflow entry *never* changes *except* when we add to the end of the file. In that case, both the highest key value in the cylinder index and

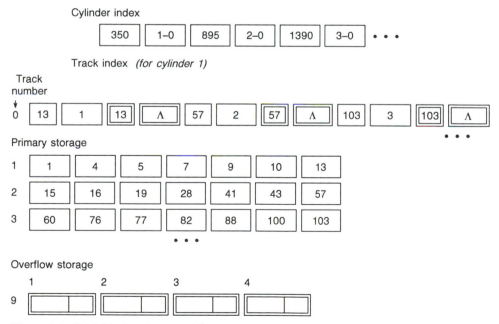

Figure 4.1 Initial indexed sequential structure.

the overflow entry of the track index would change. In this example, the overflow records are located on track 9 in consecutive locations. Normally the overflow records would not be consecutive. An available space list would link the overflow space together as records are added and deleted from the file. After a period of time, the available space list would not contain storage in consecutive locations. The overflow pointers follow the format *z-w,* where *z* gives the track number and *w* gives the record number.

Now we are ready to make insertions. Let's first add a record with a key of 8. We initially need to determine if it already exists in the file and if not, on which track it should be placed. We begin by searching the cylinder index. A record with key 8 should be located on cylinder 1 since 8 is < 350. The cylinder index sends us to the track index on track 0 of cylinder 1. We compare 8 with the highest key on the first track and determine that since 8 is < 13, it should appear on track 1. We do a sequential search of the primary records of track 1 and discover that a record with a key of 8 does not appear in that track. To insert it into its proper lexicographic position, we need to move *all* of the records that appear after that position. In this example, those would be the records with the keys 9, 10, and 13. Since no space is available for adding another record to the primary storage area for track 1, we need to use the overflow space. Which record is placed into the overflow area? Not the record that we are inserting, the one with key 8, but instead we displace the record with key 13 from the primary area when we insert the new record. The available space list would tell us that the next available space is track 9, record 1. We insert 13 in that position and *update* the entries for the highest key in the primary storage area, which becomes 10, and the overflow pointer to 9-1. Since the overflow record is at the end of the chain for track 1, we place a Λ pointer in its link field. The indexed sequential storage then appears as

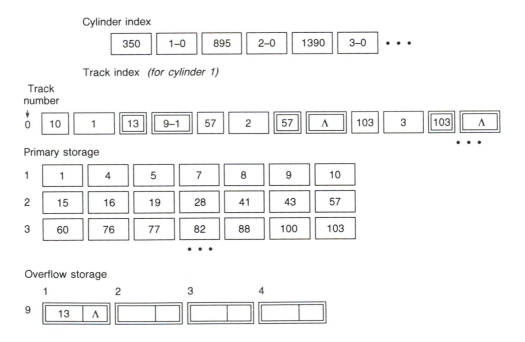

Next we insert a record with a key of 99. The index sends us to track 3. Assuming that it is not a duplicate, when inserting it into its proper position, we move the record with key 100 and place the one with key 103 into overflow storage. This insertion is similar to the previous one except that the values are different. The resulting structure is then

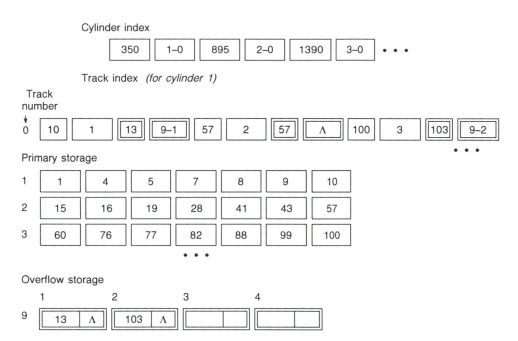

The next record to insert has a key of 14. Assuming that this record does not already exist in the file, the first question to answer after determining the proper cylinder is what track should it be inserted into. The answer is track 2. That necessitates moving *all* of the records stored in the primary area of the track. Wouldn't it be better to insert it onto the end of track 1? In that way it would be placed in the overflow area and *no* records would need to be moved in the primary area of track 2. What do you think? It sounds appealing doesn't it. We will not opt for this alternative because it would add complexity to the algorithm. The alternative would require that we search two tracks on each insertion and then choose between them. It would require more insertion effort but would provide no gain on the average for retrieval. Also, the circumstance that we have in this insertion would not appear that frequently; thus on the average, considering a single track is a preferable solution. We want to *keep the algorithm simple.* If we could have improved the retrieval performance, a more complex algorithm may have been justified, for example, as we observed previously with the binary tree collision resolution vs. linear quotient. We then insert the record with key 14 into track 2, move all of the primary records to the right one position and move the record with key 57 into the overflow storage area. We update the track index and the structure then appears as

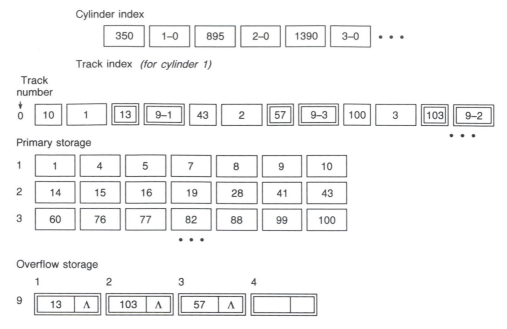

The key of the final insertion record in this example is 11. Again assuming that it is not already in the file, we determine that it should be placed in the *overflow* area of track 1. How do we know that it should be placed in the overflow area? It is a consequence of having two pairs of entries for each track. When we compare 11 with 10, the highest key of a record stored in the primary area for that track, we know that the new record will not fit into the primary area since 11 > 10. We next compare 11 with the highest key in the overflow area; since 11 < 13, to maintain order, it should be placed into the overflow area. In this case, we don't need to modify the primary storage area or the index entries associated with it. All we need to do is to place 11 into its proper position in the chain of overflow records. What we are doing here, then, is inserting into a linked list. Since 11 < 13, the new record should appear before the record with key 13. We adjust the links so that the track index overflow pointer points to the next available overflow storage, 9-4, and then set its link field so it points to the location of the record with key 13, that is, 9-1. The final structure appears in Figure 4.2.

DISCUSSION

Data Access

To sequentially process the data in an indexed sequential file entails stepping through the indexes to locate track after track of primary and overflow records. We cannot simply begin at the first record in the file and then proceed as we would with a sequentially organized file since there may be intervening tracks containing overflow records or other information, possibly from a different file. We also need to use the

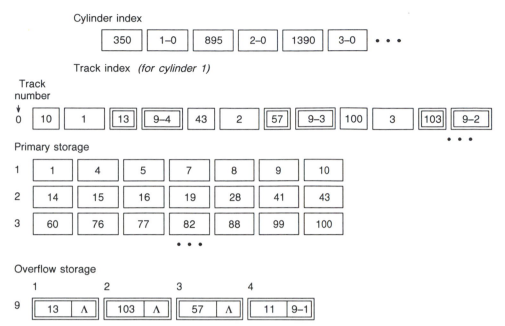

Figure 4.2   Final indexed sequential structure.

index when we move from one cylinder to another since the cylinders may not be contiguous. The index, however, does order the records for sequential processing in contrast to a directly organized file. You can then understand why sequential processing of an indexed sequential file is faster than for a directly organized file but slower than for a sequentially organized file.

To directly access a single record, we proceed in much the same manner as with an insertion. We search the cylinder index first to determine the proper cylinder. That entry in the cylinder index points to the track index. We then search the primary and overflow entries of the track index to locate the proper area that a record must be stored in if it is to be in the file. For example, if we retrieve the record with key 13, we find that it should be on cylinder 1. We check the track index and note that the highest key in the primary area of track 1 is 10. Knowing that means we *do not* need to search the primary area of track 1. We then compare the desired key of 13 with the highest key in the overflow area; they are the same, so we know that if the record exists in the file, it must be in the overflow area of track 1. We follow the link field of the overflow entry to a record with key 11. That is not a match, but it has a nonnull link; so we follow it to the record with key 13 and the search is ended.

Note.   We need to search *either* the primary or overflow area of a track but *not both,* since each track has *two* pairs of entries associated with it. The index in an indexed sequential file directs us to an individual record faster than a search of a sequentially organized file, but since we need to access several index entries and possibly many records, the retrieval time is greater than for a directly organized file.

Which takes more time: a successful search or an unsuccessful one? In the case of a sequentially organized file and a directly organized file, an unsuccessful search takes more time. With an indexed sequential file, however, both require the *same* amount of time. Why is that? Since the records in an indexed sequential file are *ordered,* we can end a search when we move beyond the position in which the sought record should have occurred.

Performance

Note that the overflow records are always unblocked; that is, only one record appears in each block or unit of storage in the overflow memory. Having a blocking factor of one simplifies the insertion algorithm, but it can reduce the efficiency of search if we have many records in the overflow area for a track. For example, consider the earlier diagram *after* several more records have been added to group 7:

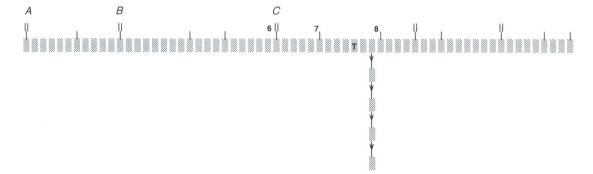

We now have four overflow records, which means that we may have to make as many as four separate accesses of auxiliary storage to locate a record. And you remember from Chapter 1 how relatively expensive an access of auxiliary memory is. This mechanism that the indexed sequential file organization uses to manage overflow records is its greatest drawback, for the chain of overflow records can grow until all the storage available for the file has been exhausted. Such a long chain would degrade the performance, for both subsequent insertions and retrievals. If you consider the entire structure, index and data, as a single tree structure, when we insert a record into an overflow area, we are adding to the depth of the subtree that we must search on retrieval. The tree structure grows from the *top down.* In Part Three, we consider a tree structure called a B-tree which has largely supplanted the indexed sequential organization for new applications since the B-tree grows from the *bottom up* and eliminates exceptionally long probe chains on retrieval.

To correct the imbalance that is likely to occur with an indexed sequential file over time, it is advisable to *reorganize* the file periodically, that is, to unload the records, allocate more space for primary records, and then reinsert the records back into the file. Since such a reorganization is likely to take considerable time, it is usually done at nonpeak times of computer usage, for example, weekends or holidays. Some businesses have a regular schedule for reorganizations. Another technique for reducing the effect of long overflow chains is to include unused space or dummy records in the primary storage area so that records inserted later can occupy those spaces instead of having to move or force a move to an overflow chain.

Deletion

The action taken when a record is *deleted* from an indexed sequential file depends upon whether it is located in the primary or overflow memory. If the record is in primary memory, it is merely removed from the file, and the space that it occupied is then available for a subsequent insertion. Removing a record may be accomplished by replacing it with a dummy record or marking its storage location as being available, à la a tombstone. It would usually not be worthwhile to move records to reclaim that space immediately since in most situations an insertion would soon occur. If the record is on an overflow chain, that chain would be relinked such that its predecessor record would then point to its successor record and the space occupied by the record would be returned to the available space list.

Overflow

The *overflow* area may be implemented in either one of two ways—cylinder overflow or independent overflow. With cylinder overflow, the overflow area is on the *same* cylinder as the primary storage area. With an independent overflow area, several cylinders of primary storage share a separate (independent) cylinder for overflow records. Figure 4.3 depicts these two methods for handling overflow records. The advantage of the cylinder overflow is that the read/write head on an auxiliary storage disk does not need to be repositioned to another cylinder to access an overflow record. In some time-sharing environments, though, the advantages of a cylinder overflow area may be lost since another user's processes may intervene between the system's accessing the primary area and the overflow area. To lessen such an adverse effect, some operating systems schedule input/output to minimize seek time.

The advantage of the independent overflow area is that several primary cylinders may *share* a single overflow region so that the amount of overflow space that needs to be reserved per cylinder can be lessened. An auto parts store does not need to have as many of a particular part on hand if there is a distributor nearby. For example, rather than each of 10 auto parts stores having two of an item on hand, it may only keep one on hand; the nearby distributor may retain three of the items from which it can quickly replenish a store. Instead of needing 20 parts, only 13 are required. The same principle holds here. We don't need to reserve more overflow tracks per cylinder than we are likely to need because we can share storage with other cylinders. The likelihood of *all* cylinders needing overflow storage in unusual amounts is very low, comparable with

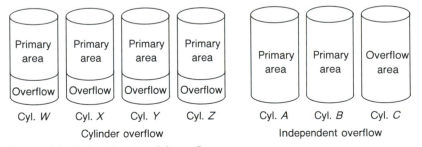

Figure 4.3   Indexed sequential overflow areas.

that of each of the auto parts stores needing the distributor to send it a particular part at the same time. It could happen, but it is not likely.

Using Indexed Sequential

Even though the indexed sequential file organization as we have described it in this section is rarely being used in new applications, it has been used with such frequency in the past that you are still likely to encounter it in practice today. Therefore, it is important to understand it. As with much software, understanding its basic structures allows us to use it more effectively. Let's assume that we are inserting many records into an *existing* indexed sequential file in which the primary storage area is filled. Why would it help if we sorted the records into descending order before insertion? On the surface, sorting the records would appear to give us poorer performance since we have added a step to the process. As an example, in column A following, we have the unordered keys of the records to be inserted. In column B are the same keys but sorted into descending order

A	B
39	72
47	68
72	47
17	39
22	36
68	22
36	17

When the primary area is *full,* the indexed sequential insertion procedure performs an insertion sort on the items in an overflow chain of a particular group. That means stepping through the records already in the chain to locate the proper position for the insertion. Each element encountered on the chain requires an access of auxiliary storage, plus we need to do this processing for each insertion. In this example, without the presorting of keys, if there were no other records in the overflow area for the group, 39 would be inserted; then we would need to access it again to insert 47. To insert 72, we would have to access both 39 and 47, and so forth. What is the advantage then of descending order? The new record that we are inserting is always at the *top* of the chain for the group of records being inserted. That characteristic eliminates accessing a record multiple times during the insertion process for a group of records. This example also introduces an important technique which we will see again in this text. Spend additional time structuring information prior to its primary processing to reduce that processing time.

We can improve performance greatly when using an indexed sequential file by understanding how to use it effectively. Coyle [1] describes how his organization was able to increase the number of transactions per hour from 8000 to 33,000 by merely changing the way in which it used the file. One technique that he used was to place the cylinder index (or top level master index) in primary memory since all accesses of an indexed sequential file need to use it; doing so reduces the number of auxiliary memory accesses for each retrieval by one, thereby improving performance.

KEY TERMS

Access method	Indexes
Available space list	Cylinder
Cylinder	Master
Cylinder overflow	Track
File reorganization	ISAM
Independent overflow	Master indexes
Indexed sequential	Overflow area
Deletion	Tabs
Direct access	Tracks
Insertion	Tree index
Sequential processing	
Successful search	
Unsuccessful search	

EXERCISES

1. Given the index structure and key values of Figure 4.1, insert records with the keys 18, 2, and 26. Assume that records one, two, and three of track 9 are available for overflow records. **(a)** What are the track index values after the three insertions? **(b)** What is the length of the longest overflow chain?

2. Given the index structure and key values of Figure 4.1, insert records with the keys 2, 6, 8, and 12. Assume that records one, two, three, and four of track 9 are available for overflow records. **(a)** What are the track index values after the four insertions? **(b)** What is the length of the longest overflow chain? **(c)** In general, what effect does the key distribution of the added records have on the retrieval performance of an indexed sequential structure?

3. In an indexed sequential file organization, what do you know about the number of overflow records on a track when the index entries in the normal and overflow key areas for that track are the same?

4. Assuming that all the indexes and data are stored on auxiliary memory, how many accesses would be necessary to retrieve the record with key 11 in the indexed sequential structure of Figure 4.2?

5. Given the index structure of Figure 4.2, how would it appear if the records with keys 13 and 14 were deleted? Remove the key and link values to indicate an empty location.

6. **(a)** If 30 records with keys that map to the first three primary storage tracks are added to the indexed sequential file of Figure 4.2, what would be the greatest number of records possible in an overflow chain? **(b)** the smallest number possible in an overflow chain? **(c)** an average number?

7. What are the disadvantages of placing dummy records in the primary storage areas of an indexed sequential file?

8. **(a)** When initially loading an indexed sequential file, why should the records be presorted? **(b)** Why should the procedure for initially loading an indexed sequential file differ from that for inserting a single record?

REFERENCE

Coyle, Frank T., "The Hidden Speed of ISAM," *Datamation,* 1971, pp. 48–49.

PART SUMMARY

In this part you have been introduced to the three primary ways of structuring or organizing records in a file: sequential, direct, and indexed sequential. All three have their place depending upon the type of access that you intend to make on the file: sequential, direct, or both. In addition to learning about these file organizations, you have also been presented with several techniques and principles that are applicable in a wide variety of software development situations. Let's review these important concepts.

TECHNIQUES

- Compute information when there are storage limitations (e.g., the address values in computed chaining).
- Subdivide a search space into many units so that only a portion of them need to be considered when searching (e.g., multiple chains in computed chaining).
- Spend additional time structuring the information during insertion to reduce the retrieval time; store once, retrieve often (e.g., Brent's method, binary tree collision resolution, and computed chaining).
- Use a separate data structure to control the organization of data in a storage structure (e.g., the binary tree in binary tree collision resolution).
- Store only the information that still needs to be accessed and cannot be readily computed (e.g., the bottom two levels in a complete binary tree in binary tree collision resolution).
- Keep some condensed information in primary memory to improve access time to the actual data (e.g., maintaining a bit string in primary memory to identify occupied locations for linear quotient insertions).
- Order data to improve search efficiency (e.g., the binary search).

PRINCIPLES

- Identify deficiencies in a procedure and eliminate them (e.g., binary tree collision resolution vs. Brent's method).

- Understand how a data structure operates even if you will not be implementing it, for your understanding of it may assist you in using the structure efficiently (e.g., indexed sequential).
- Sharing available storage among several data structures may require less storage overall than allocating a fixed amount of storage to each one (e.g., an independent overflow area in indexed sequential organization).
- Continue searching for improvements to a technique even though it has been widely used for some time (e.g., EISCH vs. LISCH).
- Look for data structures that make the difficult or impossible achievable (e.g., the binary tree for moving records from secondary and subsequent probe chains).
- Look for special situations in organizing data that may lead to storage or retrieval efficiencies (e.g., the complete binary tree in binary tree collision resolution).

PART EXERCISES

1. Which file organization, sequential, direct with LISCH, or indexed sequential, would be the simplest one to implement? Which one would be the most complicated?

2. Order the three primary file organizations from the easiest for sequential access to the most complicated.

3. Order the following three file organizations from the least amount of total space (that needed for the records plus that needed for the organization's overhead) to the greatest: indexed sequential, direct with linear quotient, and adaptive sequential with a count field. What assumptions are you basing your ordering on?

4. Explain how the frequency of access of data and the number of records in a file would affect the choice of a file organization.

5. What are the tradeoffs in using an indexed sequential file organization vs. an ordered (sorted) sequential file to store the information about your collection of 124 record albums? Assume that you would want both to access the information about a single album and to list your entire collection in sorted order.

6. What is the difference between a blocking factor and a packing factor?

PART TWO

BIT LEVEL AND RELATED STRUCTURES

PREVIEW

As computer systems have evolved, the user has moved to higher and higher levels of interfaces for communicating with them. This evolution has usually meant that the user has gradually moved away from a hardware level of interaction with the computer. For the most part, this evolutionary process has been beneficial since it has enabled more people to access computers, which in turn has increased their productivity. In this part, though, we want to move counter to that trend and consider lower-level or hardware-influenced structures for representing information. At the most basic physical or hardware level, information is represented by *bits* or binary digits, zeros and ones (0s and 1s). In many situations, a single bit is all that is needed to represent a piece of information, such as "yes" or "no," or "on" or "off." In other cases information can be encoded into a binary or bit form so that it may be searched more readily. One of the fundamental principles of this part is that

- If information is not readily processible for the task at hand in its given form, *convert* it into a form that is more readily processible.

We do not need to be content with the status quo for the information representation, especially when storage and performance costs are important; we can convert it into another form. This other form may frequently be a bit level representation or one influenced by such a representation.

The primary advantages of bit level or binary representations are that they:

- Require less storage; if we only need a bit to represent the information, why use a byte?
- Allow faster searching; most computers have hardware instructions for com-

paring or searching for bit patterns. Much information may be dealt with in a single instruction. And that means better performance.

At this point you may be saying to yourself that you do not want to program at the assembly level. But before you decide to skip this part, take note. Many higher-level programming languages have facilities for storing and accessing information in a bit form. Second, in many situations, it is the bit *concept* that is important to the algorithm development. The algorithms do *not require* a bit level implementation.

Chapter 5

Bits of Information

BINARY ATTRIBUTES

> A **binary attribute,** as the name implies, is an attribute that may take
> on only one of two values, for example, *on* or *off; yes* or *no;* or *has* or
> *doesn't have.*

Typically, a field value of a record takes on one of *many* possible values and provides a *measure* of characterizing the associated entity, such as the grade point average of a student. In many cases, as mentioned in the preview of this part, information may be represented in a binary form. When we are only interested in whether an object or person has a certain characteristic and *not* in the degree or *measure* of the characteristic, representing that characteristic as a binary attribute has advantages. For example, what if we are only interested in whether a student has at least a 3.0 grade point average, and not in its specific value? Representing that information as a binary attribute, for example, a "1" if the student has a 3.0 grade point average, and a "0" if he doesn't, requires *less* space than storing the entire grade point average, *and* it is *faster* to search for a "1" in a certain position than to compare each grade point against a 3.0. By using a binary attribute, we save not only storage but also search time.

We could use binary attributes to represent the characteristics or information about a person. In this case, the fields of a record might be the names of traits often used to describe a person, such as "tall," "dark," "handsome" ("beautiful"), etc. If a person possesses a feature, a "1" is stored as the value in the field, otherwise a "0". The record describing Hobo Joe might appear as

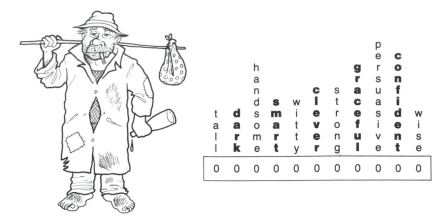

and that for Prince Charming might appear as

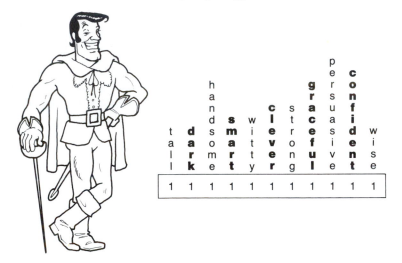

and most of the rest of us would have records somewhere in between. Since these example records have 11 fields, there are a total of 2^{11}, or 2048, different classifications for people. If we have a file of such records, in which we might think of the individuals' names being along one axis of the table and the characteristics along the other axis such as

Characteristics

0 0 0 1 1 1 0 0 1 0 1
1 0 0 0 1 0 1 0 1 0 0
• • •
0 1 1 0 1 0 0 1 1 1 0
0 1 1 1 0 0 0 1 0 1 0
1 0 1 0 0 1 1 0 0 1 0
• • •

People

File of records with
binary attributes

it is then straightforward to search for all tall people or all wise people or all people who have both attributes. We merely need to search for a "1" in the column corresponding to the desired attribute; when we find a "1", we have located a person with the desired characteristic. For an "and" or "both" condition, both columns must have a "1" in them. We could also use such a file of records with binary attributes to perform *matching* operations. For example, we could find the person or persons who had the greatest (or fewest) number of traits in common with another individual. If we added a gender field, which is also binary, we could locate the most similar person of the opposite gender.

There are many such cases in which we could use binary attributes to represent information for subsequent searching and matching. Some examples include

```
                   Features
          ┌──────────────────────────
          │  0 0 0 1 1 1 0 0 1 0 1
          │  1 0 0 0 1 0 1 0 1 0 0
          │          • • •
 Houses   │  0 1 1 0 1 0 0 1 1 1 0
          │  0 1 1 1 0 0 0 1 0 1 0
          │  1 0 1 0 0 1 1 0 0 1 0
          │          • • •
          │
```

a real estate file in which the records contain information on houses for sale. A feature of a house, for example, could be whether it has a swimming pool. It is then an easy matter for a prospective buyer to locate all the houses with swimming pools. Another feature might be having a garage. Again it is an easy matter to find all homes for sale with both a swimming pool and a garage.

Another example is a document retrieval file in which documents appear along one axis and keywords along the other as

```
                   Keywords
           ┌──────────────────────────
           │  0 0 0 1 1 1 0 0 1 0 1
           │  1 0 0 0 1 0 1 0 1 0 0
           │          • • •
Documents  │  0 1 1 0 1 0 0 1 1 1 0
           │  0 1 1 1 0 0 0 1 0 1 0
           │  1 0 1 0 0 1 1 0 0 1 0
           │          • • •
           │
```

We could then locate all documents dealing with "hashing" if that were one of the keyword columns.

A third example functions as an aid in a job search. The records of the file correspond to the positions available, and the attributes are the requirements for a position. A requirement for a position could be something such as "knowledge of SQL." The file appears as

Requirements

```
              0 0 0 1 1 1 0 0 1 0 1
              1 0 0 0 1 0 1 0 1 0 0
                      • • •
Positions     0 1 1 0 1 0 0 1 1 1 0
              0 1 1 1 0 0 0 1 0 1 0
              1 0 1 0 0 1 1 0 0 1 0
                      • • •
```

A prospective employee could then perform a search for those positions which matched his interests and abilities.

Other examples of files of records with binary attributes include information about birds, landscaping plants, flowers, and stamps. Almost any file of information can be represented in binary form *depending upon the retrieval application.* If you only need to know whether or not, and *not* the degree, binary attributes are appropriate.

An Example

Let's consider in greater detail the use of binary attributes and how we process them. The example data, about *desserts,* are illustrated in Table 5.1. The ingredients of a dessert recipe constitute the binary attributes for a record.

If chocolate is our favorite ingredient and we want to make a dessert containing

TABLE 5.1 DESSERT RECIPES REPRESENTED WITH BINARY ATTRIBUTES

Desserts	Baking powder	Blueberries	Brown sugar	Butter	Chocolate	Cinnamon	Eggs	Flour	Lemon juice	Margarine	Milk	Nutmeg	Salt	Sugar	Vanilla	Water	Special ingredients
Banana sundaes	0	0	1	1	0	1	0	0	1	0	0	0	0	0	0	0	Bananas
Berry crumble	0	1	1	1	0	1	0	1	1	1	0	0	1	0	0	0	Oats
Chocolate toffee bars	1	0	1	1	1	0	1	1	0	0	0	0	1	0	1	0	Nuts
Custard	0	0	0	0	0	0	1	0	0	0	1	1	1	1	1	0	
Fresh apple pie	0	0	0	1	0	1	0	1	0	0	0	1	1	1	0	0	Apples
Glazed pound cake	1	0	1	0	0	0	1	1	0	0	1	0	1	1	1	0	Graham crackers
Peanut-fudge pudding cake	1	0	1	0	1	0	0	1	1	0	1	0	1	1	1	1	Peanut butter, vegetable oil
Southern blueberry pie	0	1	0	0	0	1	0	0	0	1	0	0	1	1	0	1	Cornmeal, cornstarch

Source: Betty Crocker's Microwave Cookbook, Random House, New York, Copyright © 1981. Adapted with permission.

it, searching the table of binary attributes to find recipes that contain chocolate is a simple task. Which recipes are they? Chocolate toffee bars and peanut-fudge pudding cake. Wasn't this much easier and faster than searching the actual recipes in a cookbook to find the ones that contain chocolate? This situation is another example of the important principle: Spend time preprocessing information to make retrieval more efficient. The justification is that we need to do the preprocessing only once, but we can benefit from it many times in retrieval efficiency.

We could also use this table to avoid recipes that contain an ingredient we do not like. For example, how many recipes do not contain lemon juice? We merely search the lemon juice column and note all the recipes that contain a zero in that column; in this example there are six. We can also form retrieval queries using other Boolean operators such as "or" and "and." If instead of a single favorite ingredient, we have multiple favorite ingredients of "chocolate" and "ice cream," we could pose a query to find all recipes that contained at least one of the ingredients. If a one appears in either the "chocolate" or the "ice cream" column, the corresponding recipe is retrieved. If we were really interested in self-indulgence, we could even locate those recipes that contain *both* of our favorite ingredients. That requires that both the "chocolate" and "ice cream" columns contain a one for an individual recipe. Peanut-fudge pudding cake is the dessert for us. We make a complex retrieval request by combining attributes with the Boolean operators: *and, or,* and *not.* We say more about multiattribute retrievals in the next chapter.

The example file has only eight recipes with sixteen basic ingredients. Even the most reluctant cook has many more recipes than that. Typically a person has hundreds of recipes that he uses or would like to use. As the collection of recipes increases, the number of ingredients also increases. Instead of having only 16 ingredients, one may have between 100 and 200 ingredients. Searching for recipes with particular attributes becomes more difficult for both a computer and a person. It is more difficult to process a record with a 100 bits than one with 16. To make retrieval processing more efficient, we want to condense the information record into a more manageable form. That leads us to the technique of superimposed coding.

Superimposed Coding

An important concept of this part is that if information is not in a form that is readily processible, *convert it* into one that is. A 1- to 200-bit record may not be manageable, whereas one of 8 bits is. By *encoding* the original information, in this case a recipe, into a *compact* form, it is more readily processible, that is,

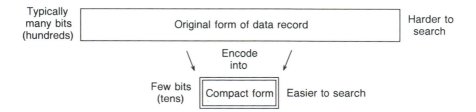

The scheme that we use for the encoding is called **superimposed coding.** To form an 8-bit encoding of a record, we *superimpose* an 8-bit code for *each* of the ingredients in the recipe. Initially, we must decide on the codes for the ingredients. Why are we using 8 bits? One reason is that we have a relatively small number of recipes in the example. But why 8? Why not 7 or 9? We need to know a little about the computer that we will be using. *Important:* Most computers have built-in hardware instructions to search bit patterns of special lengths such as the byte, word, half word, or double word. Making the code a byte size in this case should speed the computer retrieval.

We form the code by setting exactly k bits to one in an m-bit field. In this example, we have already chosen m to be 8. What should k be? We need enough different codes to represent each of the ingredients. If k were one, how many codes are possible? [1]

code

bit positions   1 2 3 4 5 6 7 8

We could place the 1 bit in eight different positions, so there are eight possible codes. What we have is a combinatorial problem of $_mC_k$, that is, the *combination* of m things taken k at a time, where

$$_mC_k = \frac{m!}{(m - k)!\, k!}$$

If we use a k value of 2, we compute that there are 28 possible codes; that is sufficient for this example. For a different circumstance, we might have needed to continue varying m and k until we found a combination that gave us at least the desired number of codes without going beyond it too greatly, unless we planned upon expanding the number of ingredients soon. Table 5.2 gives the arbitrarily assigned codes for the ingredients in the example file.

The recipes' codes are formed by *superimposing* the codes for the ingredients contained within a recipe. For example, to form the recipe code for the peanut-fudge pudding cake, we include the code for the first ingredient from Table 5.1, that is, baking powder. However, Table 5.2 does *not* contain a code for baking powder. Is this an error or oversight? No, but what is the reason? You need to know something about the application. Many, if not most, recipes contain that ingredient, so we gain little by including it in the condensed code for a recipe. We do not usually differentiate recipes based upon baking powder just as we do not use the keyword "computers" when searching for a document in a computer science database. You will notice then that "common" ingredients are not included in Table 5.2. Another reason for not superimposing codes for common ingredients is that if many codes were superim-

---

[1]The bit positions are numbered left to right starting with "1" to simplify the discussions in this part.

TABLE 5.2 INGREDIENT CODES

Apples	00100100	Ice cream	01000001
Bananas	00100010	Lemon juice	00100001
Blueberries	10000001	Nutmeg	00010001
Brown sugar	01000010	Nuts	01000100
Chocolate	10000100	Oats	00010010
Cinnamon	00011000	Peanut butter	01001000
Cornmeal	00101000	Vanilla	10000010
Graham crackers	00010100		

posed the resulting recipe code would contain primarily ones that would make it difficult to distinguish one recipe from another. The code for peanut-fudge pudding cake is

Brown sugar	01000010
Chocolate	10000100
Ice cream	01000001
Vanilla	10000010
Peanut butter	01001000
Peanut-fudge pudding cake	11001111

Superimposed coding gets its name from the manner in which the individual codes are superimposed or placed one on top of the other. If a one appears anywhere in a column, a one appears in the result. The superimposing is the same as *OR*ing the bits together.

Table 5.3 contains the superimposed codes for the eight example recipes. Let's again retrieve all recipes containing chocolate. This time, we will retrieve all recipes that match, that is, have "1"s in the same positions as the ingredient code for chocolate (positions one and six). Which recipes drop out (or match)? Chocolate toffee bars, glazed pound cake, and peanut-fudge pudding cake. Notice that when we performed the retrieval previously without superimposed codes, only chocolate toffee bars and peanut-fudge pudding cake matched. Why did glazed pound cake drop out? Does it contain chocolate? No, it is referred to as a **false drop** since it should not have dropped out. Why do we have false drops? When we condensed the coded information for a recipe, we lost something. We no longer have exact information, but we are willing to give up this property to speed retrieval. Instead of searching hundreds of bits, we only search eight. How do we avoid retrieving inaccurate information? In using superim-

TABLE 5.3 SUPERIMPOSED RECIPE CODES

Banana sundaes	01111011	Fresh apple pie	00111101
Berry crumble	11111011	Glazed pound cake	11010110
Chocolate toffee bars	11000110	Peanut-fudge pudding cake	11001111
Custard	10010011	Southern blueberry pie	10111001

posed coding for retrieval, we must always check the responses against the original file to determine false drops. This additional checking takes time, but it should be more than offset by the savings gained by using the encoded form of the information.

Discussion

The advantage of using superimposed coding with binary attributes is that a computer can search a condensed form of the information more quickly than a form that may contain hundreds of bits. Since we need to search *all* the records in the file, a small savings in the search of each record can add up. The method saves time, but does it also save space? No, the superimposed codes certainly require less space than the original form of the information, but we must *also* keep the original information to check for false drops; therefore, the technique actually uses more space since we now have added the additional superimposed recipe codes.

How could we reduce the number of false drops? One means is to increase the code size from 8 bits to 16 bits. In this way, there is a lower likelihood that a recipe code contains a pattern of bits that is the same as the code of a nonincluded ingredient. Second, we could remove ingredients such as brown sugar and vanilla from the ingredients list since several recipes contain them. Could we also reduce false drops by *increasing* the value of k so that additional 1 bits are set in a code? No, doing that might actually *raise* the number of false drops, since the code for a recipe could then have more 1 bits set that would match more possibilities on retrieval.

KEY TERMS

Binary attribute
Boolean operators
Combinatoric formula
False drop
Information conversion

Matching
Recipe
    Binary representation
Superimposed coding

EXERCISES

1. Verify the recipe codes in Table 5.3.
2. (a) Which recipes drop for the ingredient, "ice cream"? (b) Are any false drops?
3. (a) Which recipes drop for a combination of "chocolate" *and* "ice cream"? (b) Are any false drops?
4. Are there any combinations of two ingredients in Table 5.2 that produce bits in at least the same positions as the ingredient code for "chocolate"? If so, list them.
5. Would the number of false drops *increase* or *decrease* with an "and" query, that is, one that requires that the value of several attributes be set to specified values? Explain.
6. Describe an efficient procedure for use with superimposed coding to locate all recipes that *do not* contain a specific ingredient, for example, "chocolate."

7. What are the minimum values for m and k for a k-bit code in an m-bit field if m is a multiple of 8 and 72 distinct codes are needed? Minimize on m first.

8. What is the minimum field size and number of 1 bits for a code to represent 5000 attributes? Assume that no more than 15 percent of the bits should be 1s and that the field size should be divisible by 8.

TEXT SEARCHING

Originally, computers were used almost exclusively for numerical computations. But it quickly became apparent that computers were also excellent *symbol manipulators.* Today the vast majority of computer applications deal with nonnumeric or symbol manipulation processes rather than numerical calculations. Because of the original dominance of computers in numerical areas, many people are still reluctant to consider the many interesting and powerful uses of computers in nonnumerical applications. In this section, we want to consider the basic nonnumeric process of text searching, or pattern matching, that is, the searching for a pattern, a sequence of symbols, s_1, s_2, ..., s_n, in a string that is usually a much larger sequence of symbols. For example, we could search for the pattern "computer" in the following string:

string   Problem solving is a common paradigm of computer science.

pattern computer

Almost every introductory programming text contains an algorithm for pattern matching; in addition to it being an important process, it also is a fitting example of nested looping. These naive pattern matching algorithms say to compare the first symbol of the pattern with the first symbol of the string. If that comparison succeeds, then compare the next symbol of the pattern with the next symbol of the string. Continue advancing the comparison positions in the pattern and the string until either a complete match for the pattern is found or a comparison fails. On a failure of a symbol match, move the pattern one position to the right in the string and begin the symbol by symbol matching again beginning with the *first* position of the pattern.

	Number of comparisons
Problem solving is a common paradigm of computer science.	
computer	1
computer	2
computer	3
. . .	
computer	22 – 24
computer	25
computer	26
. . .	
computer	43 – 50

[The _ represents the pattern symbol being compared.]

Although illustrative, such an algorithm is not efficient. In fact, it has a worst case computational complexity of O(mn) where m is the length of the pattern and n is the length of the string. If the string were a document of any size, the processing efficiency could be very poor. Because almost all pattern matching still uses such algorithms, the pattern matching or text searching operation has gained a reputation for being a *slow* and *resource intensive* operation. Such a reputation certainly suggests that it is something to be avoided if at all possible. That is a "head in the sand argument." Just because an algorithm is widely used for a long time *does not* imply that it cannot be improved upon. Improved pattern matching algorithms have appeared, for example, the ones by Knuth, Morris, and Pratt[1] and Boyer and Moore[2].

An interesting aside of the Boyer and Moore pattern matching algorithm is that it begins the matching process of the pattern against the string at the *right* side of the pattern instead of the left. Culturally, most of us do pattern matching in a left to right scan, but that does not require that a computer algorithm must mimic the human process. Matching from the right improves the performance of the algorithm tremendously. An important principle of problem solving is to *keep an open mind.* That's what we want to do in all our data structure and algorithm design decisions. Boyer and Moore did and discovered a significantly improved pattern matching algorithm. Using the Boyer and Moore pattern matching algorithm, the previous example appears as

```
                                                                  Number of
Problem solving is a common paradigm of computer science.         comparisons
computer                                                              1
        computer                                                      2
            computer                                                  3
              computer                                                4
                computer                                              5
                  computer                                            6
                      computer                                      7 - 14

[The _ represents the pattern symbol being compared.]
```

If the symbol matched against in the string by the rightmost symbol of the pattern does not match and does not appear in the pattern, the pattern is moved beyond its current position. If the symbol does not match but does appear in the pattern, the pattern is moved so that the rightmost occurrence of this comparison symbol in the pattern is aligned with the comparison symbol in the string. The latter situation occurs in the example when the r in the pattern fails to match the leftmost m in "common" in the string. The pattern is then realigned so that the m in the pattern coincides with the m that had been matched against in "common." This repositioning is done because the m in "common" could have been, although it in fact isn't, an m in "computer." After the realignment, the pattern matching continues from the right of the pattern. (See [2] for additional details on the Boyer and Moore pattern matching algorithm.) Only 14 comparisons are needed to locate the pattern "computer" using the Boyer and Moore algorithm rather than the 50 comparisons of the naive pattern matching algorithm. This is an effective demonstration of how some thought about how to conduct an operation can have significant effects on performance. We want to continue in that tradition.

One feature common to all the pattern matching algorithms is that many or most symbols of the string or text must be examined at least once. Further improvement in string searching could be obtained if a line of text were exhaustively scanned *only* when there was a high probability of finding the sought pattern. Can text be compressed or coded so that this condensed representation could be searched initially, rather than the entire text?

When one searches a group of articles for those of particular interest, rather than looking at the entire text of each article, one looks only at a condensation. It may be in the form of a summary, abstract, or keywords. Only if that condensation appears relevant does one read the entire document.

How can the symbols of a text be condensed into an easily scanned format? One such method is that proposed by Harrison[3] and extended by Tharp and Tai[4] in which a signature is associated with a segment (e.g., a line) of text. (See Figure 5.1.) Instead of scanning each symbol of the line, only the signature needs to be examined. This comparison could be performed in a few operations on most computers. Only if the examination of the signature suggests a likely match in the associated text is the text scanned using an algorithm such as that proposed by Boyer and Moore.

In adding the text signatures to the records of a file, we are *preprocessing* the data so that the retrieval or search time will be less. **Remember:** store once, process often. We are using additional storage to keep the signatures (represented by the area enclosed in the double lines in Figure 5.1) to save retrieval time. Adding storage is another technique that we will use frequently. The use of the signatures acts as a *filtering* device. It allows us to quickly filter *out* those records which cannot possibly contain what we are searching for. This filtering allows a computer to perform the searching task *faster by doing less.* That is another important principle.

What is a signature? Each of us has one. It is something that *identifies* or *distinguishes* us from all others. In some respects it represents our essence.

> A **record signature,** or **text signature,** is an encoding of the contents of a record or a line of text to characterize its essence.

In this implementation, the text signature contains *bits* of information; its length is usually a multiple of the computer word size to allow efficient processing. The signature code is formed by *hashing* all contiguous k-symbol groups into an integer in the range

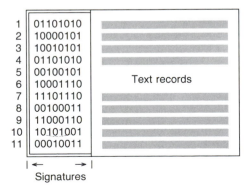

1	01101010
2	10000101
3	10010101
4	01101010
5	00100101
6	10001110
7	11101110
8	00100011
9	11000110
10	10101001
11	00010011

Text records

|← → |
Signatures

Figure 5.1 Text file with signatures added—one identifying each record.

$\{0, m - 1\}$ where m is the size of the signature. Since it is not possible to search for a pattern of length less than k, k is usually chosen to be two. The bit position corresponding to the integer result of the hashing function is then set to one. In other words, contiguous groups of symbols from a line of text set the bits in the signature for that line. For example, with the previous instance of a string

```
string    Problem solving is a common paradigm of computer science.

    if k = 2 then all 2-symbol pairs are hashed, that is,

    h(Pr), h(ro), h(ob), h(bl), h(le), h(em), h(m▓), . . . , h(e.)
        where h is the hashing function and ▓ represents a blank
        symbol.
```

If a line of text is 80 symbols, 79 symbol-pair combinations are hashed to form the signature for the line. In the example with 57 symbols, 56 symbol-pair combinations are processed. Each line of text is preprocessed to form a signature that does not need to be recreated unless the contents of the line change.

Each line of text then has an m-bit signature associated with it such as

```
00101...001001    Text signature
```

where the 1s correspond to the results of the hashing operation. Eight-bit signatures (rather than wider ones) are used only for illustrative purposes in Figure 5.1—the size of text signatures in practice is much larger.

It is now possible to perform string searching using the text signatures. If we want to search for the pattern "hash" in the example file, the first step is to form the *pattern signature* by applying the hashing function to adjacent symbol pairs, that is, **h**(ha), **h**(as), and **h**(sh). Since the pattern signature is formed using the same procedure as the text signature, it is the same length. Let's assume that it has the value 10010100 for this example. To search for the pattern "hash," its signature is compared with the signature of each line in the body of text being searched. If a text signature has 1s in *all* the same positions as the pattern signature, then that line of text is searched using an efficient pattern matching algorithm such as that of Boyer and Moore. If the comparison fails, then that pattern is *absolutely, positively not* in that line of the file. If it were in that line, then 1 bits would have been set in the same positions as the pattern signature during the creation of the text signature for that line. Hence that record *does not* need to be searched exhaustively. In *one* comparison, we can eliminate the need to perform possibly 80 or more other comparisons on a record of a file. That is a lot of improvement. Of the 11 lines of text in Figure 5.1, only line 3 needs to be searched exhaustively to determine if the word "hash" is contained therein.

If we do match the pattern signature with the text signature, that *does not* guarantee that the pattern will be found in the associated line of the file. Remember the concept of hashing collisions. We could have a collision in the formation of a signature such that two different symbol combinations set the same bit. There is more than one way in which a particular bit can be set since the hashing function does not guarantee uniqueness. If the text and pattern signatures match but a bit in the text

Algorithm 5.1
SIGNATURE MATCHING IN TEXT SEARCHING

```
1   proc search
2       repeat
3           read a line of text and its signature
4           if (¬text_signature) ∧ pattern_signature = all_zeros
5               then search the line exhaustively
6                       if found then true_drop else false_drop
7       until no_more_text
8   end search
```

where ∧ is the logical "and" operation.

signature is set by a different combination of symbols than the corresponding bit in the pattern signature, we have a **false drop,** that is, when we exhaustively search the line of text, we do not find the pattern. We discuss ways to reduce the frequency of false drops later.

A pseudocode form of the signature matching algorithm appears in Algorithm 5.1. What makes this algorithm and most of those in this part effective is that essentially all computers have special instructions built in for processing entire bit strings in a single operation. Algorithm 5.1 makes use of the machine operations (1) to complement (¬) a bit string, that is, to change all 1s to 0s and vice versa, (2) to "and" (∧) two bit strings, and (3) to check a bit string for all zeros. Using these built-in instructions is much faster than checking just one bit at a time in a text signature.

An Example of Signature Formation

For this discussion, assume that the length of the contiguous symbols input to the hashing function, that is, k, is two and m, the signature length, is 64 bits in one case and 128 bits in another. If the record length is 80 bytes prior to preprocessing, it expands to either 88 or 96 bytes after processing to include the overhead for storing the signatures. The extra storage required for the signatures is then 10 or 20 percent of the record length.

What should the hashing function be? What are the two important characteristics of a hashing function? —simplicity and providing an even distribution of values. To achieve an even distribution, the symbols of English were placed into classes based upon their frequency of occurrence (as given in Knuth[5]) since all symbols do not occur with the same frequency in English. The total frequency of occurrence for each class is approximately the same. In Table 5.4a, the symbols are placed into 8 classes and in Table 5.4b into 11 classes. The ▓ (blank) symbol, because of its high likelihood of occurrence, is placed in a class by itself; the remaining symbols are then evenly distributed among the other groups. Upper and lower cases for the same letter belong to the same class. An array (T) of 256 elements is constructed in which the index is the integer corresponding to the bit pattern for that symbol (e.g., an A is 193 in EBCDIC) and the entry is the class number to which that symbol belongs.

TABLE 5.4a   DISTRIBUTION OF TEXT SYMBOLS FOR
8-BYTE SIGNATURES

Class	Symbols
0	▓ (blank)$_{[18.6]}$*
1	$E_{\{10.3\}}$† $B_{\{1.3\}[11.6]}$ 6 : & ' " ?
2	$T_{\{8.0\}}$ $X_{\{0.1\}}$ $Z_{\{0.1\}}$ $W_{\{1.8\}}$ $G_{\{1.5\}[11.5]}$ 5 ; / * <
3	$A_{\{6.4\}}$ $F_{\{2.1\}}$ $Y_{\{1.6\}}$ $P_{\{1.5\}[11.6]}$ 4 ,) ! > ^
4	$O_{\{6.3\}}$ $L_{\{3.2\}}$ $C_{\{2.2\}[11.7]}$ 3 . (@ [—
5	$I_{\{5.7\}}$ $K_{\{0.5\}}$ $D_{\{3.2\}}$ $M_{\{2.0\}}$ $J_{\{0.1\}}$ $Q_{\{0.1\}[11.6]}$ 2 9 #] |
6	$N_{\{5.7\}}$ $V_{\{0.8\}}$ $S_{\{5.1\}[11.6]}$ 1 8 − $ {
7	$H_{\{4.7\}}$ $U_{\{2.3\}}$ $R_{\{4.8\}[11.8]}$ 0 7 + % }

*Total percentages for letters in that group.
†Percentage of occurrence for that letter.

TABLE 5.4b   DISTRIBUTION OF TEXT
SYMBOLS FOR 16-BYTE
SIGNATURES

Class	Symbols
0	▓ (blank) $_{[18.6]}$
1	$E_{[10.3]}$ 1 ! + *
2	$T_{[8.0]}$ 2 ? < @
3	$A_{\{6.4\}}$ $G_{\{1.5\}[7.9]}$ 3 . > /
4	$O_{\{6.3\}}$ $Y_{\{1.6\}[7.9]}$ 4 & (
5	$I_{\{5.7\}}$ $F_{\{2.1\}}$ $Q_{\{0.1\}[7.9]}$ 5 ; =)
6	$N_{\{5.7\}}$ $M_{\{2.0\}}$ $J_{\{0.1\}[7.8]}$ 6 : − {
7	$S_{\{5.1\}}$ $U_{\{2.3\}}$ $K_{\{0.5\}[7.9]}$ 7 — # }
8	$R_{\{4.8\}}$ $C_{\{2.2\}}$ $V_{\{0.8\}}$ $X_{\{0.1\}[7.9]}$ 8 ' ^ [
9	$H_{\{4.7\}}$ $W_{\{1.8\}}$ $B_{\{1.3\}}$ $Z_{\{0.1\}[7.9]}$ 9 %]
10	$D_{\{3.2\}}$ $L_{\{3.2\}}$ $P_{\{1.5\}[7.9]}$ 0 " $ 1

For a symbol-pair y_1, y_2, the hashing function is

$$\mathbf{h}(y_1 y_2) = \text{number\_of\_classes} * T(y_1) + T(y_2)$$

The number_of_classes for $k = 2$ is chosen to be the largest integer, n, such that $n^2 \leq m$ where m is the signature length. Therefore, the number_of_classes for a 64-bit signature is eight, which gives a hashing function of

$$\mathbf{h}(y_1 y_2) = 8 * T(y_1) + T(y_2)$$

This function maps every symbol-pair into an integer between 0 and 63. Owing to the construction of array T according to the values in Table 5.4a, all eight classes (0–7)

Algorithm 5.2
SIGNATURE FORMATION

```
1       proc set__signatures
2       repeat
3           text ← input__line
4           signature ← all__zeros
5           for i ← 1 to 79 do
6               position ← number__of__classes * T[i] + T[i+1]
7               signature (position) ← '1'
8           end
9           output ← signature || text
10      until end__of-file
11      end set__signatures
```

> where signature is a bit string of length 64 or 128,
> all__zeros is a bit string of the same length with all
> bits zero and || denotes concatenation.

have an equal (or essentially equal) likelihood of being chosen. The hashing function therefore produces numbers in the range 0 to 63 with equal probability under the assumption that the occurrence of a symbol-pair is based upon the distribution of the individual letters. In this case, since 8^2 equals 64, all 64 bits of the signature are used. The hashing function for the 16-byte signature,

$$h(y_1 y_2) = 11 * T(y_1) + T(y_2)$$

maps every symbol-pair into an integer between 0 and 120 since the corresponding array T formed from values in Table 5.4b contains 11 classes of symbols (0–10). In this case, then, since 11^2 equals 121, 7 bits of the signature area remain unused. The algorithm for setting the signatures is given in Algorithm 5.2.

The 8- and 16-byte signatures for the word "computer" are formed as follows:

Hash function	Bit position set	
	8-byte signature	16-byte signature
$h(co)$	36	92
$h(om)$	37	50
$h(mp)$	43	76
$h(pu)$	31	117
$h(ut)$	58	79
$h(te)$	17	23
$h(er)$	15	19

$h(co)$ is 36 for an 8-byte signature since $T[c]$ is 4 as a result of c being in class four, and $T[o]$ is also 4 as a result of o also being in class four. Inserting these values into the hashing function gives 8(4) + 4, which yields 36. $h(co)$ for a 16-byte signature

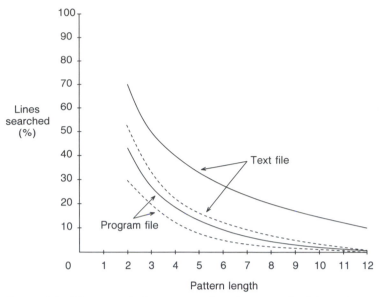

Figure 5.2 Lines searched vs. pattern length ———— for 8-byte signatures and − − − − for 16-byte signatures.

is 92 since T[c] is 8 as a result of c being in class eight and T[o] is again 4 since o is in class four.

Discussion

The effectiveness of text signatures as a filtering mechanism is illustrated in Figure 5.2. As the pattern length increases, the percentage of the lines of text that must be searched exhaustively decreases. A 16-byte signature requires fewer lines of text to be searched than an 8-byte signature. A program listing file requires less searching than does a textual file.

Does setting a pattern signature for "computer" as illustrated previously lead to any problems? Yes, for if we are searching for the string "computer," we will retrieve any string that contains "computer" as a substring, for example, "minicomputers." If we do not want to locate these larger words containing "computer" as a substring, we need to modify the search pattern with delimiters on both sides of the word that we are seeking, for example, "░computer░". The symbol groups, "░c" and "r░", are then included in setting the pattern signature. Including delimiters around patterns is even more important for strings more common than "computer," for example, "search," for otherwise many more lines of text need to be searched and many irrelevant words such as "research" and "searchlight" are located.

Encoding the record contents of text and program files into signatures accelerates the retrieval process by eliminating the need to exhaustively search many records. We can do it faster by doing less. The amount of preprocessing that this encoding requires is manageable, since once established, such a file remains relatively stable. A user typically edits or modifies only a few lines in a text or program file.

Faloutsos and Christodoulakis [6] provide additional information on the use of signature files. In [7], Faloutsos surveys a variety of methods for processing text.

KEY TERMS

Computer word size
Delimiters in pattern matching
EBCDIC
False drop
Filtering
Pattern matching
Pattern signature

Signature
   Bits
   Formation
Symbol manipulation
Text searching
Text signature

EXERCISES

1. Verify that the remaining signature bits in the example with "computer" are set properly for both 8- and 16-byte signatures.

2. **(a)** Which bits are set in an 8-byte signature for the word "cobble"? **(b)** for a 16-byte signature?

3. If a line of text contains only the words "computer science" followed by blanks, does the signature for the pattern "ten" match with it when using 8-byte signatures? With 16-byte signatures? Explain.

4. Repeat the previous problem but with the pattern "end" instead of "ten."

5. In Figure 5.2, why are fewer lines of a program file searched than for a text file?

6. Why does a 16-byte signature lead to better performance than an 8-byte signature?

7. Why do fewer lines of text need to be searched for longer patterns as compared with short patterns?

8. **(a)** What is the primary disadvantage of using a k value of three for the symbol groupings input to the hashing function? **(b)** for a k value of one?

9. Which pattern requires *more* lines of text to be searched exhaustively when using text signatures, *chop* or *cherry tree*?

10. Describe a procedure using text signatures for searching for a pattern such as "computer science" which may span two lines of text.

REFERENCES

1. Knuth, D. E., J. H. Morris, Jr., and V. B. Pratt, "Fast Pattern Matching in Strings," *SIAM J. Computing,* Vol. 6 (1977), pp. 323–350.
2. Boyer, R. S., and J. S. Moore, "A Fast String Search Algorithm," *CACM,* Vol. 20 (1977), pp. 762–772.
3. Harrison, M., "Implementation of the Substring Test by Hashing," *CACM,* Vol. 14 (1971), pp. 777–779.

4. Tharp, A.L., and K.C. Tai, "The Practicality of Text Signatures for Accelerating String Searching," *Software-Practice and Experience,* Vol. 12 (1982), pp. 35–44.

5. Knuth, D. E., *Sorting and Searching,* Addison-Wesley, Reading, MA, 1973.

6. Faloutsos, Chris, and Stavros Christodoulakis, "Signature Files: An Access Method for Documents and Its Analytical Performance Evaluation," *ACM TOOIS,* Vol. 2, No. 4 (October 1984), pp. 267–288.

7. Faloutsos, Chris, "Access Methods for Text," *Computing Surveys,* Vol. 17, No. 1 (March 1985), pp. 49–74.

Chapter 6

Secondary Key Retrieval

BACKGROUND

Until now, except for the discussions using binary attributes for retrieval, all the retrievals have been for either an *entire* file, when performing sequential processing, *or* for a *single* record, when performing direct access. But how do we manage situations in which we need information on some number of records between the extremes of one and everything? We may want to find which employees work in the accounting department, or which employees are programmers, or even which programmers work in the accounting department. Usually more than one person meet these requirements. When we performed direct access for a single record, we used the primary key field to identify the proper record. Since a primary key must uniquely distinguish a record from all others, obviously a primary key retrieval retrieves only a single record. You will recall that all the fields that are not or cannot be used to identify a record are the *secondary keys.* Fields such as *title* and *department* in an employee record are examples of secondary key fields. When we perform a secondary key retrieval, then, we want all the records that have a certain value, or values, for one or more of the secondary key fields. Boolean operators may be used to combine the secondary key values to qualify a request. This type of retrieval is also called *multiattribute* retrieval since we usually specify a value for more than one attribute. A *conjunctive query* is one in which multiple attributes must have specified values; that is, the attribute values are "AND"ed together, hence the name conjunctive. An example of a conjunctive query is to find all employees who both work in the data processing department *and* earn more than $45,000. If the attribute values are "OR"ed together, that is a *disjunctive query.* Of course, both "AND"s and "OR"s can be combined in a single query that is referred to as a *mixed query.* The form in which the user specifies the desired attribute values is dependent upon the *query language* used for retrieval. End-user-oriented information management systems usually have a specific query language associated with them, for example, DL-1 with the IMS database management system and SQL and QBE with the DB2 database management system. For the present, we focus on the mechanisms for retrieving information and not on the format of specifying what to retrieve. See [1] for additional information on database query languages.

How can we retrieve the records with a certain secondary key value? One method is to scan *all* the records in the file and check the attribute value of *each* record to determine if it matches the desired value. Obviously, for a file of any size, searching *all* the records is extremely inefficient. We want to limit the search as much as possible to only those records with the desired values. We can accomplish this goal by exchanging additional storage for less processing time during retrieval. We now consider several such methods for improving the retrieval performance for secondary key retrieval.

MULTILIST FILE ORGANIZATION

One means of reducing the number of records searched during a secondary key retrieval is to add a link field for *each* secondary key field for which retrieval is likely. Such a link field chains together all the records that have the *same* value in the associated data field, for example, all programmers. Each secondary key field for which a link field has been added has an array of list heads associated with it; each array contains as many entries as there are values for that field in the data records in the primary file. A list head points to the initial record in the primary file with a certain value. The link field in that record in turn points to the next record, which contains the *same* secondary key value. The format of the employee record, for example, might be modified to appear as

Employee record

Employee name	Number	Title	t _ link	Dept.	d _ link	Manager	. . .

in which the t___link (title link) points to the record of another employee with the same title and the d___link (department link) points to the record of another employee who works in the same department (most likely a different record from that pointed to by the t___link). The records of the primary file then have multiple lists or multilists running through them. Although this method improves secondary key retrieval performance, a disadvantage is that it uses a large number of links that require storage and must be maintained whenever a record is moved. A multilist organization is therefore not appropriate with collision resolution methods in which the records are moved. Another potential disadvantage of the multilist organization is that the user must decide which secondary key fields to link prior to creating the primary file.

INVERTED FILES

> **Inverted files,** which are not to be confused with inverted matrices, are merely additional files that are *indexes* into the primary file of data.

The primary file may be a direct-access file constructed using the procedures discussed in Chapter 3. We again use preprocessing of the data to create these indexes. An inverted file may exist for *every* field or attribute on which we are likely to retrieve records with specific secondary key values. In a *fully inverted* file, every attribute is inverted on whereas in a *partially inverted* file, only selected attributes have indexes. One question that is difficult to answer is which fields should be inverted on. If we invert on too many, we waste storage; if we invert on too few, we incur greater processing time on retrieval. Each entry in an inverted file index contains a secondary key value, which occurs in the data in the primary file, and pointers to *all* the records that contain that value, for example,

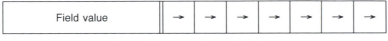

where → represents a pointer to a
record in the primary data file.

What form should these pointers take? Two possibilities exist.

- The *addresses* of the records in the primary file that contain that field value.
- The *primary keys* of the records in the primary file that contain that field value.

What is the difference? With the addresses, the record may be accessed immediately, whereas with a primary key pointer, a hashing procedure must be used to determine the actual address. Also an address may require less space than a primary key. This additional processing and storage associated with using primary keys may suggest that addresses are then preferable. But that is not always the case, for what happens when the position of a record is *moved* in the primary file, for example, in using a dynamic collision resolution technique? All the pointer addresses in the inverted files for that record need to be updated. *Maintainability* is an important consideration. It is essential that the pointer values be updated accurately. Keeping track of pointers can be a messy job. It is also harder for a person to debug and check for correctness if the pointer values are addresses such as 023987 rather than field values such as "accounting." The entry "accounting" has much more mnemonic value than "023987." If the primary file is organized for direct access using a dynamic collision resolution procedure, it is wise to use the *primary key* alternative.

Since the entries in an inverted file are likely to have a variable number of pointers, what is a space-efficient way to represent them? If they are implemented as fixed-length records in which space is allocated for the maximum number of primary file records that may have a particular attribute value, considerable space could be wasted. For example, there may be many programmers but few managers. The entries could be implemented as variable-length records *or* with two sizes of fixed-length records. In the latter case, the primary record contains (1) the field value, (2) pointers to records in the primary file that have that field value, and (3) a pointer that, when needed, points to an overflow record. An overflow record contains pointers to records in the primary file that have the specified field value and a pointer that, when needed, points to another overflow record. Figure 6.1 illustrates the use of the two fixed-length record formats to store variable-length information.

Figure 6.2 illustrates inverted files for the title and department secondary key fields of an employee file. The lines from the inverted files to the primary file indicate

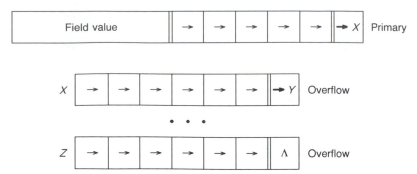

Figure 6.1 Representing variable-length information using fixed-length records.

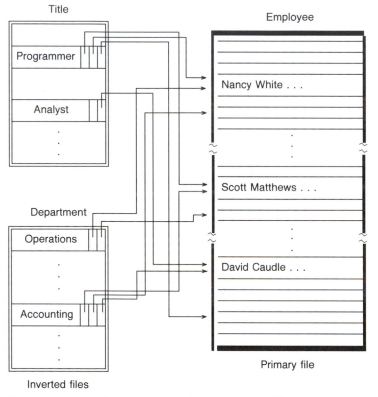

Figure 6.2 Secondary key retrieval using inverted files.

pointers that identify the relevant records in the primary file; on retrieval then, only these relevant records need to be accessed, which provides a significant savings of time.

Discussion

With the information available from the field value portions of the inverted file entries, the records in the primary file can be encoded to conserve storage. Instead of storing an actual attribute value in a record, we can store the address of the entry in the inverted file that contains the complete attribute value. For example, the record for Scott Matthews could appear as

| Scott Matthews | 0123 | x | Accounting | Melissa Jones | 0026000 | . . . |

instead of

| Scott Matthews | 0123 | Programmer | Accounting | Melissa Jones | 0026000 | . . . |

where x = address ("programmer") in the inverted file for **title.**

Less space is required to store the record in the revised format if the address requires fewer bytes than does the actual attribute value. Such savings in storage could be achieved for each field that is inverted on. The disadvantage of this technique is that the records must be reconstructed for output since the only place that the entire attribute value exists is in the entry in the associated inverted file. For example, the value "programmer" exists only in the *title* inverted file. Since reconstructing a record requires accessing additional files, and with storage costs continuing to decrease, this revised record format is of limited use. It may be appropriate, however, in situations in which records are output relatively infrequently, and using it allows the primary file to be stored in high-speed memory.

Although inverted files are widely used for secondary key retrieval, there are disadvantages associated with them. When a value of a field in a record changes, for example, when an employee is promoted from programmer to analyst, the appropriate inverted file, in this case "Title," must be updated also. The greatest disadvantage of inverted files is the amount of storage that may be required to store them; if most or all of the secondary key fields are inverted upon, then more storage may be required for them than the original primary file. In addition, in a conjunctive query, as the number of retrieval attributes increases, although the number of records retrieved will likely decrease, the number of inverted fields that must be accessed increases. In the next section, we consider an alternative technique for handling secondary key retrieval that attempts to overcome some of these shortcomings of inverted files.

KEY TERMS

Addresses
Fixed-length records
Fully inverted
Inverted file
Maintainability
Multiattribute retrieval
Multilist
Partially inverted
Pointers

Primary key
Primary key retrieval
QBE
Query
    Conjunctive
    Disjunctive
    Mixed
Query language

| Secondary key | SQL |
| Secondary key retrieval | Variable-length records |

EXERCISES

1. Given the file layout

Student number	Major	Class	Hometown	Expected Graduation Date

where student number is the primary key, what is the maximum number of inverted files that could be associated with the file?

2. How do the *number* of records retrieved differ between a primary key retrieval and a secondary key retrieval?

3. What are two methods of implementing pointers in inverted files?

4. Give an example of a situation in which a multilist organization is appropriate.

5. How many list head entries are necessary for a multilist organization of the title field with the data in Figure 2.1?

6. What are two advantages of an inverted file organization over a multilist organization?

7. Create one inverted file for presidential party and one for the state name using the PRESI-DENT file illustrated below. Assume that the president's name is the primary key.

Pres. name	Birthdate	Deathdate	Pres. party	Spouse	State
Eisenhower	10/14/1890	03/28/1969	Republican	Mamie	Texas
Kennedy	05/29/1917	11/22/1963	Democrat	Jacqueline	Mass.
Johnson	08/27/1908	01/22/1973	Democrat	Claudia	Texas
Nixon	01/09/1913	Λ	Republican	Patricia	Calif.
Ford	07/14/1913	Λ	Republican	Elizabeth	Mich.
Carter	10/01/1924	Λ	Democrat	Rosalyn	Georgia
Reagan	02/06/1911	Λ	Republican	Nancy	Calif.

PARTIAL MATCH RETRIEVAL WITH SIGNATURE TREES

Rather than having an index for each secondary attribute as with inverted files, **partial match retrieval** forms a single binary encoding for all the information contained within a record. This binary encoding may be considered a *signature* or a descriptor for a data record that is similar in purpose to the text signatures that we considered earlier. These signatures capture the essence of each data record such that instead of searching the actual records upon secondary key retrieval, we only need to search the signatures. For the same reasons that it is more efficient to search the text signatures than the actual text data, it is more efficient to search the record signatures than the actual data records in a primary file.

A basic difference between record signatures and text signatures is the manner in which they are formed. Text signatures are formed using *superimposed coding* in which any portion of the text line may set any bit in the signature. The procedure for forming a record signature, proposed by Pfaltz, Berman, and Cagley [2], uses *disjoint coding* in which each field of the record sets bits in an area of the signature that is *disjoint* from the areas set by the other fields, for example,

Employee record

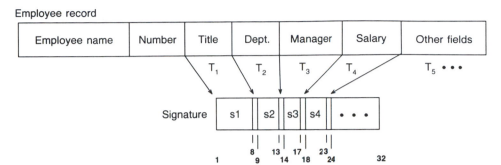

A function, specifically a hashing function, is used to convert the *value* in a specific field of a record, for example, "programmer," into a bit pattern in a portion of the signature, for example, 00001000 (for bit positions 1 through 8). Typically, each data field has a *different* hashing function associated with it for converting a data value into a bit pattern. The output of a hashing function designates which bit positions in the corresponding signature field are set to "1". For example, we could apply hashing functions that

- Set a bit in s_1 (bit positions 1 through 8) using

  T_1 = employee.title **mod** 8 + 1

- Set a bit in s_2 (bit positions 9 through 13) using

  $$T_2 = \begin{cases} 1, \text{ if employee.department's name begins with A–E} \\ 2, \text{ if employee.department's name begins with F–J} \\ 3, \text{ if employee.department's name begins with K–M} \\ 4, \text{ if employee.department's name begins with N–R} \\ 5, \text{ if employee.department's name begins with S–Z} \end{cases}$$

- Set a bit in s_3 (bit positions 14 through 17) using

  T_3 = employee.manager **mod** 4 + 1

- Set a bit in s_4 (bit positions 18 through 23) using

  $$T_4 = \begin{cases} 1, \text{ if employee.salary} \leq \$10,000 \\ 2, \text{ if employee.salary is between } \$10,000 \text{ and } \$20,000 \\ 3, \text{ if employee.salary is between } \$20,000 \text{ and } \$25,000 \\ 4, \text{ if employee.salary is between } \$25,000 \text{ and } \$30,000 \\ 5, \text{ if employee.salary is between } \$30,000 \text{ and } \$45,000 \\ 6, \text{ if employee.salary is } > \$45,000 \end{cases}$$

For x to be *between* y and z, $y < x \leq z$. The bit position set by a particular hashing function is offset by an appropriate amount to position it correctly in the record signature. Obviously, to ensure an even distribution of the bit settings, the hashing functions are designed with respect to the range of values expected to occur for the associated field. Since T_1 and T_3 take alphabetic data as input to a **mod** function, and since the conversion process from an alphabetic form to an integer one is not obvious, you will not be able to verify the output from those transformations.

Retrieval

After the record signatures have been formed by preprocessing the file, retrieval of records with certain secondary key values is possible. Using the sample hashing functions, if we want to retrieve employees in the data processing department, a retrieval signature (or search signature or query signature) is formed with a "1" bit in position 9. T_1 sets bits in positions 1 through 8 of the retrieval signature. T_2 sets bits in positions 9 through 13; the output values of T_2 are offset by eight positions to properly position them in the signature. Since "data processing" begins with a "d," T_2 yields a one, which when offset by eight, stipulates that bit position 9 should be set to "1". To locate the desired records then, a comparison is made between the retrieval signature and each record signature. If a match occurs, the record drops out; that is, it is identified for further processing. For another example, if we are interested in all employees earning more than \$45,000, the retrieval signature would have a "1" in bit position 23. If we want to retrieve all of those employees who *both* work in the data processing department *and* earn more than \$45,000, the retrieval signature would contain a "1" in positions 9 and 23. For a match to occur, both corresponding bit positions in a record signature need to match. Although each record signature must be examined, as was noted with text signatures, most computers have hardware instructions which can compare all the bits of the two signatures in a single, efficient operation.

What further processing is necessary if a record signature matches the retrieval signature? If a match occurs, we are not guaranteed that we have found a desired record. Why? Because we used hashing functions to set the signature bits, collisions could happen, which means that more than one field value could set a particular bit. In the employee example, both the data processing and accounting departments set bit 9. We then need to verify that a record that drops out does indeed have the desired value. If it does not, it is referred to as a **false drop.** We can minimize the number of false drops by ensuring that the hashing functions evenly distribute the secondary key values over the corresponding signature segment and/or by increasing the width of the signature segment. Increasing the width of a signature segment is the same as reducing the packing factor in a hashed table to reduce the number of collisions.

Signature Trees

To further improve the efficiency of searching the record signatures, we add an *index* into them. Just as the indexed sequential file organization allowed us to speed the direct-access retrieval performance, this index into the record signatures is intended to do the same with secondary key retrievals by reducing the number of signatures that must be searched. Because of the similarities with indexed sequential file organization, French [3] refers to this technique of partial match retrieval with signature trees as the indexed descriptor access method or IDAM. The signature tree is formed by grouping the record signatures for some number of contiguous addresses. A "supersignature" is formed by *superimposing* or ORing these record signatures to capture the meaning of all the associated records. As a result of this condensation process, there are fewer supersignatures than record signatures. We can carry this grouping and compaction idea to as many levels as we desire. We could group several supersignatures together to form a "super duper signature" by superimposing the associated supersignatures. The lengths of the supersignatures and super duper signatures are the same as the original record signatures because they are formed by superimposing. The general structure of a signature tree is shown in Figure 6.3. Level 1 signatures are the record signatures; level 2 signatures correspond to the supersignatures; level 3 signatures correspond to the super duper signatures; etc.

An Example

Let's assume that we have organized a file of employee records for direct-access retrieval on the primary key of the employee. To create a signature tree for the partial match retrieval of secondary keys, the first step is to form the record signatures using disjoint coding. We concentrate on only the four secondary key fields of *title, department, manager,* and *salary*. The transformations given earlier in this section will be used to form the record signatures. Figure 6.4 illustrates the signatures formed by grouping the record signatures of five contiguous *locations* (not records) to form supersignatures (or level 2 signatures) and joining three supersignatures to form super duper signatures (or

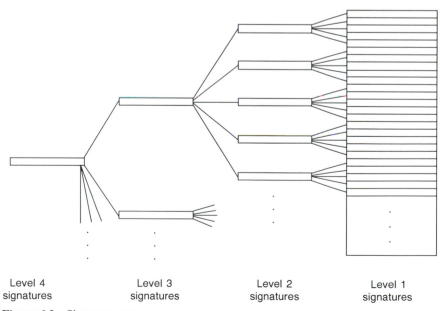

| Level 4 signatures | Level 3 signatures | Level 2 signatures | Level 1 signatures |

Figure 6.3 Signature tree.

level 3 signatures). The resulting signature tree is of depth 3. Note that records 2 and 18 have signatures of all zeros. These result from the corresponding file locations being empty. These empty locations occur unless the file has a 100 percent packing factor.

To retrieve the records of employees in the data processing department, the first step is to construct the retrieval signature. We build that by hashing the sought value through T_2. That yields a "1" in position 9 of the retrieval signature. We then begin a complete search of the signatures in the highest level of the tree; in this example that is level 3. If a level 3 record signature has "1"s in the bit positions corresponding to where the retrieval signature has "1"s, the signatures on the next lower level must be searched. If the 1 bits do *not* match, then *no further searching of that signature subtree* or its associated records is necessary. This tree pruning operation can then yield better performance by reducing the number of records that must be searched exhaustively. In the example, position 9 of the first level 3 record signature contains a "0". So there is no match. That means that this one comparison eliminated possibly 18 other signatures and 15 records from further processing.

We continue then by comparing the retrieval signature against the next level 3 record signature. In this circumstance, position 9 of the record signature does have a "1" in it. That says that we must search the level 2 record signatures, which are offspring of the matched level 3 signature. There are two such level 2 signatures, and both have a "1" in bit position 9. Therefore, we must continue searching all 10 of the

Level 3				Level 2				s_1	s_2	s_3	s_4	Address
								00010000	01000	1000	010000	0
								00000010	00001	0010	010000	1
				10010010	01011	1010	010011	00000000	00000	0000	000000	2
								10000000	00010	0010	000010	3
								00010000	00010	1000	000001	4
								00010000	00001	0010	000010	5
								10000000	00010	0010	000001	6
10110010	01011	1011	111111	10010010	01011	1010	000111	00010000	00010	0010	000010	7
								00000010	01000	1000	000100	8
								00010000	01000	1000	000001	9
								00000010	00010	0010	000001	10
								10000000	00010	0010	100000	11
				10110010	01010	1011	101001	00010000	00010	0010	000001	12
								00100000	01000	0001	000001	13
								00010000	01000	1000	001000	14
								00000010	10000	0010	010000	15
								00000100	00010	0010	001000	16
				00010110	10010	1010	011001	00010000	10000	0010	010000	17
								00000000	00000	0000	000000	18
00110111	11011	1110	011101					00010000	10000	1000	000001	19
								00000010	10000	0010	000100	20
								00000001	00010	0010	001000	21
				00110011	11011	1110	011101	00010000	00010	0010	000001	22
								00100000	01000	0100	010000	23
								00000010	00001	1000	010000	24

Level 3 Level 2 Level 1

Figure 6.4 Record signature tree.

level 1 signatures, which are their offspring. The level 1 record signatures corresponding to file locations 15, 17, 19, and 20 match the retrieval signature. Does that suggest that four employees work in the data processing department? No, because there is still the possibility of false drops. We then retrieve the four corresponding records to determine if those employees do indeed work in data processing. If the signature tree is stored in primary memory, only four accesses of auxiliary storage are necessary to retrieve these four records. Four accesses is certainly an improvement over 25.

How do we locate all the highly paid employees, that is, those earning in excess of $45,000? Again the first step is to build the retrieval signature. In this case, we apply T_4 to the retrieval value to obtain a "1" in bit position 23. What is the retrieval signature for employees earning more than $35,000? Both bit positions 22 and 23 would contain a "1" but on retrieval, if either bit position matched in a record signature, that would signify a match. In the previous case of only bit 23 being set to "1", we find that both level 3 record signatures match as do all the level 2 record signatures. The signatures corresponding to file locations 4, 6, 9, 10, 12, 13, 19, and 22 then match. Do some of the eight employees earn $45,000 or less? Are there any false drops? In this instance, no, since the range of the query is the same as the range of the hashing function. But rather than checking for this special case that occurs infrequently, the search procedure retrieves all the records, which it has to do anyway, but it *unnecessarily* checks for false drops. It may be preferable to do that than to complicate the algorithm.

What if we want to retrieve the records of all those employees who both work in the data processing department *and* earn more than $45,000? For this conjunctive query, the retrieval signature has "1"s in both positions 9 and 23. For it to match a record signature, both positions must match. With the signature tree of Figure 6.4, only the second level 3 record signature matches. Therefore, only two level 2 record signatures need to be searched but both of them match; so we need to search their offspring. Of these 10 level 1 record signatures, only the one corresponding to file location 19 matches; so at most there is only one well-paid employee in data processing. If upon actually checking the record in that position, we find a false drop, then there are no such employees.

The filtering mechanism of the signature tree can be especially useful on unsuccessful searches. For example, if we want to retrieve information about the poorly paid people performing in the accounting department, the retrieval signature has "1"s in positions 9 and 18, assuming that poorly paid means earning less than $10,000. Since neither record signature at level 3 matches the retrieval signature, we know *absolutely, positively* at that point that there are no poorly paid individuals in the accounting department; so additional searching is unnecessary.

Discussion

Encoding the secondary attributes of records into signatures or descriptors provides us with a filter to quickly discard those records which are not relevant to our current information needs. We have accelerated the retrieval performance by reducing the amount of information that has to be processed. This is another illustration of the principle: *do it faster by doing less.*

Partial match retrieval with signature trees has advantages over inverted files for secondary key retrieval.

- As the number of attributes ANDed together to form a query increases, the amount of information processing actually decreases. With inverted files on the other hand, as the "and" complexity increases, the number of inverted files that have to be processed increases and so overhead processing time increases.
- Different attributes may be weighted differently in determining the amount of storage overhead to be associated with each. With inverted files, each decision is binary; either an attribute is inverted on or it is not. With signatures, the number of bits assigned to a segment of it can be determined by the relative importance of the associated field. The more important the field, the greater is the number of bits associated with it in the signature.
- The amount of storage overhead may be less. Partial match retrieval has overhead percentages for storage that typically range from 5 to 40 percent, depending upon how the index is formed, whereas inverted files may have overhead percentages that are greater than 100 percent.

When the value of a field in a record changes, the signature for that record needs to be changed as do all its ancestor signatures at the higher levels in the signature tree. If only one field of a record changes, this updating of the signatures may take more time than the updating of the indexes required with inverted files. As more fields are updated at one time, though, the updating costs for the signatures remain essentially constant, whereas with inverted files, the costs increase proportional to the number of fields updated.

For a practical implementation, the signatures are typically in the 100- to 200-bit range, and the blocking factor, the number of signatures combined to form a higher-level signature, is in the order of 100 or more. For better performance, as much of the signature tree index as possible should be stored in primary memory; it is not necessary that the index be stored with the records in the primary data file.

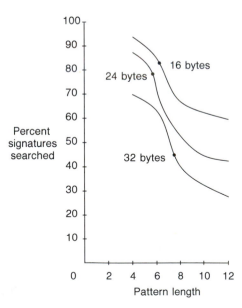

Figure 6.5 Effect of signature length; signatures searched vs. pattern length; text file.

Signature Trees and Text Searching

The concept of using an *index,* in the form of a tree, into a set of signatures to accelerate searching can also be applied to the use of text signatures. By building a signature tree, the number of signatures that must be examined can be reduced. What we are adding with a signature tree is another level of *filtering.* We can accelerate the search by doing less. Adding a signature tree to text searching also exemplifies the principle of *applying familiar techniques to new contexts.* A goal of this text is to encourage the reader to apply the techniques described here to novel situations. Figure 6.5 displays the reduction in signatures searched for various pattern lengths and signature sizes when using a signature tree.

KEY TERMS

Blocking factor	Record
Signature tree	Retrieval
IDAM	Search
Partial match retrieval	Super
Signature tree	Super duper
Signature	Text
Query	

EXERCISES

1. Using the signature tree in Figure 6.4, which records are accessed when seeking the records of the employees who work with the sanitation department?

2. Describe a procedure using partial match retrieval with signatures for responding to the query, "List the names of the employees who do *not* work in the sanitation department."

3. Build a signature tree for the records in Figure 2.1 from Chapter 2. Form the signatures using only the fields: title, department, and salary. Use the following hashing functions.

   - Set a bit in s_1 (bit positions 1 through 3) using

$$T_1 = \begin{bmatrix} 1, \text{ if employee.title begins with A–E} \\ 2, \text{ if employee.title begins with F–L} \\ 3, \text{ if employee.title begins with M–Z} \end{bmatrix}$$

   - Set a bit in s_2 (bit positions 4 through 8) using

$$T_2 = \begin{bmatrix} 1, \text{ if employee.department's name begins with A–E} \\ 2, \text{ if employee.department's name begins with F–J} \\ 3, \text{ if employee.department's name begins with K–M} \\ 4, \text{ if employee.department's name begins with N–R} \\ 5, \text{ if employee.department's name begins with S–Z} \end{bmatrix}$$

   - Set a bit in s_3 (bit positions 9 through 14) using

$$T_3 = \begin{cases} 1, \text{ if employee.salary} \leq \$10,000 \\ 2, \text{ if employee.salary is between } \$10,000 \text{ and } \$20,000 \\ 3, \text{ if employee.salary is between } \$20,000 \text{ and } \$25,000 \\ 4, \text{ if employee.salary is between } \$25,000 \text{ and } \$30,000 \\ 5, \text{ if employee.salary is between } \$30,000 \text{ and } \$45,000 \\ 6, \text{ if employee.salary is } > \$45,000 \end{cases}$$

Use a blocking factor of three to form the higher-level signatures.

4. What is the difference between superimposed coding and disjoint coding?

5. How is hashing used in partial match retrieval?

6. Is superimposed coding used in partial match retrieval? Explain.

7. What are the advantages and disadvantages of a multilist organization when compared with partial match retrieval using a signature tree?

PARTIAL MATCH RETRIEVAL WITH PAGE SIGNATURES

The implementation of partial match retrieval in the preceding section used a signature tree as an index into the record signatures. The signature tree acted as a *filter* to reduce the number of record signatures that had to be examined. An alternative filtering mechanism is to group the data records (and signatures) into pages or buckets based upon the values of the secondary attributes. Records with the same attribute values are stored on the same page. As a result, rather than having to search the entire file, only those portions of the file with records containing characteristics similar to the desired characteristic need to be examined. We want to use a *divide and conquer* search strategy.

In addition to organizing the data records based upon attribute values, **partial match retrieval with page signatures** [4] associates a *signature* or descriptor with each page, which captures the essence of all the records on a page. It is formed by *superimposing* or *OR*ing the record signatures for the records on that page. Upon retrieval, only those signatures of pages likely to contain the sought attributes need to be checked. If the page signature of a candidate page does not match the retrieval signature, then none of the record signatures on that page need to be searched.

The primary extension and difference of partial match retrieval with page signatures from partial match retrieval with tree signatures is the mapping from the secondary attributes into storage pages. Not only is the information of a record condensed to form a record signature, but it is *also* condensed to form the page number of the page on which the record is to be stored. By employing both processes, we have *two* filtering mechanisms to reduce the search space. By having less to search, we can do it faster. For retrieval, the two-step filtering process may be viewed as

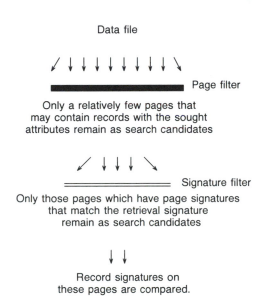

Data file

Page filter

Only a relatively few pages that
may contain records with the sought
attributes remain as search candidates

Signature filter

Only those pages which have page signatures
that match the retrieval signature
remain as search candidates

Record signatures on
these pages are compared.

A process using disjoint coding, similar to the process used for forming the record signatures, is used to classify the records. Transformations or hashing functions are used to convert the secondary attributes to a page number, for example,

Employee record

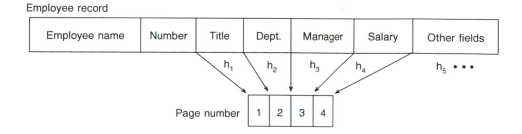

A hashing function, h_i, converts the ith attribute value into the ith section of the page number containing d_i digits. In the preceding example, all the page number sections are one digit in length to preserve the simplicity of the example. In general, though, the segments would be larger; overall the number of digits in a page number is considerably less than in a record signature. Let's use the following hashing functions to map an employee record to one of 16 possible pages, numbered from 0000 to 1111.

$$h_1 = \text{employee.title } \textbf{mod } 2$$

$$h_2 = 0, \text{ if employee.department's name begins with A–M}$$
$$1, \text{ if employee.department's name begins with N–Z}$$

$$h_3 = \text{employee.manager } \textbf{mod } 2$$

$$h_4 = 0, \text{ if employee.salary is } \leq \$25{,}000$$
$$1, \text{ if employee.salary is } > \$25{,}000$$

The records of employees with salaries $\leq$ \$25,000 will then be located on even-numbered pages.

The second filtering mechanism forms the record signatures just as in the previous implementation of partial match retrieval. A transformation function, T_i, converts a secondary attribute value into a disjoint segment of a record signature,

Employee record

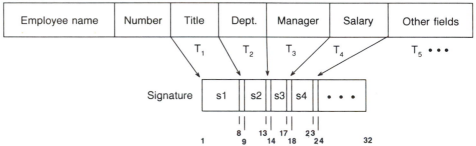

After the record signatures are formed, the page signatures are formed by *superimposing* the individual record signatures on each page.

An Example

We again use the employee record example with the record signature data given in Figure 6.4. First, the records are classified into the 16 pages using the h_i. The 23 records, not the 25 locations, of Figure 6.4, are distributed into 13 of the 16 possible pages. Since we do not know the numeric equivalents for the keys processed by h_1 and h_3, we cannot verify the placement of records within a page. The page signatures are formed by superimposing the record signatures on each page. For example, the page signature for page 2 is formed by superimposing the record signatures of its four records, that is,

	00010000 01000 1000 001000
Record	00100000 01000 0100 010000
signatures	00010000 10000 0010 010000
	00000010 10000 0010 010000

Page signature   00110010 11000 1110 011000   *OR*ing

The complete set of page and record signatures is illustrated in Figure 6.6.

Retrieval

To retrieve the records of the employees in the data processing department, we first determine on which pages such records might appear by applying h_2 to "data processing." The result is a 0 in the second digit of the page number and ? (don't cares) in the other digits, since we do not have any values specified for them; the page number must then have the template ?0??. Out of the 16 possible pages, only 8 have a 0 in the second bit position, that is, pages 0–3 and 8–11. The page signatures of these pages are compared against the retrieval signature. The retrieval signature is formed in the same manner as with partial match retrieval with tree signatures. T_2 is applied to "data processing," yielding a "1" in bit position 9. Since no records appear yet on pages 8 or 10, only six page signatures need to be compared. Matches occur with the page signatures for pages 2, 3, and 9. We then match the retrieval signature against each of the record signatures on these three pages. A total of four record signatures match; we are not yet finished, for the corresponding records in turn must be checked for *false drops.*

How do we locate all the highly paid employees, that is, those earning in excess of $45,000? Again, the first step is to build the page retrieval template. $h_4(>\$45,000)$ yields a "1" in digit four; the page retrieval template is ???1. The page signatures of the odd-numbered pages must be compared against the retrieval signature formed by applying T_4 ($>\$45,000$) which yields a "1" in bit position 23. The page signatures match for pages 1, 3, 5, 7, and 15. Only the record signatures on these five pages require additional processing. The eight records on these pages that drop then need to be retrieved to check for false drops.

If we want to retrieve those employees who *both* work in the data processing department *and* make more than $45,000, the page retrieval template is ?0?1. In this case, the page signatures of only 4 of the 16 pages need to be compared against the retrieval signature that has a "1" in *both* bit positions 9 and 23. Only the page signature for page 3 matches. The record signature on page 3 is then checked, and one record drops. Only a total of five signatures had to be compared in this case. We have reduced the retrieval effort by filtering out those *pages* which are not applicable and then filtering out those *records* which are likewise not applicable. The use of the page filtering is particularly effective in reducing the retrieval effort.

If we want to retrieve information about the poorly paid people performing in the accounting department, the page retrieval template is ?0?0 and the retrieval signature would have "1"s in positions 9 and 18, assuming that poorly paid means earning less

Page signatures				Record signatures				Page	number
00010000	01000	1000	010000	00010000	01000	1000	010000	0000	0
00110000	01000	1001	000001	00010000	01000	1000	000001	0001	1
				00100000	01000	0001	000001		
00110010	11000	1110	011000	00010000	01000	1000	001000	0010	2
				00100000	01000	0100	010000		
				00010000	10000	0010	010000		
				00000010	10000	0010	010000		
00010000	10000	1000	000001	00010000	10000	1000	000001	0011	3
00000000	00000	0000	000000						4
00010000	00011	0010	000011	00010000	00001	0010	000010	0101	5
				00010000	00010	0010	000001		
00000010	00001	0010	010000	00000010	00001	0010	010000	0110	6
10000010	00010	0010	000001	10000000	00010	0010	000001	0111	7
				00000010	00010	0010	000001		
00000000	00000	0000	000000						8
00000010	10000	0010	000100	00000010	10000	0010	000100	1001	9
00000000	00000	0000	000000						10
00000010	01000	1000	000100	00000010	01000	1000	000100	1011	11
10000011	00011	1010	111000	10000000	00010	0010	100000	1100	12
				00000010	00001	1000	010000		
				00000001	00010	0010	001000		
10001000	00010	0010	000010	10000000	00010	0010	000010	1101	13
				00001000	00010	0010	000010		
00000100	00010	0010	001000	00000100	00010	0010	001000	1110	14
00010000	00010	1010	000001	00010000	00010	0010	000001	1111	15
				00010000	00010	1000	000001		

Figure 6.6   Page and record signatures.

than $10,000. The page signatures of pages 0 and 2 need to be compared. No records exist yet for the other two pages matching the template, pages 8 and 10. Neither page signature matches the retrieval signature; so we know *absolutely, positively* that no such employees exist in the file after only two comparisons. Contrast that effort with a sequential scan of all the records in the file.

Discussion

The page and record signatures *do not* need to be stored with the actual records. If they are stored in primary memory, the filtering operation can be performed quite efficiently.

An advantage of using page descriptors instead of a descriptor tree is that when a record is inserted, deleted, or updated on a page, only the page signature for that page needs to be modified. It is not necessary to *propagate* that change up through the signature tree, that is, to modify the superimposed ancestor signatures at the higher levels of the signature tree. This feature is valuable in applications in which the location of a record may change periodically throughout the life of the file, for example, (1) with a collision resolution method using dynamic positioning of the records, or (2) with expandable hashed files, which we consider in Part Three on tree structures.

The primary disadvantage of page signatures occurs when the records on which we intend to perform secondary key retrieval have previously been inserted into a direct-access file for primary key retrieval. In that case, we *cannot physically move* those records to different locations based upon the secondary key values, for if we did we could no longer retrieve them for primary key retrieval. Therefore, to be able to perform both primary and secondary key retrieval with a single copy of each record, we need to construct a series of *pointer pages* as illustrated in Figure 6.7. These pointer pages map the records as they would be grouped based upon their secondary attributes into their actual storage locations. The difficulty with needing the pointer pages is that they require *extra storage*. The m secondary key fields set n bits that specify the page addresses of the pointer pages for grouping on those fields. The m fields also set p bits

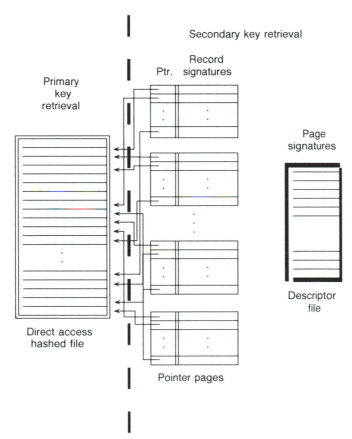

Figure 6.7 Pointer pages for connecting primary key retrieval with secondary key retrieval.

for the record signatures for a pointer page which in turn are superimposed to form the page signatures. The structures to the left of the vertically dashed line are needed for primary key retrieval—structures such as we saw in Chapter 3 on direct-access file organization—and those to the right of the dashed line are needed for secondary key retrieval.

Page signatures are not effective for queries over a wide range of values, for example, in this case, employees earning $\leq$ \$45,000 since they would not eliminate many pages from being searched.

A question that arises for partial match retrieval with both signature trees and page signatures is how does one determine the number of bits to associate with each segment of a signature. Moran [5] has investigated the optimal determination of such parameters.

KEY TERMS

Disjoint coding   Signature
Page retrieval   Page
Partial match retrieval   Record
    Page signatures   Template
Pointer page

EXERCISES

1. (a) In Figure 6.6, which page signatures are compared if we want to retrieve all employees in the training department? (b) How many records actually drop?

2. (a) Which page signatures are compared if we want to retrieve all employees in the personnel department who earn more than \$45,000? (b) How many records actually drop?

3. If a page has a large number of records mapped into it, its page signature could become quite dense, that is, have a large percentage of "1" bits. A dense signature is not effective in eliminating records from subsequent searching since it is likely that the 1s of the search signature would match its "1" bits. Describe how a signature tree might be associated with a page of records and used in addition to a page signature for the purpose of improving secondary key retrieval. Contrast the tradeoffs between going to a larger signature size vs. using a signature tree.

4. Build page and record signatures for the records in Figure 2.1 from Chapter 2. Map the records to eight pages using the following hashing functions on the fields: title, department, and salary.

   • Set page bit position 1 using

   h_1 = 0, if employee.title begins with A–L
        1, if employee.title begins with M–Z

   • Set page bit position 2 using

$$h_2 = \begin{array}{l} 0, \text{ if employee.department's name begins with A–M} \\ 1, \text{ if employee.department's name begins with N–Z} \end{array}$$

- Set page bit position 3 using

$$h_3 = \begin{array}{l} 0, \text{ if employee.salary is } \leq \$25,000 \\ 1, \text{ if employee.salary is } > \$25,000 \end{array}$$

Form the record signatures using the same fields as those used to determine the page mappings but with the following hashing functions.

- Set a bit in s_1 (bit positions 1 through 3) using

$$T_1 = \begin{array}{l} 1, \text{ if employee.title begins with A–E} \\ 2, \text{ if employee.title begins with F–L} \\ 3, \text{ if employee.title begins with M–Z} \end{array}$$

- Set a bit in s_2 (bit positions 4 through 8) using

$$T_2 = \begin{array}{l} 1, \text{ if employee.department's name begins with A–E} \\ 2, \text{ if employee.department's name begins with F–J} \\ 3, \text{ if employee.department's name begins with K–M} \\ 4, \text{ if employee.department's name begins with N–R} \\ 5, \text{ if employee.department's name begins with S–Z} \end{array}$$

- Set a bit in s_3 (bit positions 9 through 14) using

$$T_3 = \begin{array}{l} 1, \text{ if employee.salary } \leq \$10,000 \\ 2, \text{ if employee.salary is between \$10,000 and \$20,000} \\ 3, \text{ if employee.salary is between \$20,000 and \$25,000} \\ 4, \text{ if employee.salary is between \$25,000 and \$30,000} \\ 5, \text{ if employee.salary is between \$30,000 and \$45,000} \\ 6, \text{ if employee.salary is } > \$45,000 \end{array}$$

5. What might be done to obtain a more even distribution of the records among the pages in the previous problem?

REFERENCES

1. Date, C.J., *Introduction to Database Management Systems,* 4th Edition, Addison-Wesley, Reading, MA, 1986.
2. Pfaltz, John L., William J. Berman, and Edgar M. Cagley, "Partial-Match Retrieval Using Indexed Descriptor Files," *CACM,* Vol. 20, No. 9 (September 1980), pp. 522–528.
3. French, James C., *IDAM File Organizations,* UMI Research Press, 1985.
4. Ramamohanarao, K., John W. Lloyd, and James A. Thom, "Partial-Match Retrieval Using Hashing and Descriptors," ACM *TODS,* Vol. 8, No. 4 (December 1983), pp. 552–576.
5. Moran, Shlomo, "On the Complexity of Designing Optimal Partial-Match Retrieval Systems," *ACM TODS,* Vol. 8, No. 4 (December 1983), pp. 543–551.

Chapter 7

Bits and Hashing

SIGNATURE HASHING

When we considered direct-access files in Chapter 3, a primary goal of the many addressing techniques was to locate the desired record in a *single access of auxiliary storage*. The motivation for this goal is that auxiliary storage accesses are an order of magnitude more expensive than main memory accesses. To optimize the retrieval performance then, we want to minimize the number of auxiliary storage accesses. One reason we want to study *signature hashing* is that it does just that. Only *one access* of auxiliary memory is needed to retrieve a record. And as the name implies, it involves hashing; so the address space may be much smaller than the key space. The previous techniques for locating a record in a single access required the address space and the key space to be the same, which wasted considerable storage.

A second reason, perhaps even more important, for considering signature hashing is that it illustrates the important principle of

combining known techniques to create new and improved means of organizing and processing data.

We are constantly striving, in all our endeavors, not just in file organization, to improve the status quo. That is one reason we are now able to use computers in so many novel applications. To continue developing innovative uses, we must continue developing better data structures. One reason for considering such a variety of data structures in this text, some of which may not seem immediately applicable, is to provide a basis for discovering new structures. The richer one's experience, the better able one should be to invent a better "mousetrap."

Larson and Kajla [1] introduced signature hashing. As the name suggests, *signature hashing* combines the techniques of using *signatures* and *hashing* to improve retrieval performance by reducing the number of auxiliary storage accesses. Signatures provide a means for encoding and compressing the key information to such an extent that it can then be stored in primary memory. By doing so, we can search the signatures in main memory much more quickly than if they were on auxiliary storage. At most, only a single access of auxiliary storage is needed to verify that we have found the desired record (akin to checking for false drops) and to retrieve the information within the record. Instead of processing the key information in the form given, by using signatures, we are *converting it into a form that is more readily processible.* That is an important concept also. If it seems that your data structure is a lemon, convert it to lemonade. Make it into something useful. You do not have to be restricted to using information in the form that you receive it. Don't be limited in your thinking.

The Method

Signature hashing is another means for resolving collisions. Normally, the order of the records within a probe chain is based upon either (1) the order of insertion or (2) the movement of records resulting from subsequent insertions. With signature hashing, the order of the records on a probe chain is determined by signature values. Two types of signatures are involved. One type, a position-key signature value or **p-k signature** value, is associated with each {probe position, record key} combination. That is, as we move through a record's probe chain, the p-k signature will change with each position, for example,

Probe chain	p-k signature
p_1	1011
p_2	0101
p_3	1100
⋮	
p_i	0010
p_{i+1}	1001
⋮	

An explanation for actually constructing a p-k signature will be deferred until the discussion of an example. The other type of signature, a **separator signature,** is associated with each table location. A separator signature value for a location is the minimum of the p-k signature values for those records stored *after* the current position in the probe chain.

Two processes integral to signature hashing are (1) determining the location in the table for attempting to insert or retrieve a record, and (2) setting the separator signature for a location when storing a record into it. In determining an insertion or retrieval location, the p-k signature value is compared with the separator signature associated with each potential location. If the p-k signature value for the record being inserted or retrieved is less than the separator signature, the relevant location has been found. If the comparison is not less than the separator signature, then the p-k signature value is matched with the separator signature for the next location on the record's probe chain.

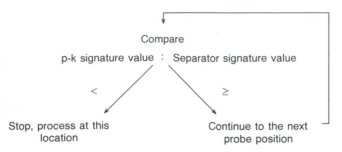

Inserting a record and setting separator signatures are intertwined. Initially, a separator signature value larger than any p-k signature value is associated with each table location. Such a separator signature value ensures that during insertion, the p-k signature value will always compare less than if the location is empty. A less than comparison does not, however, mean that the location *is* empty. When inserting a record, if the potential storage location is empty, the record is placed there and the initial separator signature remains unchanged. Otherwise, the p-k signature value for the record stored at the inspected location and p-k signature value for the incoming record are compared. The record with the *lower* p-k signature value is placed at the inspected location, the other record is considered for insertion at the next position on its probe chain, and the signature for the inspected position is revised.

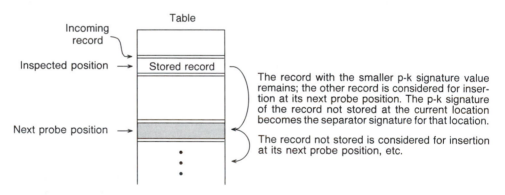

The algorithm for inserting a record into a table using signature hashing appears in Algorithm 7.1.

This process of comparing a p-k signature value against separator signature values is similar to the process with indexed sequential structures of comparing the key of the desired record against the key values in a cylinder or track index. Ordering the separator signatures guides us to the location where the record we are seeking must be stored. Instead of searching the actual locations in a file, which would typically require accesses of auxiliary storage, we search a table of separator signatures stored in high-speed memory. Signature hashing then requires extra storage for a separator table. Because we have *condensed* the key information into the form of separator signatures, it is *feasible* to store it in high-speed memory. Providing this space for the separator signatures guarantees that only a single access of auxiliary storage will be needed upon retrieval.

Algorithm 7.1
SIGNATURE HASHING INSERTION

  I. Hash the key of the record to be inserted to obtain the initial probe address.

 II. While a record remains to be inserted.

/* Search the signature table in primary memory to determine the location in the data table for attempting an insertion. */

    A. While the p-k signature for the record ⊀ separator signature for the probe address

        1. Compute the next probe address.

        2. If that probe address is the home address for the record being inserted, terminate with a "full table" message.

/* Process the data table stored in auxiliary storage. */

    B. If the location corresponding to the current probe position is empty, insert the record and terminate successfully, else,

        1. Compare the p-k signatures for the incoming record and the stored record.

        2. The record with the lower p-k signature remains at the current location.

        3. The separator signature for the current position is set to the higher p-k signature value.

        4. The record with the higher p-k signature value becomes the record to be inserted.

An Example

The best way to better understand how signature hashing operates may be through an example. Let's again use the example records from Chapter 3 on direct-access files, those with the key values of

$$27, 18, 29, 28, 39, 13, \text{ and } 16$$

We also use

$$f(\text{key}) = \text{key } \textbf{mod } 11$$

as the hashing function and

$$i(\text{key}) = \text{Quotient}(\text{key } / 11) \textbf{ mod } 11$$

as the incrementing function. The signature generation function will be slightly more complicated. We need a function that will return different values as we proceed through a probe sequence. Since we will be using a signature size of 4 bits, we use the generating function of

TABLE 7.1 EXAMPLE POSITION-KEY SIGNATURE VALUES

Key	Probe chain position			
	1	2	3	4
27	12(1100)	6(0110)	9(1001)	0(0000)
18	3(0011)			
29	14(1110)	0(0000)		
28	13(1101)	2(0010)	9(1001)	
39	9(1001)			
13	13(1101)			
16	1(0001)			

$$s_1(\text{key}) = \text{key mod } 15$$

and

$$s_i(\text{key}) = (s_{i-1}(\text{key}) + 1) \, {}^* \, \text{key mod } 15$$

The signature sequence for 27 is then 12 (1100), 6 (0110), 9 (1001), 0 (0000), ...
Fortunately, for this example, we will not need to compute many signature values since
the probe chains will be relatively short. The signature generating function given above
is only for explanatory purposes. In practice, a pseudo-random-number generator is
used to generate signatures where successive random numbers could be computed as

$$r_{i+1} = (ar_i + d) \text{ mod } 2^{32}$$

with constants $a = 3141592653$ and $d = 2718281829$; $r_0 = \text{key}$. See Knuth [2] for
additional details. The desired number of signature bits would then be selected from
the high end of the random number generated by this function. Since such a function
is not easily computable by people, we have resorted to a simpler, but less random,
generator for illustrative purposes. Table 7.1 gives the necessary computed signature
sequences for the example keys.

We begin by inserting 27 into its home address of 5. The record storage table on
auxiliary storage and the separator table in main memory would appear as

	Separators		Records	
			Key	Other fields
0	1 1 1 1	0		
1	1 1 1 1	1		
2	1 1 1 1	2		
3	1 1 1 1	3		
4	1 1 1 1	4		
5	1 1 1 1	5	27	
6	1 1 1 1	6		
7	1 1 1 1	7		
8	1 1 1 1	8		
9	1 1 1 1	9		
10	1 1 1 1	10		

15 (1111) is an acceptable value for the initial separator signatures since it is larger than any p-k signature value produced by the generation function for this example. In inserting 27, the p-k signature of 12 (1100) is less than the separator signature for location 5, that is, 15 (1111). The less than comparison tells us that we should attempt to insert the record with key 27 at location 5 in the table. Upon accessing location 5, we find an empty location; so the insertion can be completed without additional processing.

The record with key 18 is inserted into location 7 in a similar fashion. When we insert the record with key 29, its p-k signature value of 14 is less than the separator signature value for its home address of location 7. The less than comparison tells us to try to place 29 into location 7. But that cell is occupied. So we need to order the records on the probe chain based upon their p-k signatures. The p-k signature of 0011 for 18 is less than the 1110 for 29, so 29 is not stored at location 7 but an insertion attempt is made at its next probe position, location 9. Since 29 was not stored at location 7, the separator signature is set to the higher p-k signature value, in this case, 1110. The separator and record tables become

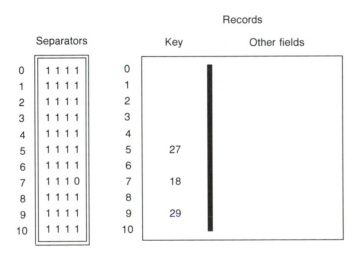

To retrieve 29, only *one* probe of auxiliary storage is necessary. We first compare the p-k signature value of 29, 1110, with the separator signature value of its home address of seven, 1110. Since the comparison is not less than, we proceed to the next position on the probe chain, which is location 9. The new p-k signature value of 0000 for that position is less than the separator signature value of 1111 for location 9, which tells us that if 29 is in the table, it must be found in location 9. Upon retrieving the record at location 9, we confirm that it is the record with key 29. Although we had to examine two separator signatures, we had to make only *one* access of an auxiliary storage location. Since the separator table is stored in main memory, comparisons against it are much, much faster than accessing auxiliary storage.

The record with key 28 is inserted into location 6 without obstacle. When we begin the insertion process for 39, its p-k signature value of 1001 is less than the separator signature value of 1111 at its home address of 6; we then attempt to place 39 at location 6. But 28 is stored there. Which one should be placed elsewhere? We decide based upon p-k signature values. Since 39 has the lower p-k value of 1001 vs. 1101 for 28, the record with key 39 is stored at location 6 and 28 is removed for possible reinsertion at location 8, the address of its next probe position. The separator signature at location 6 is then set to the higher p-k signature, or 1101. The tables then become

Records

	Separators		Key	Other fields
0	1 1 1 1	0		
1	1 1 1 1	1		
2	1 1 1 1	2		
3	1 1 1 1	3		
4	1 1 1 1	4		
5	1 1 1 1	5	27	
6	1 1 0 1	6	39	
7	1 1 1 0	7	18	
8	1 1 1 1	8	28	
9	1 1 1 1	9	29	
10	1 1 1 1	10		

Only one access of the record file is necessary to retrieve 28.

The record with key 13 is inserted into location 2 without difficulty. The final record with key 16 hashes to a home address at location 5. Its p-k signature value of 0001 is less than the separator signature of 1111; so we attempt to insert 16 there. Upon retrieving the contents of location 5, we observe that a record with key 27 is stored there. Since it has a p-k signature value of 1100, which is greater than the corresponding value of 0001 for 16, it is removed for possible reinsertion at the next position on its probe chain, that is, location 7. The separator signature value for location 5 is then revised to the higher p-k value of 1100. The record with key 16 is inserted at location 5. At location 7, 27's new p-k signature value of 0110 (for the second position on the probe chain) is less than the separator signature value for that location; therefore, we attempt to insert 27 at position 7. But that cell is occupied. Since its p-k signature value of 0110 is greater than that of the p-k signature value of 0011 for the record stored in that location, that is, 18, we should then attempt to insert 27 at the next location on its probe chain, that is, location 9. The separator signature value for location 7 is revised to the higher p-k value, that is, 0110. The record with key 27 continues to cascade down its probe chain. At location 9, the new p-k signature value of 1001 is less than the separator signature value of 1111; so we attempt to insert 27 at location 9. That position is filled with a record with key 29. We then compare the new p-k signature value of 1001 for 27 with that of 0000 for 29. We continue the reinsertion process for 27 at the next location on its probe chain, location 0, and we change the separator signature value at location 9 to 1001. Since location 0 is empty, 27 may then be reinserted at that location. The final tables then appear as

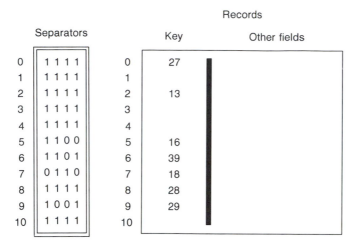

Retrieval

What is certainly interesting about signature hashing is that the *average number of auxiliary storage retrieval probes* needed to retrieve each record once is only *one*, assuming that the separator table can be stored in primary memory. Compare that with the average of 1.9 for linear quotient or 1.4 for coalesced hashing, or 1.7 for Brent's method. We really can't do any better than that. What we give up is the additional storage for the separator table and a more complex insertion process. The retrieval process is only slightly more complicated.

To retrieve any record in the table requires only a single access of auxiliary storage. For example, to retrieve the record with key 27, we initially compare its first p-k signature value of 1100 with the separator signature value of 1100 for location 5, the home address for 27. Since it is not a less than comparison, we compare the next p-k signature value of 0110 with the separator signature value of 0110 associated with the next probe position. Again, we do not have a less than comparison so we continue by comparing the p-k value of 1001 for the third probe position with the corresponding separator signature value, that is, 1001. As before, we do not have a less than compare. We finally compare the next p-k signature value of 0000 with the separator signature value of 1111. This then tells us that if 27 is in the table, it must be stored in location 0. We then access auxiliary storage and retrieve the record at location 0 to verify that it does contain the record with key 27.

We retrieve 18 in a similar fashion. Its initial p-k signature value of 0011 is less than the corresponding separator signature value of 0110 for its home address of 7. We retrieve the record at location 7 to verify that it has a key of 18.

All searches of this table require only a single auxiliary storage access—even unsuccessful ones.

Discussion

The insertion effort per record could be reduced by increasing the bucket size.

Deleting a record is achieved by placing a tombstone beside its storage location rather than actually removing the record from the file. The latter alternative would be too costly, for it would require changing the separator table and possibly moving some records.

What is fascinating about this method is that it uses a level of indirection in the search. Instead of searching the actual data located on auxiliary storage, we search an encoding of the data kept in high-speed memory. Thus, we can speed the search significantly.

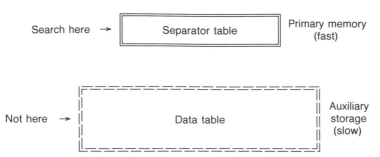

A *general principle* demonstrated by this method is: if data are stored or structured in such a way that processing it is not efficient, encode it into a form that is efficient. Can you think of any examples in your activities that require you to use a level of indirection? For example, instead of visiting five colleges you might want to consider for graduate studies, it is much faster (and less expensive) to look at the college catalogs of those five schools in your current college library. You might then be able to narrow your selection to a single school for an actual visit. Instead of going directly to the campuses to search for a graduate school, you visit them indirectly through your library.

One reason for having a sequence of signature values instead of only one associated with a key is to handle the situation in which the current record stored at a location and the incoming record both have the same signature. If that occurs, we move *both* records. If they had the same signature values forever, there would be no way to separate them. With a signature *sequence,* the two records would likely have *different* signatures at the next probe position, which would enable them to be separated.

Signature hashing as it is currently defined does have a serious difficulty in that the cascading of records to new locations during the insertion process may not terminate. In practice, then, we must place a limit on how much cascading to allow. But placing a limit may prevent an insertion from occurring. What suggestions do you have for proceeding should that happen in practice? We will give one suggestion in the next section after we discuss another data structure.

Signature hashing is another example of triple hashing: first to obtain the home address, second to get the increment, and third to obtain the p-k signature value.

KEY TERMS

Position-key signature   Separator table
Pseudo-random-number generator   Signature hashing
Separator signature   Triple hashing

EXERCISES

1. **(a)** What constraint is placed upon the initial values for the separator table? **(b)** What role does the initial value in a separator table play in determining whether the corresponding auxiliary storage location is empty?

2. Insert a record with key 38 into the final table in the discussion on signature hashing. Note changes in the location of any records in the data table and the signatures in the separator table.

3. Insert a record with key 49 into the final table in the signature hashing example. Note changes in the location of any records in the data table and the signatures in the separator table.

4. Reinsert the records for the example but in the reverse order of that given.

5. Is computed chaining a space-efficient collision resolution technique to use with signature hashing? Explain.

6. Complete the p-k signature values for Table 7.1.

7. How many probes of auxiliary storage are required for an unsuccessful search?

BLOOM FILTERS

Usually, when we search for a desired piece of information, we expect to retrieve it. Most of the discussion of searching and retrieving thus far has dealt with successful searches. But there are, nevertheless, many situations in which we *do not* expect to find what we are searching for. In those cases, we anticipate an *unsuccessful search.* As we have noted many times before, an unsuccessful search is usually lengthier than a successful one because we continue searching until we have exhausted all possibilities. With a successful search, we normally stop looking when we find what we are seeking.

What are examples of circumstances in which a search is likely to be unsuccessful? Have you ever waited behind a person in a checkout line at a supermarket or a department store who is paying for her purchases with a credit card? Depending upon the policy of the store, the sales associate may search some data to determine if the credit card is "bad" in a sense, for example, lost or stolen. Often this searching involves comparing the credit card number against a microscopic listing of such "bad" numbers. Depending upon the length of the list, the search efficiency, and the eyesight of the sales associate, it may take what seems to be an eternity to complete this search. Other stores require the sales associate to take the credit card and additional identification to a store guru located a great distance from the payment area. This guru usually has a sixth sense such that he can just look at the number and identification and tell whether it is "good" or not. Maybe people who use "bad" credit cards have a distinguishing visible trait that only this guru can identify. In any event, it usually seems to take a significant amount of time for the sales associate to visit the guru and return. Fortunately, the vast majority of people who use credit cards do not use "bad" ones. So it is usually expected that the search will be *unsuccessful;* it would, however, be helpful if the search did not take so long. The same type of activity often accompanies a purchase by check. Is the purchaser a bad check writer? Again, usually not. That then is another example of an unsuccessful search. We later describe other examples of unsuccessful searches. What we need is a technique that will allow us to conduct an unsuccessful search in a relatively efficient manner. For computer searches at least, a Bloom filter [3] is such a technique.

> A **Bloom filter** is an encoding of the primary keys of a file into a single bit string. As the name suggests, it is a *filter* for quickly eliminating nonmatches (those records which do *not* exist in an associated stored file) from further consideration; only rarely will even a single access of auxiliary storage be necessary.

It is a technique for improving the performance of unsuccessful searches. We consider the technique in this section because the filter itself is a loo . . . ong bit string, for example, with m bits represented as

0 $\hat{m-1}$

As we insert the records such as those of "bad" credit card or "bad" check users, we encode information about their primary keys in the Bloom filter. We do so by employing the now familiar technique of hashing. We apply a battery of n hashing functions to the primary key of each record, for example, the checking account number or credit card number, as it is inserted into the file, for example,

$$h_1(\text{key}) \rightarrow (0 - (m - 1))$$
$$h_2(\text{key}) \rightarrow (0 - (m - 1))$$
$$h_3(\text{key}) \rightarrow (0 - (m - 1))$$
$$\cdot$$
$$\cdot$$
$$\cdot$$
$$h_n(\text{key}) \rightarrow (0 - (m - 1))$$

The output of each hashing function is a number within the range of bits in the Bloom filter. The bit in the position corresponding to each output value is set to *one*. As you can understand, the Bloom filter needs to be relatively large to be effective. What makes it practical is that the filter information is encoded in bits and not bytes. A Bloom filter may be thought of as a signature for the entire file.

When we attempt to retrieve a record, for example, that of a potential "bad" check writer, the primary key for that record is fed through the same battery of hashing functions. The output of each function is compared against the corresponding bit position in the Bloom filter. At the first occurrence of a *zero* in the filter, we know *absolutely, positively* that there is no way for that record to be in the file. For if it were in the file, all the bits in the positions corresponding to the output of the hashing functions would be one since they would have been set during the insertion process. If we do encounter a *zero* bit in matching against the Bloom filter, we will not have made a single access of auxiliary storage in attempting to locate the record. But what if all the bit positions checked have ones in them? Should we immediately arrest the person? No, for we could have a *false drop*. If a search is not terminated by the filter, we must continue and actually perform a normal search of the file for the "bad" record. But as we said before, having to do that—search the file—is the definite exception.

The great advantage of a Bloom filter is that it *filters* out most unnecessary accesses to auxiliary storage in cases of unsuccessful searches. And that improves performance greatly.

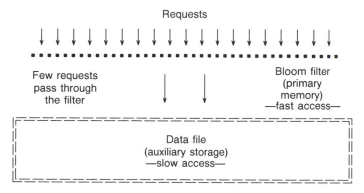

An Example

Let's assume that records with the keys

27, 18, 29, and 28

are the "bad" records that we want to store in a file of five locations using a hashing function of

$$f(\text{key}) = \text{key } \textbf{mod } 5$$

and, for simplicity, progressive overflow for collision resolution. If we use the following two hashing functions to set a Bloom filter of 16 bits:

$$h_1(\text{key}) = \text{key } \textbf{mod } 16 \text{ and}$$

and $$h_2(\text{key}) = \text{key } \textbf{mod } 14 + 2$$

after insertion, the file would appear as

	Key	Data
0	28	
1		
2	27	
3	18	
4	29	

and the Bloom filter would be

	1	1			1					1	1	1		1	
0	1	2	3	4	5	6	7	8	9	10	11	12	13	14	15

Notice how many costly probes of auxiliary storage are eliminated in attempting to locate the records with the following keys:

Key	Retrieval probes
85	2
22	5
51	1
26	1
34	3
19	3

For example, to locate a record with key 22, five probes of the file are needed before we know that the record is not there. The home address is location 2; we then consider the consecutive locations 3, 4, 0 and only when we encounter the empty location at one do we know for certain that the record is not in the file. How many of the above keys would be filtered out by the use of the Bloom filter? $h_1(85)$ is five; since bit five is zero, we do not need to apply the next hashing function or perform any other checking. We know definitely that 85 is not in the file. Only the record with key 51 is a false drop. $h_1(51)$ is 3 and $h_2(51)$ is 11; since both corresponding bit positions are set to one, we must check the file. But the home address, $f(51)$, is location 1, which is empty. After one probe we know that 51 is a false drop. Fourteen probes are saved by using a Bloom filter to accelerate the search for the six records in this example.

Other Examples

Recently, I was asked to evaluate a leading word processing system from an "ease of use" perspective. Since the product had been receiving excellent reviews, I was expecting to be impressed with its user friendliness. Overall I was pleased with what I saw, but I was struck by the user *un*friendliness of one of its features. That feature dealt with hyphenating words to create an attractive output page. What disturbed me was that I was asked by the system to hyphenate words, many of which I could barely spell. I didn't like doing this for numerous reasons, among them

- Since I didn't know how to hyphenate most of the words, I had to spend considerable time looking up each word in a dictionary to determine its hyphenation.
- The hyphenation process was prolonged and tedious. I felt that the computer system was controlling me rather than vice versa.

I reported my concerns to the system developers, and they explained their reasons for implementing the hyphenation system as they had. They said that automatic hyphenation algorithms were not always accurate and they did not want the system to insert incorrect hyphenation. They stressed that the amount of hyphenation required in a document could be reduced by increasing the number of superfluous blank characters permitted on a line. They referred to this as increasing the size of the hyphenation zone or "hot" zone that determined whether a word should be hyphenated. The permissible range was from two to nine characters. When set to nine, there would be little hyphenation required.

The designers' solution to the unfriendliness of hyphenation was to change the documentation for the system. They included a suggestion, printed with a different font and enclosed in a thick black border, that the user set a large "hot" zone to reduce the number of hyphenation requests. That was a simple modification from the designers' point of view, but what would have been a better solution? And how does that relate to Bloom filters? The reason that the system designers had not used a hyphenation algorithm was that there were too many exceptions to the rules and looking up those exceptions would require either a large amount of primary storage or a considerable amount of processing to locate them on auxiliary storage. Since most words follow the rules for hyphenation, there would be only a small percentage of exceptions. So most of the time when the system would want to know if a specific word were an exception, the answer would be "no." It would be an unsuccessful search. The exceptions could be stored in a file, just like the information on the "bad" check writers, with a Bloom filter created in primary memory to assist in reducing the number of accesses to the file when determining whether a word was an exception. If a word to be hyphenated did not pass through the Bloom filter, the automatic hyphenation algorithm would be applied. If a word did pass through the filter, it would need to be looked up. If it were an exception, the correct hyphenation would be determined by the lookup; otherwise, in the case of a false drop, the hyphenation algorithm would be applied.

Another example of the use of Bloom filters, described by Gremillion [4], deals with the processing of updates to a file in real time. Since accessing auxiliary storage is relatively expensive in terms of time, it may be preferable *not* to perform the updates in real time so that the system's performance will not be degraded. The updates would be written to a difference file for further processing in a batch mode at a later time, for example, at a nonprime time when the system was not as heavily loaded. But that leads to another problem. Either the difference file needs to be accessed for each record retrieved *or* out-of-date information will be given. How often have you wanted to know your current account balance or whether a certain transaction had occurred and been told that the computer system that the organization was using would not allow access to the information? It could only give the information as of some prior date. That is not a satisfying solution. Can we do better? Yes, create a Bloom filter of those records in the difference file. Then a record could be retrieved in real time. Only if a record's key passed through the Bloom filter would the difference file have to be accessed. And the number of false drops could be minimized by the proper choice of the length of the Bloom filter and the associated hashing functions. Since only a fraction of the records in a file are likely to be accessed multiple times during a day, most accesses to a record in the primary file would not require looking at the difference file also.

Gremillion [4] also studied the effects of filter size and the number of hashing functions (transformations) on the rate of false drops. His results are graphically displayed in Figure 7.1. Overall, as the size of the filter and the number of transformations is increased, the filtering error rate (or false drops) decreases. By choosing a proper size for the filter and an appropriate number of hashing functions, one can usually get as low an error rate as one needs for an application. Observe, however, the number of false drops when the filter size is small and the number of transformations is increased. The number of false drops actually *increases*. What is an explanation? When the filter size is small, as the number of transformations is increased, the density of the one bits in the filter increases. That means more matches and more false drops. The size of a filter is similar to the size of a hashed table. As the packing factor of a table increases, the number of collisions increases. If the size of a filter is small compared with the number of bits being set, we have the equivalent of a high packing factor and there will be many collisions and hence false drops.

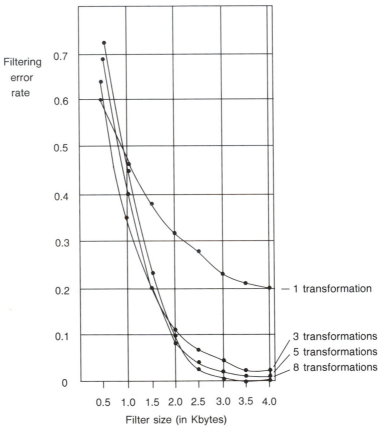

Figure 7.1 False drops as a function of filter size and the number of transformations. (*Source:* Association for Computing Machinery, Inc., Copyright © 1982, reprinted by permission.)

Filter Design

An important question that arises when implementing a Bloom filter is what should be its length and how many hashing functions should be used. Figure 7.1 gives experimental guidelines for making these choices. Mullin [5] provides a mathematical analysis to assist in making these decisions. This topic could have been entitled "theoretical analysis," but that might have encouraged many readers to skip it. Since the analysis is straightforward and useful, we include it here. We use the following definitions:

- m, the number of bits in the filter.
- n, the number of hashing functions.
- k, the number of records in the file associated with the filter.
- α, the percentage of records in the file compared with all possible records, for example, the percentage of "bad" check writers out of all check writers.

The probability of a single bit in the Bloom filter of size m being set to "1" by a hashing function applied to the key of one record is

$$P_{\text{set}} = \frac{1}{m}$$

if we assume that the hashing function has an even distribution. The probability of a single bit not being set by the hashing function then is one minus the probability of it being set or

$$P_{\text{unset}} = 1 - \frac{1}{m}$$

The probability of a single bit remaining unset after applying the n hashing functions is

$$P_{n.\text{unset}} = \left(1 - \frac{1}{m}\right)^{n}$$

assuming that the hashing functions are independent. Since there are k records in the file, each record must pass through the battery of hashing functions during the creation process for the filter. The likelihood of a bit remaining unset after the processing of k records is then

$$P_{nk.\text{unset}} = \left(1 - \frac{1}{m}\right)^{nk}$$

The probability of a single bit of the filter being set is then one minus the probability of it not being set, or

$$P_{nk.set} = 1 - P_{nk.unset}$$

On retrieval, each of the positions checked as a result of applying one of the n hashing functions to the primary key must be a one to require that the corresponding record be retrieved from auxiliary storage. The probability then of *all n* of these bit positions being *checked* for containing a one is

$$P_{allset} = P_{nk.set}^{n}$$

By substituting for $P_{nk.set}$ we obtain

$$P_{allset} = (1 - P_{nk.unset})^{n}$$

and finally by substituting for $P_{nk.unset}$, we arrive at

$$P_{allset} = \left[1 - \left(1 - \frac{1}{m} \right)^{nk} \right]^{n}$$

The probability of a false drop is

$$P_{false\ drop} = (1 - \alpha)P_{allset}$$

The theoretical results for four different filter sizes and two transformations are compared in Table 7.2 with the experimental results from [4]. The experimental values were obtained from a simulation with 100,000 transactions, 30,000 unique values accessed and 7000 unique key values updated. Since a value for α was unknown, it was assumed to be zero when computing the theoretical values, and as a result they are larger than one would expect. Notice, however, that the experimental results are higher than the theoretical ones. How might you account for this? One explanation is that the theoreti-

TABLE 7.2 FALSE DROP RATES; THEORY VS. PRACTICE

Filter size	Simulation		Theory	
	Number of transformations		Number of transformations	
	4	6	4	6
24,576	0.356	0.431	0.214	0.302
32,768	0.210	0.252	0.109	0.142
49,152	0.078	0.082	0.036	0.036
65,536	0.035	0.029	0.014	0.011

Source: Association for Computing Machinery, Inc., Copyright © 1983, reprinted by permission.

cal analysis assumed a perfectly even distribution for the hashing functions. As we know, that is basically impossible. With imperfect hashing functions, we would expect more false drops.

Another major assumption made in the theoretical analysis is that all records are equally likely to be accessed. In practice, this is not usually true; certain records are more apt to be retrieved than others. This assumption suggests a lower rate of false drops in practice than in theory. The fact that the actual rates were higher for the experimental situation suggests that the evenness of the hashing function distribution is the more significant factor.

One benefit of the theoretical analysis is that it can be used to assess the effectiveness of the hashing functions used in creating the filter. A program can be easily written to produce the theoretical values. Those values can then be compared against the experimental ones. If there is a great difference, that might suggest that the hashing functions be modified to give more even distributions.

One More Application

Since the Bloom filter is still an underused data structure, we apply it just one more time, in this case to a situation that does *not* involve the retrieval of records. Perhaps the diversity of the applications that we are considering may suggest other applications to you. In many situations we prefer not to perform a costly operation. A Bloom filter can be used as a decision aid. Rather than proceeding in all cases with the operation, such as a search of auxiliary storage, we can check the filter and in most or many cases eliminate the need to perform the operation. An operation that appears frequently in relational query and retrieval languages is one to eliminate duplicate tuples or records in the output. Eliminating duplicates is an expensive operation, for its usual implementation involves a relatively slow sorting operation. The records are sorted and then adjacent records are compared for equivalence.

Because of the slowness of the operation, a few relational query languages, for example, SQL, make the default specification the inclusion of duplicates. The user must take special action, that is, including the keyword DISTINCT, to eliminate the duplicates. Since in the majority of situations, the user does not want duplicate tuples included in the result, such a requirement is not user-friendly. An alternative implementation procedure is to use a Bloom filter in the process of checking for duplicates. Initially the Bloom filter is empty

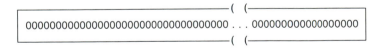

As each record is retrieved, a check of the Bloom filter is made to determine if the positions corresponding to the results of applying the hashing functions to the retrieved record have been set to "1." If all have been set, a sort operation is then required. If not, these positions are set to "1." The Bloom filter would change throughout the process.

```
                                ─────( (─────────────
  00000100000000000010000000000010000 . . 000000000000000000
                                ─────( (─────────────
```

Over time, as more records are added, the Bloom filter would become more dense.

```
                                ─────( (─────────────
  010001001000101011001010101011111010 . . . 010101001011101010
                                ─────( (─────────────
```

It is important then to make the filter large enough so that we do not often need to perform an unnecessary sort. Usually, if the records in the query result do not contain a duplicate, it would not be necessary to invoke a sort procedure.

Signature Hashing and Bloom Filters

When we considered signature hashing earlier in this chapter, we noted the concern that the cascading of records during the insertion process may not terminate. In implementing signature hashing, we definitely need to resolve this infrequent but possible situation. One solution is to create an overflow grouping or subfile of all those records which remain uninserted after some number of attempts. This overflow record area would contain few if any records. Therefore, since we are unlikely to find the record that we are looking for in that overflow area, a search of it is likely to be unsuccessful. The Bloom filter is a prime candidate to minimize the performance effect of unsuccessful searches. We could create a Bloom filter of all those records in the overflow area. Before doing a retrieval, we could check the overflow area for the record. Is it one of these infrequent exceptions? We would do that checking first by searching the Bloom filter. Only if the request passed through the filter would we need to access the overflow area. There would be few cases of false drops. By checking the overflow area first, we eliminate a considerable amount of searching on the separator table for those records in the overflow area. Of course, we would want to have a check to determine if there were any records in this overflow area. If there weren't, then we would want to eliminate the search of the overflow area altogether. Also if few records had been placed in this overflow area we may want to perform the check of it *after* the normal search rather than before it. We could make that decision dynamic. If fewer than some threshold number of records were stored in the overflow area, we would search for the overflow records at the end of the normal process; otherwise we would check the area before the normal searching process.

KEY TERMS

Bloom filter   Filter size
Duplicate records   Hyphenation
False drop   Real time updates
Filter design   Unsuccessful search

EXERCISES

1. (a) Given the following hashed file,

	Key	Data
0	85	
1	51	
2	22	
3	26	
4		

formed using the hashing function of key **mod** 5 with progressive overflow for collision resolution, set the bits for a Bloom filter where $h_1(\text{key}) = \text{key} \bmod 16$ and $h_2(\text{key}) = \text{key} \bmod 14 + 2$.

0 1 2 3 4 5 6 7 8 9 10 11 12 13 14 15

(b) Which of the following keys would cause false drops on lookup when using the Bloom filter?

$$81, 27, 46, 23, 51, 39$$

2. How many probes of auxiliary memory would be saved by using the Bloom filter to retrieve the six records with keys 81, . . . , 39 in Problem 1?

3. (a) Given the following hashed file,

	Key	Data
0	10	
1		
2	27	
3	83	
4	19	

formed using the hashing function of key **mod** 5 with progressive overflow for collision resolution, set the bits for a Bloom filter where $h_1(\text{key}) = \text{key} \bmod 15 + 1$ and $h_2(\text{key}) = \text{key} \bmod 14 + 2$.

0 1 2 3 4 5 6 7 8 9 10 11 12 13 14 15

(b) Which of the following keys would cause false drops on lookup when using the Bloom filter?

63, 81, 83, 30, 12, 95

4. How many probes of auxiliary memory would be saved by using the Bloom filter to retrieve the six records with keys 63, . . ., 95 in Problem 3?

5. Describe another application of how a Bloom filter might be used to speed processing.

6. What happens to the number of false drops when using a Bloom filter as the number of hashing functions is increased?

7. For which filter sizes in Table 7.2 does an *increase* in the number of hashing functions lead to an *increase* in the rate of false drops? Since such a result is contrary to what is expected, explain.

8. What effect would using a perfect hashing function have upon the construction and use of a Bloom filter? Would it be worthwhile?

CLASSIFICATION HASHING

Several times before we have stated that it is better to focus a search on a subset of the data rather than all the data. It is faster to search one room of a house for a lost item than to search the entire house. We have called this strategy **divide and conquer.** We divide the data into several groups so that we can subsequently conquer (locate) the desired item more quickly. We did this with multiple synonym chains in collision resolution, and with partial match retrieval with page descriptors. Classification hashing is a generalization of those instances. Rather than search all the data, we organize or classify them based upon the data values of the fields in each record. Then on retrieval, we need only look at those groupings which can contain the sought value. All the other classifications may be eliminated from further searching.

> **Classification hashing** is the partitioning of data records by hashing the field values of the record. On retrieval, only classifications that can contain the sought values need to be searched.

Each classification has a distinct m-bit identifier associated with it, where the size of m determines the number of classifications. For a file of records with r fields, that is,

Field 1	Field 2	Field 3	Field 4	$\cdots$	Field r

the classification identifier for a record is constructed by applying a hashing function, h, to each of the field values, yielding an output value in the range of the m-bit identifier.

$$h(\text{field}_1) \rightarrow (0 - (m - 1))$$

$$h(\text{field}_2) \rightarrow (0 - (m - 1))$$

$$h(\text{field}_3) \rightarrow (0 - (m - 1))$$

.
.
.

$$h(\text{field}_r) \rightarrow (0 - (m - 1))$$

The bit positions corresponding to the outputs from the hashing function are set to "1" and superimposed to form the classification identifier. As many as r bits may be set in an identifier, but fewer bits are possible since collisions may occur. All records with the same identifier are classified together. The records within a classification may be grouped together using almost any organization that we have considered in this text. Constructing a classification identifier is similar to constructing a text signature, but instead of the hashing function being applied to pairs of contiguous characters, it is applied to field values. With classification hashing, we are dividing the data records into $_mC_r$ classifications, where

$$_mC_r = \frac{m!}{(m - r)! \, r!}$$

Placing records into classifications is analogous to mapping them into pages in partial match retrieval with page descriptors. On retrieval then, only the classifications that have "1" bits in the appropriate positions of the classification identifier need to be searched.

In contrast to Bloom filters, the hashing functions in classification hashing set bits in a string to represent the items we *expect to find;* with a Bloom filter, they set bits for those items we *don't expect to find.* A Bloom filter is used for primary key retrieval whereas classification hashing is used for secondary key retrieval.

An Example

For the sake of computation, if we assume there are no collisions when setting the bits, and $m = 16$ and $r = 6$, there would be 8008 classifications. As an example of retrieval in classification hashing, let's arbitrarily decide that we wish to locate all records with specific values for *three* fields, for example, *programmers* in the *accounting* department who have *Melissa Jones* for a manager. To find these records, we would need to search only $_{13}C_3$ or 286 or ≈ 3.5 percent of the classifications. We narrowed the search considerably by merely hashing the desired values for the three fields. Why was it $_{13}C_3$? The three fields for which we knew the values would set 3 of the 16 bit positions. That would leave 13 positions, which could be set by the other three fields. How many classifications would need to be searched if we wanted the records with specific values for *two* fields instead of three? We would then need to search $_{14}C_4$ or 1001 or 12.5 percent of the classifications. The two fields for which values are specified would set 2 of the 16 bit positions. That would leave 14 positions to be set by the other four fields. Does it appear reasonable that the percentage of classifications that would need to be searched would *increase* as the number of fields designated is *decreased*? Yes, because with fewer restrictions on the type of records that we are searching for, more records will meet those restrictions. For example, in a student database, if you wanted to locate all blond students with a grade point average greater than 3.0, a certain number of records would be retrieved. But if you were only interested in blond students, you would expect more records to match.

Classification hashing is useful for secondary key retrieval and for document retrieval in which you want to retrieve all records with certain keywords. It is a technique that allows us to narrow the search considerably with relatively little effort.

KEY TERMS

Classification hashing   Classification identifier

EXERCISES

1. **(a)** How many classifications would need to be searched if we needed all records with *four* field values using the example of this section? **(b)** Does this answer differ in the direction that you would expect compared with the results given for the examples?

2. Assume the following classification hashing scheme for storing documents. Each document possesses up to five keywords. The classification identifier for each document is a 12-bit code with five positions set to "1". A bit position, i, in the classification code is set to "1" if and only if $h(\text{keyword}_j) = i$ for some i and j where h is the hashing function. If fewer than five bits are set to one by the hashing function, additional bits are randomly set to give a total of exactly five "1" bits. If a request is made to retrieve documents with keywords A_1 and A_2, what percentage of the document classes would need to be searched?

3. List two differences between the formation of a classification identifier and a record signature.

4. Contrast and compare classification hashing to partial match retrieval with page signatures.

CHECK HASHING

Yes, you did read the name of this technique correctly; it is check hashing and *not* check cashing. Check hashing is another example of encoding information into a form to make it more readily processible, in this case, into a condensed form so that it may be stored in primary rather than auxiliary storage. By storing the information in primary memory we are able to achieve an order of magnitude increase in retrieval performance.

> **Check hashing** is the encoding of information, often key information, into a compact form to permit storage in primary memory.

As the name of the technique implies, hashing is used to encode the information. Instead of storing the actual information, we store a condensed, encoded form of it. Where is such a technique useful? In some instances, as we noted with Bloom filters, we only need to know if something exists or not. Is a person a "bad" check writer or not? Is a person's identifier on a list? Bloom filters are an effective technique when most of the searches are unsuccessful, but what about cases in which most retrievals *are* successful. That's where check hashing is useful. A spelling checker (not a corrector) is an instance in which we merely want to determine if a word is on a list of correctly spelled words. Since most words are spelled correctly, this is a case in which most searches will be successful. If we were to store the entire word, we may need a field width of 20 to 30 bytes. If we *encode* the information, we may be able to store it in as few as 2 bytes. Whereas it might be impossible to store a dictionary with 20-byte entries in primary memory, it may be possible with 2-byte entries. For a 10,000-word dictionary, we need 200,000 bytes if we store the words in their entirety, but only 20,000 bytes if we store the words in their check hash form. Obviously there is a significant savings in storage.

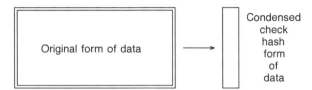

Check hashing can be considered a form of **triple hashing.** We noticed triple hashing earlier when considering signature hashing and computed chaining with multiple chains. With computed chaining, one hashing gave us the home address, another the increment, and a third the chain number. In the case of check hashing

h_1 (key)   gives the home address of the record.

h_2 (key)   gives the increment for collision resolution.

h_3 (key)   gives the check hash value for the record.

Just as cyclic check bytes are a hashed encoding of the information contained in a record, so too is the check hash value stored in a table. The primary advantage of check hashing is that it takes considerably less space to store the check hash value than the actual information. The encoded information stored in primary memory can be accessed faster than the original information stored on auxiliary storage.

An Example

To gain a better understanding of check hashing, let's work through an example. For simplicity, we use two-digit decimal numbers, but in practice the key values could be alphabetic and would normally be much longer. The keys of

23, 51, 48, 11, 73

are placed into a seven-element table using the hashing functions of

$$h_1 (\text{key}) = \text{key } \textbf{mod } 7$$

$$h_2 (\text{key}) = \lfloor \text{ key } / 7 \rfloor \textbf{ mod } 7$$

and
$$h_3 (\text{key}) = (\text{ key } \textbf{mod } 5) + 1$$

h_2 is used with the linear quotient collision resolution method. The check hash value for 23 is 4 which, since there is no collision, is placed at its home address of location 2. The table of check hashes is then

	Check hash	Data (optional)
0		
1		
2	4	(23)
3		
4		
5		
6		

The portion of the table enclosed within double lines ‖ is optional and is used in this case to store the complete key for purposes of reader comprehension. Obviously, in practice, we would not store the actual key, for that would defeat the purpose of using a check hash. There are circumstances in which we may want to store additional data in a check hash table, for example, in a table lookup situation in which we want to retrieve a value associated with a key.

In this example, the next key of 51 hashes to location 2, which is not empty; so we check for a duplicate entry.

Note. **When we are checking for a duplicate entry, we are comparing check hash values, and *not* the actual key values.**

Since the check hash for 23 is 4 and that for 51 is 2, there is not a duplicate. To find a proper storage location for 51, we use h_2 in conjunction with the linear quotient collision resolution method. Since the increment is zero, we set it to one. We then store the check hash value 2 in location 3. The check hash value of 4 for 48 is stored in location 6 and the check hash value of 2 for 11 is stored in location 4. The table is then

	Check hash	Data (optional)
0		
1		
2	4	(23)
3	2	(51)
4	2	(11)
5		
6	4	(48)

When we insert 73, its home address of 3 is occupied. We match the check hash value for 73, which is 4, with the check hash value stored at location 3, which is 2. We next look in location 6, since the increment is three. Then we compare the check hash for 73 with the check hash stored at location 6. *They are the same!* What does that mean? For all we know, 73 has already been inserted into the table and thus the insertion process is terminated. This last insertion attempt does illustrate a drawback in using check hashing; that is, we may *lose* some information in the encoding process, which may lead to inaccuracies.

Discussion

Because we have incomplete information in a check hash value, we may experience a few errors when searching. For example, in the case of a spelling checker, if the check hash of a misspelled word coincides with that of a correctly spelled word, we would assume that an incorrect word was actually correct. However, the error rate is relatively small and can be adjusted by the size of the check hash. As a check hash gets larger, there are fewer chances for collisions, which means fewer errors. For a 10,000-word dictionary in a table of 12,000 nodes, the error rate is

$$P_{\text{error}} = 8 \times 10^{-5} \quad \text{for a 16-bit check hash}$$
$$P_{\text{error}} = 4 \times 10^{-7} \quad \text{for a 24-bit check hash}$$

With a 16-bit check hash value for the 10,000-word dictionary, if we looked up 100,000 items, on the average eight would be incorrect. For a 3-byte or 24-bit check hash, the expected number of errors would be 0.04, much less than one. That may not be too many if we have an application in which speed is the most important consideration.

But what about applications in which accuracy is essential? Check hashing was introduced by Dodds [6] to assign Canadian postal codes to addresses. Since the volume of addresses was so large, it was important to have superior performance. For that reason, it was important to keep the table of addresses and associated postal codes in primary memory. Since this assignment was being done on a small minicomputer without extensive primary memory, it would have taken too much memory to store the entire street and community address as the key portion of the table lookup record containing the appropriate postal code for that address. Therefore, the address key information was stored as a check hash. But what about accuracy? Surely, the post office would not tolerate addresses being assigned to incorrect postal codes. To prevent errors, when the table was constructed, the synonyms were removed from the table. That is, if two or more different addresses mapped to the same table location, then *all* of those records were removed from the table. That means that for a table lookup, no collisions would occur but sometimes a record would not be found. About 85 percent of the addresses were assigned postal codes in this manner. The other 15 percent required another technique such as storing the assignment table on auxiliary storage. But check hashing was an effective means to quickly reduce the amount of data that required processing. No correct addresses were assigned incorrect postal codes, but incorrect addresses could be assigned postal codes. Would it matter, though, where a letter with an incorrect address would be sent?

KEY TERMS

Check hashing
Error rate
    Check hashing

EXERCISES

1. (a) Build a check hash table for the keys 32, 24, 17, 11, and 26 using the same hashing functions for h_1 and h_2 as used in the example but use $h_3(\text{key}) = \text{key} \bmod 3 + 1$. (b) How many bits are needed for the check hash in this case?

2. (a) Repeat Problem 1, but use the hashing function $h_3(\text{key}) = \text{key} \bmod 5 + 1$. (b) How many bits are needed for the check hash in this case? (c) How did the number of collisions compare with those of Problem 1? (d) How do you account for these results?

3. (a) Repeat Problem 2 but use $h_3(\text{key}) = \text{key} \bmod 5$. (b) How many bits are needed for the check hash in this case? (c) How did the number of collisions compare with those of Problem 2? (d) How do you account for these results?

4. Would it be reasonable to use a Bloom filter for all the keys that were *not* in the check hash table in the postal code example? Explain.

5. How would it be determined in the postal code example whether another technique would be needed to assign a postal code?

REFERENCES

1. Larson, Per-ake, and Ajay Kajla, "File Organization: Implementation of a Method Guaranteeing Retrieval in One Access," *CACM,* Vol. 27, No. 7 (July 1984), pp. 670–677.
2. Knuth, D. E., *Seminumerical Algorithms,* 2d Edition, Addison-Wesley, Reading, MA, 1981.
3. Bloom, Burton H., "Space/Time Trade-offs in Hash Coding with Allowable Errors," *CACM,* Vol. 13, No. 7 (July 1970), pp. 422–426.
4. Gremillion, Lee L. "Designing a Bloom Filter for Differential File Access," *CACM,* Vol. 25, No. 9 (September 1982), pp. 600–604.
5. Mullin, James K., "A Second Look at Bloom Filters," *CACM,* Vol. 26, No. 8 (August 83), pp. 570–571.
6. Dodds, D. J., "Reducing Dictionary Size by Using a Hashing Technique," *CACM,* Vol. 25, No. 6 (June 1982), pp. 368–370.

PART SUMMARY

The unifying theme of the data structures and techniques contained in this part was the use of the bit. They all made use of a bit representation of information. Another important theme was the conversion of information from a form that was not readily processible into one that was—which in the examples of this part involved the use of bits. Computers can process bits quickly even though people cannot. The use of bits allowed us to save space or time or in some cases both. As in the previous part, in addition to learning about specific data structures and techniques, you have also been introduced to or reacquainted with several general techniques and principles that would be applicable in other situations. Let's review these ideas.

TECHNIQUES

- Use additional storage to save processing time (for example, text signatures, inverted files, partial match retrieval, Bloom filter).
- Subdivide a search space into many units so that only a portion of them need to be considered when searching (e.g., classification hashing, partial match retrieval with page signatures).
- Spend additional time structuring information prior to its use to reduce subsequent processing time; preprocess once, process often (e.g., text signatures, partial match retrieval, inverted files).
- Keep condensed information in primary memory to improve access time to the actual data (e.g., signature hashing, Bloom filter, check hashing).
- Search an encoded form of the information rather than the actual information (e.g., signature hashing, Bloom filter, check hashing).

PRINCIPLES

- Do it faster by doing less (e.g., partial match retrieval, text signatures).
- Apply familiar techniques to new contexts (e.g., signature trees with text searching).

- Combine known techniques to create new and improved means of organizing and processing data (e.g., signature hashing).
- Continue searching for improvements to a technique even though it has been widely used for some time (e.g., Boyer and Moore pattern matching).
- Keep an open mind to nontraditional ways for processing data (e.g., Boyer and Moore with pattern matching from the right instead of the left).

PART EXERCISES

1. Matching—choose the *most* appropriate technique to apply to each situation listed.

Situation	Technique
____ i. Primary memory is at a premium—a few errors are tolerable.	a. Bloom filter
	b. Difference file
____ ii. Credit card validation.	c. Disjoint coding
____ iii. Updating is allowed—real time access is important.	d. Check hashing
____ iv. Search time is more important than storage.	e. Text signatures
	f. Superimposed coding

2. Insert *(increase, increases, decrease,* or *decreases)* in the following questions:
   (a) As the packing factor of a file increases, the average number of probes _____.
   (b) As the number of bits in a check hash increases, the probability of an error _____.
   (c) As the length of a text signature increases, the number of lines in the associated file that must be searched exhaustively _____.
   (d) As the size of a Bloom filter increases, the number of false drops _____.
   (e) As the pattern length increases, the number of lines (with text signatures) that must be searched exhaustively _____.

3. Give two examples of *triple* hashing.

4. (a) Which *page* signatures would need to be examined if we wanted to retrieve employees in the safety department who earn between $25,000 and $30,000? Use the data in Figure 6.6.
   (b) How many *record* signatures would need to be examined to answer the query in part a? Use the same data. (c) How many signatures (at all levels) would need to be examined to answer the query in part a using the signature tree in Figure 6.4?

5. Name a technique that uses both disjoint and superimposed coding.

PART THREE

TREE STRUCTURES

PREVIEW

One of the fundamental and important data structures of computer science is the **tree.** Its importance results from its:

Flexibility A tree allows stored data to be processed in different ways relatively efficiently, for example, directly or sequentially. It is a fitting data structure for "maybe" situations in which you do not know how the data might ultimately be processed. It is also useful when you know in advance that the data will be processed in different ways. A tree may *not be the best* data structure for a single purpose, but its flexibility often makes it the best for multiple purposes.

Search Efficiency A tree aids in the efficient search of data, for when a *branch* of a tree may be pruned in a search, that means that all the data in that branch do *not* need to be processed further. A tree structure can often be used as a filtering mechanism in locating a desired item of information. As we proceed through the tree structure, we filter out unwanted data.

Naturalness of Representation A tree is often a natural way to represent information. What we are describing may have *parent, offspring,* and *sibling* relationships that fit easily into a tree structure. A problem is more readily solved if we have a straightforward representation for the data.

Power of Representation A tree sometimes makes an application possible where otherwise it would not have been. (Recall the binary tree collision method.)

Trees have always been an important data structure, and they are likely to continue to be so for the reasons mentioned.

In this part, we build upon what you already know about trees. You have previously seen tree structures used in this text; we used them in the binary tree collision resolution method and the indexed sequential file organization. It will be helpful if you know how to store a tree structure in a sequential and linked fashion. Also, knowing the standard orders for traversing a binary tree will come in handy.

We examine ways to build upon the basic tree structures to provide greater flexibility and performance in storing and retrieving data. We consider better ways to construct binary trees to retrieve data more efficiently. We look at alternatives to the indexed sequential file organization method that we discussed earlier. The alternative methods provide better retrieval performance and allow the files of information to grow and shrink without major reorganization. We see hashing again—this time with methods that allow variable size tables. Plus we look at a way to represent keys on a character by character basis rather than as an entire unit. And we look at two more methods for secondary key retrieval. But this is not all; we also look at applications and many variations of the methods and structures.

Again, one of the goals of this part is to continue the emphasis on *problem solving*. How can you put known methods together in novel ways to solve current problems? We want to expose you to a wide variety of techniques so that you will have a full bag of tools when you encounter problems on your own. Some of the ideas may not appear particularly useful today, but who knows how you may be able to apply them in the future?

Chapter 8

Binary Tree Structures

BINARY SEARCH TREES

Many of the reasons mentioned in the preview for studying trees are applicable to binary search trees. A binary tree is the simplest tree structure. As the name implies, it has at most two branches emanating from a node or juncture. To make a binary tree a binary search tree, we follow a convention for constructing and processing it. A node in a binary search tree is used both to store data and to act as a decision marker in locating information. In storing (or retrieving) information, the first record becomes the root node of the tree. In inserting subsequent records, a comparison is made between the key of the record to be inserted and the key of the record stored in the binary tree. If the keys are identical, we have a duplicate record (or we have found the record that we are looking for on retrieval). If the keys do not match, we then proceed. If $key_{insert} < key_{stored}$, we take a left branch. If $key_{insert} > key_{stored}$, we take a right branch. If the branch is null, we have found the place to make the insertion (or we know that the record is not in the file if we are retrieving). If we do not encounter a null branch, we repeat the process with the entry at the next level and compare for equal, less than, or greater than. The process terminates upon locating a null branch.

> A **binary search tree** is a binary tree in which the nodes are used both to store information and to provide direction to other nodes. The information is organized such that all key values less than that of the current node are in its left subtree and those greater are in its right subtree.

An algorithm for inserting into a binary search tree appears in Algorithm 8.1.

An Example

Let's build a binary search tree by inserting the names of favorite cities of a recent group of students. The cities and their insertion order are

Dallas, Oakland, Pittsburgh, Miami, Chicago, Denver, Boston, and Spivey's Corner

Algorithm 8.1
BINARY SEARCH TREE INSERTION

I. Set a tree traversal pointer, p, to the root node of the binary search tree. Initialize a trailing pointer, q, equal to p.

II. While p is not null, compare the key of the record to be inserted with the key of the record in the node pointed to by p.

    A. If lexicographically equal, terminate with a "duplicate record" message.

    B. If lexicographically less than, set $q := p$, then set p to the left offspring node.

    C. If lexicographically greater than, set $q := p$, then set p to the right offspring node.

III. If p was last updated to a left offspring, insert the record as a left offspring of the node pointed to by q; otherwise insert it as a right offspring.

(For those unfamiliar with Spivey's Corner, it is the home of the National Hollerin' Contest held each year.) The binary search tree after the insertions is shown in Figure 8.1.

Discussion

The depth of a binary search tree, as for trees in general, is the maximum level of any node contained in it, relative to the root node. The root is level one and the level of each of the remaining nodes is recursively defined to be one greater than the level of its parent. The depth of the example binary search tree is four since "Denver" and "Spivey's Corner" are at level four. Knowing that the depth of the binary search tree is four tells us that we are able to retrieve a single record directly without making more than four probes. The average number of probes to retrieve each record once is 2.75. But note that this average is considerably *more* probes than we made with a typical hashing scheme. What is the advantage? *Flexibility.* We can process the data sequentially by using one of the standard orders for traversing a binary tree. Do you know what that standard order is? It is **inorder.** We can then process data directly or sequentially with relative ease. Where have we previously seen a similar data structure? Yes, an indexed sequential file uses a tree mechanism to organize data for both direct

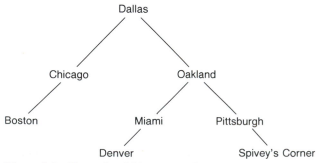

Figure 8.1 Example of binary search tree.

and sequential processing. What are the differences? With indexed sequential, the tree is used for index purposes only, and the data is stored in the leaf nodes; whereas with a binary search tree, the tree is used both as a directory and as a storage structure.

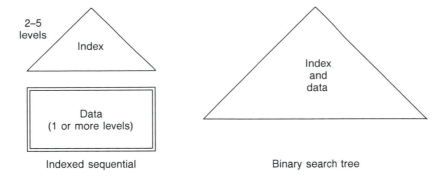

Indexed sequential   Binary search tree

Which takes less effort to build? Which is the simpler data structure? The binary search tree. But which is the less efficient structure if you have large amounts of data to store? Which tree structure has greater depth? The binary search tree; so it is less efficient when performing direct access of a single item. What about sequential processing? Again the binary search tree is less efficient, since moving up and down the tree takes more effort than moving through the indexed sequential index structure. Why is the indexed sequential structure more efficient for large amounts of data? Notice the branching factors. The branching factor of a binary tree is two, whereas that of an indexed sequential structure is much greater. This difference implies that the binary search tree will have greater depth and therefore will take more processing time for both sequential and direct processing. For small amounts of data and/or data that is not to be processed frequently, the binary search tree has advantages.

Notice that the structure of a binary search tree is *dependent upon the order of insertion.* What would the structure of the example binary tree have been if the records had been inserted in the alphabetic order of the keys?

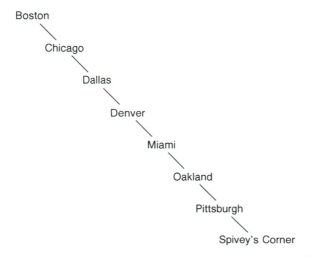

What is the maximum depth of this tree? What is the maximum number of probes needed to retrieve an item? What is the average number of probes needed to retrieve an item? Notice that the structure is a degenerate tree that is equivalent to a sequential linear list. As we know, such a structure is not efficient for performing direct access. Since its depth is eight, the maximum number of probes needed to retrieve an item is eight. The average number of probes to retrieve a single item is 4.5. With the original binary search tree, only 4 and 2.75 retrieval probes are needed, respectively. We obviously prefer to have performance less dependent upon the order of insertion. The next tree structure that we consider is intended to remove this drawback. If at any time the binary search tree is not approximately in balance, we perform additional processing to bring it back into balance. This is a technique that we have used before, that is, to spend additional time processing the data at insertion, so that we may spend less time for retrieval. The justification again is that we usually insert an item into a table once but retrieve it many times. This variation of the binary search tree which will be presented next is called an AVL tree or a height-balanced tree.

KEY TERMS

Binary search tree	Node level
Index	Storage structure
Inorder tree traversal	Tree depth

EXERCISES

1. Insert the names of the months of the year into a binary search tree using calendar order.

2. (a) What is the average number of probes needed to retrieve each record in Problem 1 once? (b) What is the tree's depth? (c) What is the maximum number of probes needed to retrieve a record? (d) What is the minimum depth needed to store 12 records in a binary tree?

3. Insert the names of the months of the year into a binary search tree using calendar order beginning with "April" and continuing through "March."

4. (a) What is the average number of probes needed to retrieve each record in Problem 3 once? (b) What is the tree's depth? (c) What is the maximum number of probes needed to retrieve a record?

5. Compare the results of Problems 1 and 3. What do they tell you about the effect of the order of insertion upon the performance of a binary search tree?

6. (a) Devise an expression in terms of n, the number of nodes, to give the maximum depth of a binary search tree. (b) Do the same for the minimum depth.

7. Write an algorithm in pseudocode to delete a node with a specified key value from a binary search tree. Assume a linked node structure of

lchild	record	rchild

The resulting structure should be a binary search tree. The algorithm should allow the deleted node to be in any position in the binary search tree.

AVL TREES

The AVL tree gets its name from its originators, Adelson-Velskii and Landis [1].

> An **AVL tree** is a binary search tree in which the subtrees of a node are maintained at approximately the same height (or depth). Transformations on the binary tree keep its height balanced.

For that reason, it is also called a height-balanced tree. As we noted in the previous section, a deficiency of a binary search tree is that the insertion order of the records may cause it to become unbalanced. The AVL tree allows us to maintain a balanced tree independent of the insertion order.

What is meant by a height-balanced tree? First, we need to define the term *balance factor* for a node. The **balance factor** of a node is the height of the right subtree minus the height of the left subtree. As before, the root node is at level one and each of the other nodes is the level of its parent plus one. The height (or depth) of a subtree is the maximum level of any node in that subtree relative to its root. In the binary tree:

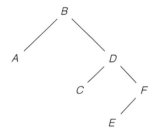

node E is at level four and the height of the subtree with F as its root is two. Since nodes A, C, and E are terminal, they all have balance factors of 0. F has a balance factor of −1, D is +1, and B is +2. Node D has a balance factor of +1 since the height of its right subtree is two and its left subtree is one. An **AVL tree** is a binary search tree in which *all the nodes have a balance factor of 0 or +1 or −1*. The tree in the example is not balanced since node B has a balance factor of 2. If node E were removed, the resulting tree would be an AVL tree.

Caution: The balance factor of a node is determined by the *height* of its left and right subtrees and *not* by a count of the number of nodes in them.

Inserting a record into an AVL tree begins just as with a binary search tree; then, the balance factors for all the nodes on the path from the inserted node up through the root node are computed. If the resulting tree is *not* balanced, one or two rotations are performed to bring the tree back into balance. Fortunately, there are only two circumstances in which an AVL tree may become out of balance (or four, if you count the two symmetric equivalents). These two (or four) situations are shown in Figure 8.2 in which X and Y are nodes and A, B, and C are subtrees whose heights are given in terms of h. Identifying which case arises is a pattern matching operation. We move from the inserted node back to the root to try to match one of the out-of-balance templates.

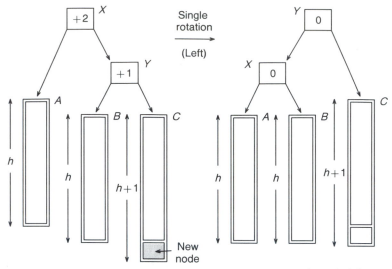

Figure 8.2a   AVL tree—case I, single rotation—tree unbalanced right.

When a match occurs, we then perform the associated transformation. The template matching is relatively easy if you note

> that in case **I** situations the signs on the balance factors of nodes X and Y are the *same,* whereas in case **II** the signs are *different.*

Only the balance factors associated with nodes X and Y need to be considered when determining which case is applicable. Can the rotations be just any rotations? No, recall

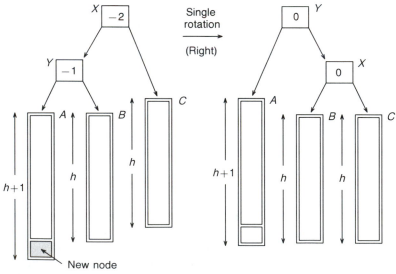

Figure 8.2c   AVL tree—case I, single rotation—tree unbalanced left.

that one advantage of the binary search tree is that if we traverse such a tree in *inorder,* we have an alphabetical ordering of the keys. In performing a rotation, we need to preserve that ordering. In Figure 8.2a, if the tree is traversed in *inorder,* the nodes and subtrees are visited in the order *A, X, B, Y,* and *C.* Since the first case I tree is out of balance to the right, we perform a rotation to the left to bring it back into balance. After the rotation, an *inorder* traversal of the tree yields *A, X, B, Y,* and *C,* thus preserving the traversal order. Case II is a little more complex. In looking at Figure 8.2b, you will

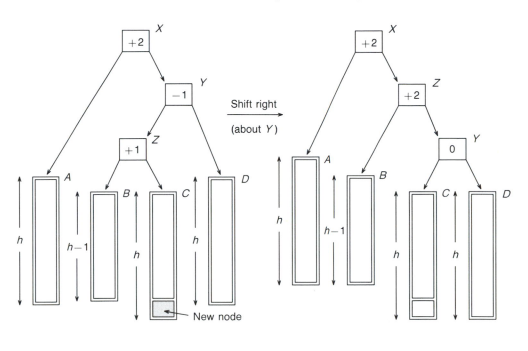

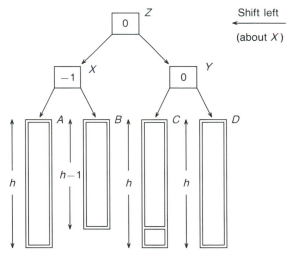

Figure 8.2b   AVL tree—case II, double rotation—unbalanced right.

note that the entire tree is out of balance to the right, and that the subtree Z makes node Y unbalanced to the left. In this case, we need *two* rotations to bring the tree back into balance. First, we perform a right shift around node Y to bring that subtree into better balance. Then a left rotation around node X completes the transformation, bringing the entire tree into balance while preserving the *inorder* traversal order. In performing a template match, the left and right subtrees of node Z may be the same height or they may be interchanged from those shown in Figure 8.2b. The English

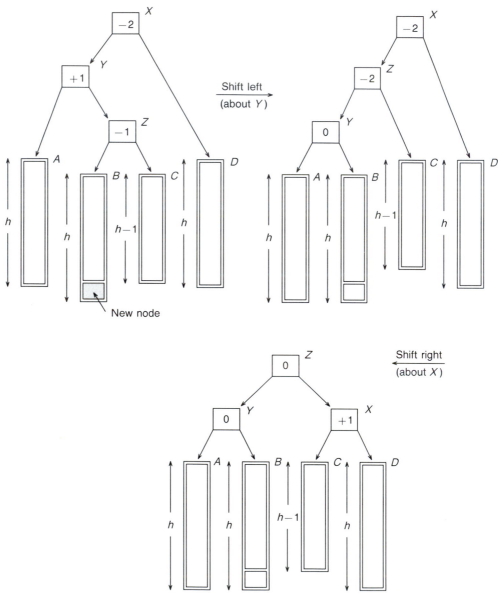

Figure 8.2d   AVL tree—case II, double rotation—tree unbalanced left.

Algorithm 8.2
AVL INSERTION

I. Insert the record into an AVL tree using the binary search tree insertion procedure.

II. Adjust the balance factors for all the nodes on the path from the parent of the newly inserted record to the root node.

III. If all the adjusted balance factors are 0, or ± 1, then terminate, else ascertain whether it is a case I or case II situation and then perform the appropriate transformation.

description of the algorithm appears in Algorithm 8.2. **Remember** that the AVL insertion algorithm applies only to the insertion of a record into an *existing* AVL tree.

An Example

To better understand the procedure, let us work through an example using the favorite cities that we considered in the binary search tree example in Figure 8.1, that is,

Dallas, Oakland, Pittsburgh, Miami, Chicago, Denver, Boston, and Spivey's Corner

Dallas is inserted first. After Oakland is inserted, the tree is still balanced. However, when we insert Pittsburgh, the tree becomes out of balance.

Is this a case I or a case II situation? Since the signs on the nodes corresponding to nodes X and Y in the diagram are the same, it is case I. We then do a left rotation to bring the tree into balance.

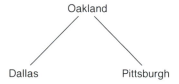

Notice that if we step through the resulting tree in *inorder,* we still have an alphabetical ordering. Next we add Miami. The resulting tree is still balanced. (You have to know your alphabet well or you will have trouble with these insertions.) Then Chicago is added. We are still OK. Denver is added next, and the tree becomes out of balance to the left.

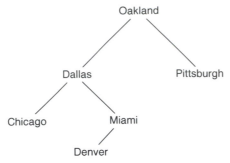

Is it a case I situation or case II? Since the signs of the nodes corresponding to X and Y are different, it is case II. So we perform a double rotation giving us the tree:

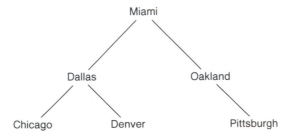

After we add Boston, the tree is still balanced. Finally, we add Spivey's Corner. The resulting subtree with the node containing Oakland as its root is out of balance to the right with a case I situation. **Remember that we check the balance factors beginning with the parent node of the inserted record and working back to the root node.** The tree is rebalanced with a single rotation to the left giving:

A check, but not a guarantee, for correct insertion is to process the data in the nodes of the resulting tree in *inorder* to notice whether an alphabetical ordering is obtained. If one is not, there is definitely an error. If one is, we still have the possibility that we may have done something incorrectly, but we have greater certainty of correctness than before the check.

Discussion

At most, only *one* transformation, either a single or a double rotation, is needed for each insertion of a node into an AVL tree. We perform the transformation if necessary after *each* insertion rather than waiting until all the nodes have been inserted and then transform the tree. The extra processing that we are performing at insertion time to balance the tree will pay dividends at retrieval time with better performance. Also notice that we have a bounded worst case performance of $O(\log_2 n)$ for direct access, which is an advantage over the indexed sequential file organization of Chapter 2.

For ease of implementation, as we traverse the AVL tree to determine the proper position for inserting a record, we should maintain two pointers in addition to the one for traversing the tree and its trailer. We should keep pointers to the most recent (bottom-most) node on the insertion path with a balance factor of $+1$ and its parent. The node with such a balance factor identifies a potentially unbalanced subtree. We know that only one transformation of an unbalanced tree is needed to rebalance it. When we insert a record, a balance factor may change at most by $+1$ (since we are only adding one record). All situations of subtree imbalance are indicated by a $+2$ balance factor of the root node of the unbalanced subtree. Therefore, only a node that had a balance factor of $+1$ prior to insertion is a candidate for the root node of an unbalanced subtree. By noting the most recent node with such a balance factor, we limit the portion of the tree that must be checked for imbalance. If there does not exist a node on the insertion path prior to insertion with a $+1$ balance factor, then we know that we can insert one record without encountering an unbalanced tree. The trailing pointer for the node with the $+1$ balance factor is needed to identify the node whose link field must be modified to join the transformed subtree into the remainder of the tree since the transformed subtree may have a different root node after the transformation.

Do we need to update the balance factors of all the nodes in an AVL tree after an insertion? No, fortunately, the region of modification is localized to only the subtree that is out of balance. Why is that? Notice that all four (or two) of the transformations *reduce* the depth of a subtree by one. When we insert a node such that it leads to an imbalance, we *increase* the depth of that subtree by one. So after the transformation, the net effect is that depth of the subtree has not changed as a result of the insertion; thus, none of the other balance factors are affected.

We do need extra storage for keeping a balance factor field in each node.

KEY TERMS

AVL tree   Height-balanced tree
Balance factor   Inorder

EXERCISES

1. Insert the following list of favorite cities, compiled by students from several years ago, into an AVL tree:

Jefferson, Hampton, Greenville, Ahoskie, Gurnsey, Boone, Gusey, and Hamilton

2. Insert the seven words, *happy, days, will, soon, be, here,* and *again* into an AVL tree.

3. Repeat Problem 2 replacing *again* with the pair of words, *once* and *more* (for an eight-word total).

4. Construct an AVL tree for the eight records with the keys inserted in the following order: *Fall, is, the, best, time, of, year,* and *unequivocally.*

5. Does an AVL tree of *n* nodes have the minimum number of levels for a binary tree of *n* nodes? Prove or give a counterexample.

6. Insert the names of the months of the year into an AVL tree using calendar order.

7. **(a)** What is the average number of probes needed to retrieve each record in Problem 6 once? **(b)** What is the tree's depth? **(c)** What is the maximum number of probes needed to retrieve a record?

INTERNAL PATH REDUCTION TREES

What is the purpose of the transformations associated with the AVL trees? It is to reduce the number of probes required to retrieve a record. Can a binary search tree be transformed to even further reduce the number of retrieval probes? If the average depth of a node in a binary search tree could be reduced, we would achieve the desired result, since the number of retrieval probes for a node is equivalent to its depth. Even though the AVL tree structure is relatively mature, that does not suggest that it can't be improved upon. Recently, another form of a height-balanced tree was introduced by Gonnet [2] called the **internal path reduction (IPR)** tree. This structure often provides better performance than an AVL tree *and* it uses a parameter that is easier to ascertain than a balance factor for checking whether a binary tree is balanced. A disadvantage of the method is that more than one transformation of the tree may be necessary for a single insertion. The parameter used for determining the balance is the **internal path length** of a tree, which is defined to be *the sum of the depth of all the nodes* in the tree. The binary tree:

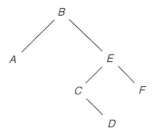

has an internal path length of 15 (since the root level is one). The *internal path length* of a tree is equivalent to the number of probes needed to retrieve each record in the tree once. A transformation occurs whenever we can reduce the internal path length of a tree. By *reducing* the internal path length of a tree, we are also *reducing* the number of retrieval probes and therefore the time necessary to retrieve a record.

> An **IPR tree** is a binary search tree in which the subtrees of a node are maintained at approximately the same height by manipulating their internal path lengths. Performance with an IPR tree is no worse than an AVL tree, and in many cases it is better.

Just as with an **AVL** tree, an **IPR** tree has four possibilities for transformation. The two in Figures 8.3a and b may be applied when the tree is **MR,** that is, *more nodes to the right;* the other two in Figures 8.3c and *d* are possible when the tree is **ML,** that is, *more nodes to the left.*

What makes the computation of the internal path length simple is that it reduces to a count of the nodes in a subtree. Counting is something that computers can do readily. If we transform the tree in Figure 8.3a, we reduce the internal path length of the tree by $n_c - n_a$, where n_c is the number of nodes in subtree c (represented by a triangle in the figure) and n_a is the number of nodes in subtree a. The internal path

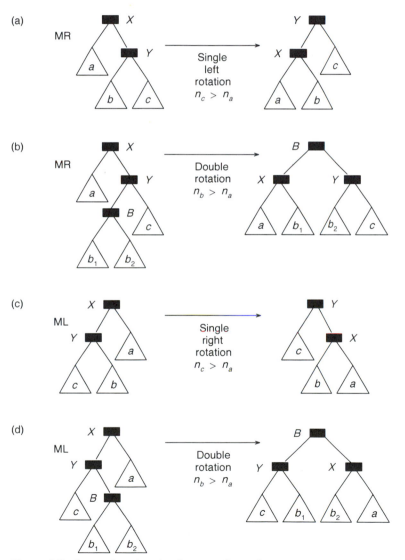

Figure 8.3 Internal path reduction transformations.

length of the tree before the transformation is:

$$I_a + n_a + I_b + 2n_b + I_c + 2n_c + 3$$

I_a, I_b, and I_c are the internal path lengths of the subtrees a, b, and c, respectively. The term n_a is included since an extra level is traversed when accessing any node in subtree a through node X. The term n_b is multiplied by two since there are two levels from the root of subtree b to node X. The term $+3$ accounts for the internal path of the nodes X and Y. In applying this expression to the example binary tree, we get the same result of 15 for the internal path length. After the rotation, the internal path length is

$$I_a + 2n_a + I_b + 2n_b + I_c + n_c + 3$$

The difference between the internal path length after the transformation and before is

$$n_c - n_a$$

thus the internal path length will be reduced if $n_c - n_a > 0$, or equivalently, if

$$n_c > n_a$$

The binary tree structure shown prior to a single rotation in Figure 8.3a is the *same* binary tree as that shown before the double rotation in Figure 8.3b. A given binary tree structure is checked for both types of rotations. What appears to be a difference in the two structures is actually an expansion of the subtree b in Figure 8.3a

into the structure

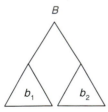

for Figure 8.3b. Accordingly n_b is equal to $n_{b1} + n_{b2} + 1$. The reason for the greater detail in Figure 8.3b is that subtree b is nearer to the middle of the structure than subtree a. As a result, subtree b_1 is rotated in one direction and subtree b_2 is rotated separately in another direction during a transformation.

Repeating the analysis for the binary tree in Figure 8.3b, we obtain the internal path length of the binary tree before rotation to be:

$$I_a + n_a + I_{b1} + 3n_{b1} + I_{b2} + 3n_{b2} + I_c + 2n_c + 6$$

The internal path for the tree after the transformation is:

$$I_a + 2n_a + I_{b1} + 2n_{b1} + I_{b2} + 2n_{b2} + I_c + 2n_c + 5$$

The difference between the internal path after and before the transformation is

$$n_{b1} + n_{b2} - n_a + 1$$

The criterion for performing a double rotation is then

$$n_{b1} + n_{b2} - n_a + 1 > 0$$

If $n_{b1} + n_{b2} + 1$ is replaced by n_b, the criterion can then be written as $n_b - n_a > 0$, or equivalently,

$$n_b > n_a$$

If both a single and a double transformation are possible at a node in the tree, the single transformation is performed since it is simpler. The balance criteria are checked at each node in the binary tree from the parent of the inserted node to the root node of the entire binary tree. Note again that unlike AVL trees, the IPR trees may require *more than one transformation*. If a transformation does occur, we still check for an unbalanced tree at the parent of the root of the previously rotated subtree and continue checking until we reach the root node of the entire tree. We also need to check the IPR balance of the immediate left descendant of the root of the rotated subtree after a single left rotation, the immediate right descendant after a right rotation, and both immediate descendants for a double rotation. The IPR insertion algorithm appears in Algorithm 8.3.

Note. A significant difference in determining the imbalance in an IPR tree vs. an AVL tree is that an IPR tree uses a *count* of nodes in subtrees whereas an AVL tree uses the *height* of subtrees.

This IPR insertion algorithm can be implemented recursively or with a stack to record the locations of nodes requiring additional processing.

An Example

To better understand the IPR method, let's consider an example using as data the list of cities in the first AVL exercise:

Jefferson, Hampton, Greenville, Ahoskie, Gurnsey, Boone, Gusey, and Hamilton

Algorithm 8.3
IPR INSERTION

I.   Insert the record into an IPR tree using the binary search tree insertion procedure.

II. For each node on the path from the parent of the newly inserted record to the root of the tree,

    A. Determine if the associated subtree has more nodes to the right or left (or neither). [to know whether to perform a left or right rotation]

    B. If $n_c > n_a$, then

        1. Perform a single rotation in the proper direction.

        2. After a left (right) rotation, check the subtree of the immediate left (right) descendant of the root of the rotated subtree for balance. If unbalanced, rotate and recursively test for and perform any additional rotations.

    C. Else if $n_b > n_a$, then

        1. Perform a double rotation in the proper direction.

        2. Check both immediate descendants for an imbalance and recursively perform any needed rotations.

As with the AVL tree, insertion is performed in the same manner as with a binary search tree until an imbalance arises. Jefferson and Hampton are inserted without requiring a transformation. We do not check for an imbalance until we have at least three levels in the tree. Greenville is added and the IPR tree becomes:

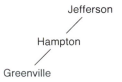

The resulting binary tree is a case ML tree, *more nodes to the left,* with Jefferson corresponding to node X and Hampton to node Y in the template in Figure 8.3c. Since n_c, which is one, is $> n_a$, which is zero, the tree is out of balance and should be rotated to the right once to bring it into balance as

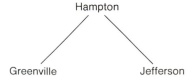

The right descendant of Hampton is balanced; so no further rotations are necessary. When Ahoskie is added, the binary tree becomes

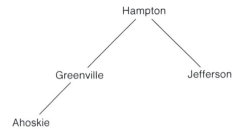

This tree matches the template for an ML tree, but since neither n_c nor n_b is $> n_a$, no rotation is needed. Then Gurnsey is added. The resulting binary tree of

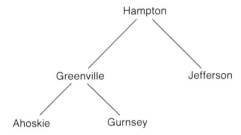

is also an ML tree, and for the same reasons as before, no rotation is needed. When Boone is added, the binary tree becomes

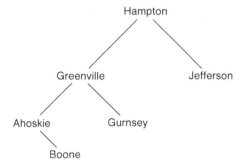

We begin the testing for imbalance at the subtree with Ahoskie as its root, but since it contains only two nodes including the newly inserted Boone, it is obviously not out of balance. We then need to test for imbalance at the subtree with Greenville as its root. This subtree is ML and does *not* require a rotation. But in addition we need to check for imbalance at the root of the tree (at Hampton). Here too the tree is ML. But this time, since n_c is $> n_a$, a single rotation to the right is performed giving the structure

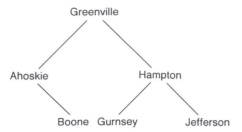

After checking the immediate right descendant of Greenville for balance, we continue by adding Gusey and Hamilton, yielding the tree

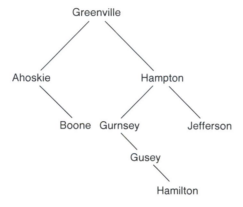

We begin working our way up from the most recently inserted node with a key of Hamilton. We first check the balance of the subtree beginning with Gusey, then Gurnsey as the root. This latter subtree is **MR**, *more nodes to the right*. Since n_c is $> n_a$, we perform a single rotation to the left to bring that subtree back into balance, giving

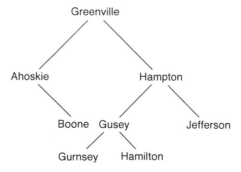

Since the subtree with Gurnsey as its root is balanced, no additional rotations are needed for that subtree. Note that at this point the result is the same as the AVL insertion of this data in the exercise. (The AVL tree is given in the Answers to Selected Exercises at the back of the book.) But we are not finished with this insertion. We still need to check the balance at the remaining nodes on the path to the root. The subtree with Hampton at the root is an ML situation which *is* balanced. But with the full tree, which is MR, the tree is unbalanced since $n_b > n_a$. A double rotation is performed, first right, then left, yielding the tree

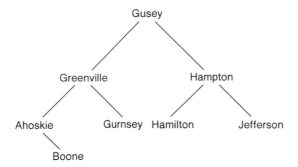

A check is made of both descendants of "Gusey" to determine if they are IPR balanced, which they are in this case. Note the *improvement* over the AVL tree. We have a better average performance with the IPR tree than with the AVL tree because we have a smaller internal path length. What we are doing in performing a rotation is moving a subtree with *more* nodes *up* in the tree, while concurrently moving a subtree with *fewer* nodes *down* in the tree. Since the subtree going up has more nodes than the tree going down in the tree, the overall internal path length is being *reduced*.

Discussion

How does the performance of an IPR tree compare with an AVL tree? From Table 8.1 (with values from 2), we see that the height of an IPR tree is never worse than an AVL tree with the same data. This means that the worst case performance for both an IPR and an AVL tree are the same. But the internal path length (for the worst case) is 10 percent better. That means that we would need 10 percent fewer probes on the average to retrieve each record once with an IPR tree vs. an AVL tree.

As with binary search trees and AVL trees, IPR trees are not particularly efficient when the number of nodes, n, in the tree becomes large. As you can observe from Table 8.1, the worst case path length (or the worst case number of retrieval probes) is proportional to $n \log n$.

In addition to the quantitative advantages of IPR trees, they also have the qualitative advantage of greater *flexibility*. The degree of balancing desired can be regulated by the user. The framework for checking imbalances allows the method to be extended easily to form what are called *k*-balanced trees. We can vary the degree

TABLE 8.1 SUMMARY OF WORST CASE COMPLEXITIES

	AVL	IPR
Height	$\leq 1.4402 \log_2 n$	$\leq 1.4402 \log_2 n$
Internal path length	$\geq 1.2793\, n \log_2 n$	$1.0515\, n \log_2 n$

of imbalance that we are willing to accept by comparing the subtree difference counts against a general parameter, k, instead of against zero.

$$n_c - n_a > k$$

and

$$n_b - n_a > k$$

The greater the value of k, the greater will be the allowable imbalance, but that in turn will reduce the insertion effort.

   We can also modify the insertion process so that the imbalance decision is based upon how large the tree is. For instance, with a tree of a thousand nodes, a loss of one in the internal path length may not be worth the processing time required to IPR balance the tree. A tree formed in this manner is called an ϵ-balanced tree; its criterion for a single rotation is

$$\frac{n_c - n_a}{n_a + n_b + n_c + 2} > \epsilon$$

To weight the decision of whether to rotate, the comparison criterion is divided by the total number of nodes in the binary tree (or subtree) under consideration. As the tree gets larger, the difference between n_c and n_a needs to increase for a rotation to be performed.

   The IPR algorithm may also be more readily programmed than the AVL algorithm since maintaining the node counts of the subtrees in an IPR tree is more straightforward than maintaining the balance factors in an AVL tree. In an IPR tree, if a COUNT field were included in each node, then, when an insertion occurred, only the nodes on the path from the newly inserted node back to the root would need to be incremented by one. These nodes need to be dealt with anyway during the balancing process, so adding a statement to perform the incrementing would not be difficult. However, the COUNT field in the IPR algorithm requires more storage than the 2 bits needed for the BALANCE FACTOR in the AVL algorithm.

   Improved average retrieval performance, flexibility in when to balance a tree, and programmability are the primary advantages of IPR trees. The disadvantage is the increased insertion effort. But again, we insert once and retrieve many times.

   The algorithm for inserting into an IPR tree and the one for inserting into an AVL tree both assume that the tree being inserted into is already a properly balanced binary search tree for which they maintain that balance after each insertion. What then does one do to bring an existing binary search tree that is not balanced into balance? One method, of course, is to reinsert the records of the tree using either the AVL or IPR insertion algorithm. This solution is not unreasonable if one has already coded the algorithm for accomplishing it. It is an efficient use of the person's resources even though it may take significant machine resources to rebuild the binary tree structure. An alternative is to apply one of the algorithms for bringing an entire, existing unbalanced binary search tree into balance. Chang and Iyengar [3] review such algorithms, and Stout and Warren [4] describe one that is optimal in both time and space but less complex than previous ones.

Tarjan [5] applies the concept of a self-correcting data structure, such as we considered with linear lists in Chapter 2, to an unbalanced binary search tree for the purpose of improving its balance and direct-access retrieval performance. When a record is retrieved, the binary search tree is restructured. In the operation to rearrange or splay (spread out) the tree, the retrieved record is moved to the root position (which is analogous to moving the retrieved item to the front of a self-correcting linear list). The splay operation also approximately halves the depths of all the nodes on the retrieval path while not increasing the depth of any other node in the tree by more than two. As more records are retrieved in a self-adjusting binary search tree, called a splay tree, it becomes more balanced.

KEY TERMS

Binary search tree	Internal path reduction trees
Self-adjusting	k-balanced trees
ϵ-balanced trees	ML
Internal path length	MR

EXERCISES

1. Insert the seven words *Happy, days, will, soon, be, here,* and *again* into an IPR tree.

2. Repeat Problem 1 but replace *again* with the two words *once* and *more* (for eight words total).

3. Insert the names of the months of the year into an IPR tree using calendar order.

4. (a) What is the average number of probes needed to retrieve each record in Problem 3 once? (b) What is the tree's depth? (c) What is the maximum number of probes needed to retrieve a record?

5. Construct an IPR tree for the eight records with the keys inserted in the following order: *Fall, is, the, best, time, of, year,* and *unequivocally.*

6. What is the reason that k and ϵ have different ranges of appropriate values?

7. Insert the example data into a 1-balanced IPR tree.

8. (a) Insert the nine records with the keys: *What, fun!, Just, like, play, on, April, Fools', Day* into an IPR tree. (b) How many transformations are performed during insertion? (c) What is the average number of retrieval probes for retrieving each record once?

9. (a) Insert the nine records from Problem 8 into a 1-balanced IPR tree. (b) How many transformations are performed during insertion? (c) What is the average number of retrieval probes for retrieving each record once?

10. (a) Insert the nine records from Problem 8 into a .2-balanced IPR tree. (b) How many transformations are performed during insertion? (c) What is the average number of retrieval probes for retrieving each record once?

11. (a) Describe a procedure for creating a balanced binary tree for storing records in a situation in which the data is static and known at the time of the creation of the tree. *Hint:* Do not use either the AVL or IPR procedure that processes the data one record at a time in the order of insertion. (b) What are the tradeoffs of this algorithm vs. those for AVL or IPR insertion?

12. What is the criterion for a double rotation in an ε-balanced IPR tree?

REFERENCES

1. Adelson-Velskii, G.M., and E. M. Landis, "An Algorithm for the Organization of Information," *Soviet Math,* Vol. 3 (1962), pp. 1259–1263.

2. Gonnet, Gaston H., "Balancing Binary Trees by Internal Path Reduction," *CACM,* Vol. 26, No. 12 (December 1983), pp. 1074–1081.

3. Chang, Hsi, and S. Sitharama Iyengar, "Efficient Algorithms to Globally Balance a Binary Search Tree," *CACM,* Vol. 27, No. 7 (July 1984), pp. 695–702.

4. Stout, Quentin F., and Bette L. Warren, "Tree Rebalancing in Optimal Time and Space," *CACM,* Vol. 29, No. 9 (September 1986), pp. 902–908.

5. Tarjan, Robert E., "Algorithm Design," *CACM,* Vol. 30, No. 3 (March 1987), pp. 205–212.

Chapter 9

B-Trees and Derivatives

B-TREES

The B-tree is one of the most important data structures in computer science because of its versatility and performance characteristics. It is a good choice for storing data that must be accessed both sequentially and directly. Comer [1] describes the B-tree as ubiquitous.[1] Its ubiquity may result from its utility. What does the B stand for? As Comer notes, the origin of the term has never been explained. The B could stand for balanced, or broad, or bushy. One of the originators was named Bayer (with McCreight) [2] and he worked for Boeing. The one thing that the B does *not* stand for is binary. A binary tree has a *branching factor* (the number of branches emanating from a node) no greater than two, whereas a B-tree does not have a theoretical limit. The higher branching factor is a consequence of storing multiple records per node. Therefore, in contrast with the binary search trees considered thus far, a B-tree is a *multiway* search tree.

In terms of their formation, the tree structures that we have seen thus far have grown from the top down; that is, the height of the tree has grown by inserting a new node at the bottom of the tree. The root node remains the same but the terminal nodes change as new items are added. With a B-tree, the new records are inserted into the leaf level, but it is the *root* level or top of the tree that changes if the B-tree must increase in height. An advantage of the bottom-up growth pattern is that it automatically preserves a balance to the structure *without* the rotations or transformations needed with the AVL and IPR trees. This balance places an *upper bound* on the worst case performance for accessing a single record. Since the branching factor is usually greater for a B-tree than an AVL or IPR tree, its depth is *less* for the same number of records. The shallower depth means fewer retrieval probes, which in turn means improved performance. In this chapter, we consider not only the B-tree but also two of its variants called the B#-tree and the B+-tree (the latter is the terminology from Knuth [3]). **Warning:** The terminology for B-trees is not yet uniform; so be sure to read the definitions carefully when using other sources of information on B-trees.

[1]For those who packed away their dictionaries after freshman English, "ubiquitous" means present everywhere at the same time.

221

Definition

The composition of a B-tree node is similar to a *bucket* or *block* of records in that it also contains *multiple* entries. The number of records that may be stored in a node is dependent upon its capacity order. The *capacity order* and other constraints on a B-tree are specified in its formal definition. It states that each node in a B-tree of capacity order d has

1. $d \leq$ keys $\leq 2d$ (except the root that has between 1 and $2d$ keys).

   $d + 1 \leq$ pointers $\leq 2d + 1$

   (except the root that has between 2 and

   $2d + 1$ pointers).

2. All the leaf nodes on the same level.

Each node except the root node must have a storage utilization of at least 50 percent. The pointers in a node point to offspring nodes (on the next lower level) of the B-tree; since leaf nodes do not have offspring, their pointers are NULL (represented by a Λ). The number of pointers in a nonleaf node is one greater than the number of records stored in the node. From the definition then, a B-tree of capacity **order one** has nodes with one or two keys and two or three pointers, represented pictorially as

Pointer	Key (record)	Pointer	Key (record)	Pointer

Although we only illustrate the key field in the record portion of a node, keep in mind that the entire record is stored there. When a node becomes full, it splits into two nodes and the middle record of the set of records, composed of the existing ones in the node and the one to be inserted, is elevated (bumped up) to the next higher level. The middle record is chosen based on key order. With the definition of the *capacity order* of a node, it is easy to determine the middle record, since when we attempt to add one more than the maximum number of records to a node, we always have an odd number of records. Some definitions of *capacity order* allow an even number of records at this stage, and then it may not be clear which one is the middle one. The ambiguity of which one is the middle one could cause problems if people interpret "middle" differently.

Examples

We will describe the insertion procedure in more detail through two examples of constructing a B-tree: first, a B-tree of order one with alphabetic keys and then a B-tree of order two with numeric keys. Such low-capacity orders will produce frequent splittings that will assist in explaining the mechanisms of insertion into a B-tree. Typically the capacity order is much larger, for example, between 50 and 200. The value chosen for the capacity order is dependent upon parameters such as the computer block or page size and the record size for the file. The high-level description of the insertion algorithm appears in Algorithm 9.1.

   Let's begin with the B-tree of capacity order one. For data, we will use three-letter animal names, just to keep the writing to a minimum:

Algorithm 9.1
B-TREE INSERTION

I. Navigate the B-tree as an index to locate the proper *leaf* node for inserting a new record. If $<$, go left; if $=$, the record is a duplicate; if $>$ *and* another record exists to the right, repeat the previous comparisons for $<$ and $=$ with that record; else take the branch to the right.

II. If space is available, insert the new record into its lexicographic position in the leaf node, else while an overflow condition exists

    A. If the overflowing node is a root node, create a new root node in a position parent to the current root node.

    B. Split the overflowing node into two nodes.

    C. Choose the middle record based upon lexicographic order from among the incoming record and those in the overflowing node. Elevate the middle record to its parent node and place it into its proper lexicographic position.

    D. Place the records that occur lexicographically before the middle record into one of the nodes and place the remaining records into the other node.

    E. Set the pointer fields in the parent node to which the middle record is elevated. Set the pointer at the left of the elevated record to the node with the lexicographically lower half of the divided records. Set the pointer field at the right of the elevated record to the node with the upper half of the divided records.

---

cat, ant, dog, cow, rat, pig, and *gnu.*

First we insert *cat,* giving us

Then we add *ant;* we still have room in the one node, so splitting is not necessary. We rearrange the two records within the node to maintain an alphabetical ordering. We then have

Next we add *dog.* That causes an overflow condition. What is the key of the middle record? It is *cat,* which then gets elevated to the next higher level. The one node on the leaf level is divided into two nodes and we create a *new* root node with *cat* in it. This splitting increases the height of the B-tree by one, but notice that the growth is at the *top* of the tree. The B-tree then becomes

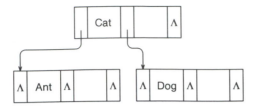

(The pointers point to the nodes and not the fields within the nodes.) The record with *cow* as its key is added next. To find the proper node for its placement, we navigate the B-tree using a procedure similar to that used in inserting into a binary search tree. We compare the key of the record that we are inserting with the key of the leftmost record in the root node. If the comparison is less than, we take a left branch. If it is equal, we have a duplicate record, or we have found the record that we are searching for on retrieval. If it compares greater than, we then compare it against the key of the next record in the node. And then we repeat the comparison process. We follow the pointer to the left of the record with the key that we are comparing against on a successful less than comparison. If we have a greater than comparison against the last (rightmost) key in a node, we take the rightmost branch emanating from that node. By following this process with *cow*, we insert it into the node containing *dog*. The key for the next record is *rat*. It maps to the leaf node with *cow* and *dog;* since that node is full, we have to split. The middle record has the key of *dog;* so it is promoted to the root level. The original leaf node is split into two nodes. Since there is room in the root node for *dog*, no further splitting is necessary. The B-tree then appears as

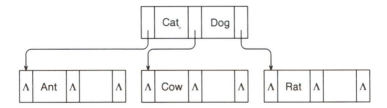

Important detail: All the records are inserted into *leaf* nodes. Only by being elevated from a lower level is a record inserted into a nonleaf node.

We next add *pig*. Since *pig* is lexicographically greater than *cat*, we then compare it against *dog*. Again we have a greater than result, but since *dog* is the last record in the node, we take a right branch that points us to the node with *rat* in it. There is room in that node; so the insertion is complete. The B-tree becomes

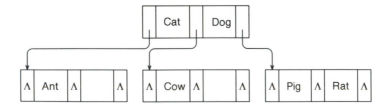

The last record has *gnu* as its key. (Even though it has been a lengthy example, we resist telling any gnu jokes.) This record maps to the leaf node with *pig* and *rat*. That node is full, so we split. *Pig* is moved to the root level, and the one leaf node is split into two nodes. Since the root node is full, there is no room for the elevated record. The root node must be split into two nodes, and a new root node (and level) is created. The level of the B-tree is then increased by one. The final B-tree appears in Figure 9.1.

Since the B-tree is such an important data structure, we include another example to strengthen your understanding of the basic mechanisms. The data for this example is numerical and will be used in subsequent examples. We use a higher capacity order, two, in this second example since we have more data. The insertion steps illustrated in Figure 9.2 should be self-explanatory.

Discussion

There are many interesting characteristics about the resulting B-trees. Notice that the tree structures are balanced—and that rotations were not required. Also, the balancing of the tree was *not* dependent upon the order of insertion of the records as it was with the binary search tree. We can retrieve a single record by stepping through the tree as we did during insertion except that when we find a match we have located the record that we are searching for. If we encounter a null pointer before we find a match, that means that the record is not in the B-tree and that circumstance terminates the search process. A B-tree is a structure for storing large quantities of information; therefore, the nodes of a B-tree are usually placed on auxiliary storage.

> **A retrieval probe** for a B-tree is defined to be an *access of a node,* and *not* a record.

When a node is accessed, its contents are copied into main memory. The subsequent searching that may be necessary to locate a record with a node is insignificant when compared with the time required to access the node in auxiliary storage. In the B-tree of Figure 9.1, three probes are necessary to retrieve *gnu,* while two probes are needed to retrieve the record with key 150 in the final tree of Figure 9.2. Because a B-tree is

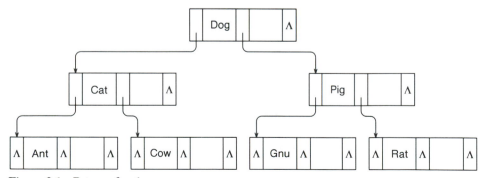

Figure 9.1 B-tree of order one.

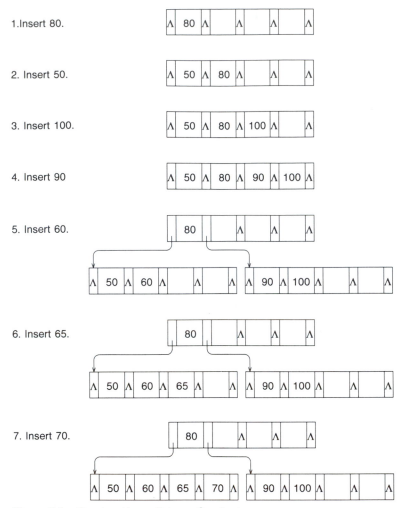

Figure 9.2 Constructing a B-tree of order two.

balanced, the worst case retrieval performance is bounded by its height, which is specified by

$$\text{Height} \leq \log_{d+1} \frac{n+1}{2} + 1$$

for $d > 1$, where n is the number of records stored in the B-tree. Since the height of a B-tree is $O(\log_d n)$, the worst case retrieval performance is then $O(\log_d n)$. A B-tree of capacity order 50 containing *one million* records requires at most *four* retrieval probes to locate a record. The worst case retrieval performance for a B-tree is less than it would be with an AVL or IPR tree since the B-tree has a higher branching factor.

We can also process the data of the B-tree sequentially by stepping through the nodes of the tree in an *inorder* traversal. Since we have more than two-way branching,

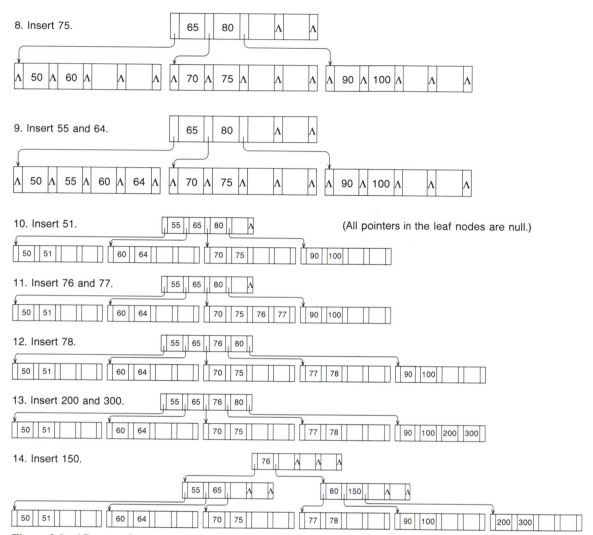

Figure 9.2 (*Continued*)

the order is left, node, pointer, node, pointer, . . . , *or* by following the lines around the perimeter of a B-tree as shown.

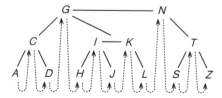

Beginning with the deepest node to the left, in this case, *A,* the nodes are processed on a path to the inside of the structure rather than a path to the outside, or in this diagram, on an upward path rather than a downward one.

Trace perimeter paths around the final structures of the two example B-trees to obtain a lexicographical ordering of the data in the nodes. Such an ordering is appropriate for sequential processing.

What is the storage utilization of the final B-tree of order one? *Storage utilization* of a B-tree is the

$$\frac{\text{Number of records stored}}{\text{Number of storage slots available}}$$

It is basically the packing factor for a B-tree. For the B-tree of Figure 9.1, the storage utilization is $\frac{7}{14}$, or 50 percent. In the B-tree of Figure 9.2, the space utilization of the final structure is even less—$\frac{17}{36}$, or 47 percent—since the root node is not even half full. The storage utilization is the lowest immediately after a splitting, especially in a case in which the splitting propagates up through the root. As additional records are added, the space utilization increases until the next splitting. One advantage of the B#-tree that we consider in the next section is that it has greater space utilization.

Applications of B-trees are discussed near the end of this chapter after we consider them and their derivatives in greater detail.

Deletion

Deletion, as you would expect, is the opposite of insertion. When we remove a record from a node, we must maintain *both* the B-tree constraints for minimum space utilization *and* the index structure for retrieving records. If the deleted record is in a leaf node and the minimum capacity constraint is maintained, no other nodes in the tree are affected by the deletion. If the deleted record, however, is in a nonleaf node, we need to move a record from a leaf node into the position of the deleted record to maintain a key to compare against when searching the index.

Note. The B-tree is being used *both* as a storage structure *and* as an index structure.

Therefore, we need to maintain the index. The record deleted from a nonleaf node is replaced by its inorder successor which is located by following the pointer immediately after the deleted record and then following the leftmost pointers in subsequent nodes as long as they are nonnull until a leaf node is reached, for example,

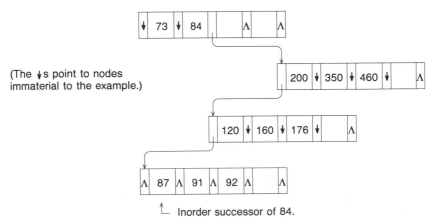

(The ↓s point to nodes immaterial to the example.)

Inorder successor of 84.

87 is the inorder successor of 84. If the minimum capacity constraint is violated when a record is either deleted from a leaf node or removed from a leaf node to replace a deleted record elsewhere, the capacity constraint is realized either by redistributing the records in the leaf node and one of its sibling nodes or by coalescing the leaf node with a sibling. An example of redistribution is

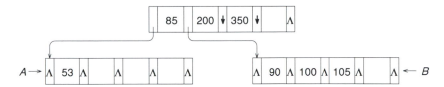

Node *A* violates the minimum capacity constraint since it is a nonroot node that is less than half filled. But its right sibling, node *B,* is filled above the minimum level. Node *A* can then borrow from node *B* to achieve the minimum capacity. The records in the two nodes together with the comparison record in the parent node are redistributed as

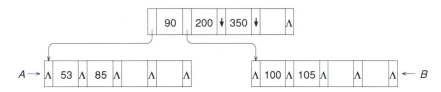

Now both nodes *A* and *B* meet the minimum capacity constraint. Notice that the comparison record in the parent node has changed; this was necessary to maintain the lexicographic ordering. If redistribution is not possible, that is, if a sibling node does not have an extra record, then the deficient node is coalesced with a sibling node. With coalescing, the records in the two nodes together with the comparison record from the parent node are combined into one node, for example, if we have

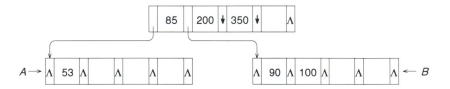

node *A* is deficient in its number of records but its sibling, node *B,* does not have an extra record. The records in the two leaf nodes and the comparison record in the parent node are then combined into one node, and the structure becomes

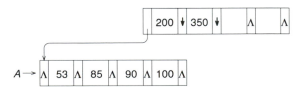

Notice that the parent node now contains one fewer record since it has one fewer offspring. If removing the record from the parent node would have caused its capacity constraint to be violated, then a redistribution or coalescing would have been performed with one of its sibling nodes. Coalescing could propagate all the way to the root node. If the root node contains only a single record prior to coalescing, when that record is removed during the coalescing process, that root node is no longer needed and the level of the B-tree decreases by one.

Deletion Example

To review, four situations are possible when deleting a record from a B-tree. Given the example B-tree of order two:

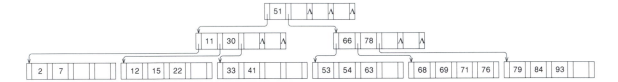

1. Deleting from a leaf node that does not violate the minimum capacity constraint.

Deleting 68 from the example tree illustrates this situation. Three records remain in its node after removing it, so the storage constraint is not violated. The tree becomes:

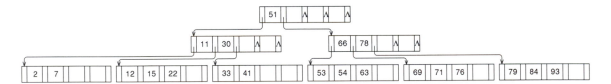

2. Deleting from a nonleaf node and replacing that record with a record from a leaf node that does not violate the minimum capacity constraint.

If 51 is deleted from the example B-tree, since it is in a nonleaf node, it is replaced by its inorder successor record of 53. Removing 53 from its leaf node does not violate the minimum capacity constraint; so the resulting tree is

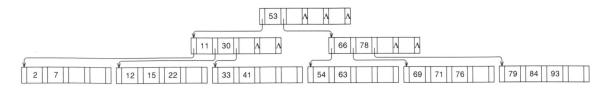

3. Deleting from a leaf node that causes a minimum capacity constraint violation that can be corrected by redistributing the records with an adjacent sibling node.

If the record with key 63 is deleted, the minimum capacity constraint for that node is violated. We first check for a redistribution. By convention, we look initially to its right sibling node; if that one does not exist or does not have an extra record, we then look to the left sibling node. In this example, the right sibling node does have an extra record, so redistribution occurs and the B-tree becomes

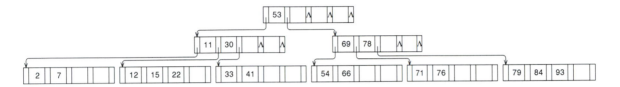

4. Deleting from a leaf node that causes a minimum capacity constraint violation that requires a coalescing of nodes.

If we delete the record with key 53 from the example B-tree as it now stands, since it is in a nonleaf node, we need to replace it with its inorder successor, 54. But removing 54 from its leaf node causes the storage constraint to be violated. Redistribution is not possible since its right sibling node does not have an extra record and it does not have a left sibling node. The record in the leaf node from which 54 was removed together with the records from its right sibling node and the comparison record in the parent node are coalesced giving the B-tree

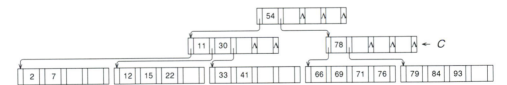

Notice that the capacity constraint for node C is violated. We must then attempt a redistribution or a coalescing to bring it back into compliance. Node C does not have a right sibling and its left sibling does not have an extra record. We then coalesce node C with its left sibling. In doing so we remove the single record from the current root node, which means that it can be removed and the level of the B-tree becomes one less. The example B-tree then becomes

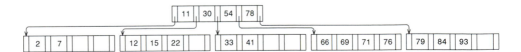

As we add records to a B-tree, its size and depth increase; but as we delete records from a B-tree its size and depth decrease. All this happens automatically as the result of the insertion and deletion procedures. And what this does is eliminate the need for file reorganizations that were necessary with the indexed sequential file structures.

KEY TERMS

B-tree
    Capacity order
    Deletion
    Retrieval probe
Branching factor
File reorganization

Index structure
Inorder
Inorder successor
Multiway search tree
Storage structure
Storage utilization

EXERCISES

1. Insert the seven words *happy, days, will, soon, be, here,* and *again* into a B-tree of order one.

2. Repeat Problem 1 but replace *again* with the two words *once* and *more* (for a total of eight words).

3. (a) Repeat Problem 1 but insert the words in alphabetical order. (b) How does the final structure compare with the result in Problem 1? (c) Is the insertion order for the records of a B-tree important to the *insertion* performance? (d) Is the insertion order for the records of a B-tree important to the *retrieval* performance? (e) Does the insertion order affect the *insertion* performance for an AVL tree? (f) Does the insertion order affect the *retrieval* performance of an AVL tree?

4. (a) What is the storage utilization of the final B-tree in Problem 1? (b) in Problem 2?

5. Under what conditions can the storage utilization of a B-tree be less than 50 percent?

6. Construct a B-tree of order one for the eight records with the keys inserted in the following order: *Fall, is, the, best, time, of, year,* and *unequivocally.*

7. Is it possible for a B-tree to contain only two nodes? Explain.

8. (a) Insert the records with the keys 6, 8, 30, 40, 50, 61, and 70 into a B-tree of order one. (b) Insert the records with the keys 12, 29, 30, 31, 32, 33, and 56 into a B-tree of order one. (c) From the results of a and b, what conclusion can you draw about the effect of key distribution on retrieval performance in a B-tree?

9. (a) Insert the data from Problem 8a into the indexed sequential structure of Figure 4.1. Assume that consecutive records on track nine are available for overflow records. (b) Repeat with the data from Problem 8b. (c) From the results of a and b, what conclusion can you draw about the effect of the key distribution of the added records on the retrieval performance?

10. (a) For a file of 100 records, what is the *minimum* depth for a B-tree of order three? (b) for a B-tree of order ten?

11. (a) For a file of 100 records, what is the *maximum* depth for a B-tree of order three? (b) for a B-tree of order ten?

12. What is the minimum depth of a B-tree of order 100 used to store 45,000 records?

13. What is the minimum number of probes for a *successful* retrieval in the B-tree in Figure 9.2?

14. What is the minimum number of probes for an *unsuccessful* retrieval in the B-tree in Figure 9.2?

15. Write an algorithm in English or pseudocode for deletion from a B-tree.

16. Write a *recursive* algorithm in pseudocode for traversing the records of a B-tree in inorder.

17. Draw the B-tree resulting from deleting *cow* from the B-tree in Figure 9.1.

18. Draw the B-tree resulting from deleting *pig* from the B-tree in Figure 9.1.

19. Draw the B-tree resulting from deleting *gnu* from the B-tree in Figure 9.1.

20. Draw the B-tree resulting from deleting 80 from the B-tree in Figure 9.2.

21. Does a B-tree or an indexed sequential structure provide better performance for a file in which a large number of records are added in real time? Explain.

B#-TREES

One observation about the B-tree insertion examples was that the space utilization was at or slightly below 50 percent after the final splitting. That relatively low percentage means that one half of the storage was not being used. Because of the minimal capacity constraints, 50 percent is the lowest space utilization for all the nodes *except* the root. As a result, the overall *minimum* storage utilization is slightly less than 50 percent; the *maximum* storage utilization is 100 percent, but that utilization would not occur often with random data. Can we improve the space utilization of the B-tree structure? Yes, one way to accomplish an increase is to *postpone splitting* during insertion since splitting *reduces* the space utilization. Splitting may also increase the depth of a B-tree, which will negatively affect retrieval performance. We call this version of the B-tree structure a B#-tree; it is a variant of the B*-tree described in Knuth [3] and introduced by Bayer and McCreight [2]. The B* variation of a B-tree also postpones splitting a B-tree, and by doing so, it often has a shallower structure than the equivalent B-tree. The primary difference between a B#-tree and a B*-tree is that the root node in a B*-tree is larger than the other nodes in the tree and may contain up to $4d + 1$ records. All the nodes of a B#-tree are the same size, so it is essentially a B-tree with a modified insertion algorithm. This algorithm, which is described in Algorithm 9.2, delays the splitting and by so doing, gives a minimal space utilization of approximately $\frac{2}{3}$. The retrieval and deletion processes function the same as with a B-tree. The basic idea of the B#-tree is to (1) postpone splitting by redistributing the records of the overflowing node into adjacent siblings and (2) when splitting is necessary, to convert *two* nodes to *three* nodes. These modifications keep the nodes more fully utilized.

> A **B#-tree** is a B-tree in which the insertion and splitting operations are processed in a manner that postpones splitting and as a result yields higher space utilization and faster retrieval performance.

An Example

Let's consider an example B#-tree of capacity order two. A node in such a tree may contain up to four records and five pointers. The data is the same as in the example of Figure 9.2, that is,

Algorithm 9.2
B#-TREE INSERTION

___

I. Navigate the B#-tree as an index to locate the proper *leaf* node for inserting a new record. If <, go left; if =, the record is a duplicate; if > *and* another record exists to the right, repeat the previous comparisons for < and = with that record, else take the branch to the right.

II. If space is available, then insert the new record into its lexicographic position in the leaf node, else

   A. While an overflow condition exists at a nonroot level, first check its right sibling, then its left for available space. If yes,

      1. Redistribute the records, including the record causing the overflow, between the overflowing node and its sibling with the space available,

   else

      2. Split the overflowing node and redistribute (1) its records, including the record causing the overflow, (2) the records of its filled right sibling node (or if none, its filled left sibling node), and (3) the comparison record in the parent node, among *three* nodes. Place the two appropriate comparison records in the parent node.
      3. Set the three pointers in the parent node.

   B. If the overflow condition exists at a root node,

      1. Create a new root node in a position parent to the current root node.
      2. Split the overflowing current root node.
      3. Elevate the middle record to the new root node and divide the remaining records between *two* nodes. Make the division based upon lexicographic key order.
      4. Set the pointer fields in the parent node to which the middle record is elevated. Set the pointer at the left of the elevated record to the node with the lexicographically lower half of the divided records. Set the pointer field at the right of the elevated record to the node with the upper half of the divided records.

___

80, 50, 100, 90, 60, 65, 70, 75, 55, 64, 51, 76, 77, 78, 200, 300, 150

We insert the first four records with keys of 80, 50, 100, and 90 without difficulty. At this stage, the B#-tree and the B-tree are the same.

| Λ | 50 | Λ | 80 | Λ | 90 | Λ | 100 | Λ |

When attempting to insert the next record with a key of 60, we have an overflow situation. Since the single node does not yet have a sibling, we cannot redistribute the records but must split the node as with a B-tree.

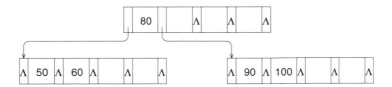

Records with keys of 65 and 70 are added next.

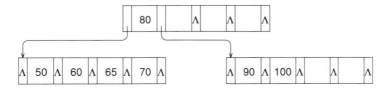

We encounter an overflow condition when we insert 75. Since space is available in the right adjacent sibling node, rather than splitting the overflowing node as we would do with a B-tree insertion, we redistribute (1) the records in the node being added to, (2) the records in the right adjacent sibling node, (3) the record being inserted, and (4) the record in the root node used for comparison purposes. We have eight records to divide among two leaf nodes and the comparison position in the root node. Which record should we choose for comparison purposes? First, we lexicographically order the records to be redistributed and then calculate the comparison record's position in this ordering by the general formula

$$\text{comparison.record} = \lfloor (r + 1) / n \rfloor$$

where r is the number of records to be redistributed and n is the number of leaf nodes that they are being distributed into.

This formula provides a *convention* for choosing a "middle" record or records when redistributing records among two or three leaf nodes. For this redistribution, the comparison record is in lexicographic position

$$\lfloor 9/2 \rfloor = 4$$

Because the record in that position has a key of 70, the records with keys that are lexicographically less than 70 are placed in the left leaf node, and the remaining records are placed in the right leaf node.

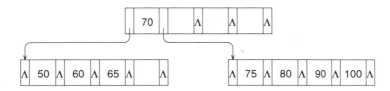

We add 55 without a problem.

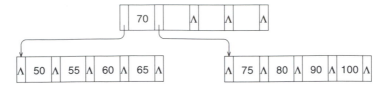

When we add 64, both the node that it would normally be inserted into and its right adjacent sibling node are full. So this case requires splitting. Unlike a B-tree, with a B#-tree, we split *two* nodes into *three*. We redistribute ten records (four from each leaf node, the one being added and the one in the root used for comparison purposes). The initial comparison record, located in lexicographic position

$$\lfloor 11/3 \rfloor = 3$$

has a key of 60. The comparison record for dividing the remaining records, located in position

$$\lfloor 8/2 \rfloor = 4 \qquad \text{(of the remaining seven records)}$$

has a key of 75. The resulting tree is

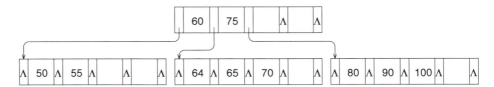

Records with keys 51 and 76 are inserted without difficulty, yielding the tree

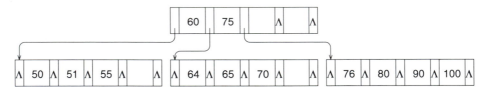

When inserting 77, we redistribute the records in the overflowing node with its left sibling. The tree then becomes

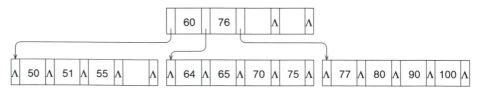

When we insert 78, we have an overflow situation in which redistribution is impossible since we limit the search for available space to the immediately adjacent sibling nodes. (To search beyond these nodes would require additional processing time and if space were located in a more distant sibling node, a more complicated redistribution procedure would be needed.) We then split the two rightmost nodes into three. In a situation in which both the right and left siblings of an overflowing node are full, *by convention,* the overflowing node and its right sibling (rather than its left) are divided into three nodes. In this example, we then have

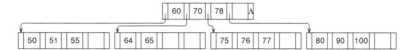

The record with key 200 is inserted without difficulty. When we attempt to insert the next record, that with key 300, we need to redistribute records with its left sibling. The resulting tree is

The final insertion of the record with key 150 causes an overflow condition in which it is necessary to combine and split. The final B#-tree appears in Figure 9.3.

Compare this result with that of the B-tree insertion. The space utilization of the B#-tree is $\frac{17}{24}$, or 71 percent, whereas the space utilization for the B-tree was $\frac{17}{36}$, or 47 percent. Not only does the B#-tree have better space utilization but it also is shallower, which means fewer probes for direct access. (See Exercise 6 at the end of this section.) This is one of the rare cases where we do not have a time-space *tradeoff* but rather a *gain* in *both* performance *and* storage; the cost, however, is a slightly more complex insertion algorithm.

Although the redistributions and splittings in this example occurred only at the leaf level, they are not limited to the leaf level but arise at any level in which an overflow situation occurs in a B#-tree.

Definition and Analysis

Since you have already seen the formal definition of a B-tree, we delayed the definition of the B#-tree until after an example so that the basis for the definition might be clearer. The formal definition of a B#-tree of capacity order d is the same as that for a B-tree of capacity order d *except* that the lower bounds for the keys and the pointers of the nodes at the intermediate levels (those not including the root, the leaves, and the

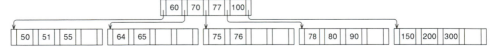

Figure 9.3   B#-tree of order two.

nodes at the level immediately beneath the root) are greater. The modified relationship for the number of keys is

$$\frac{4d - 2}{3} \leq \text{keys} \leq 2d$$

and the relationship for the number of pointers is

$$\frac{4d + 1}{3} \leq \text{pointers} \leq 2d + 1$$

How did we arrive at these greater lower bounds for a B#-tree? At the time that a splitting is necessary, the portion of the B#-tree affected by the splitting (other than a root node) has the structure:

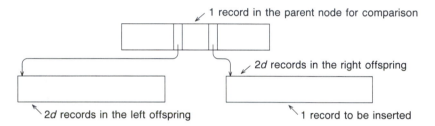

The total number of records to be redistributed after the splitting is then $4d + 2$. Since we are splitting the two nodes into three, we need two records in the parent nodes for comparison purposes. That leaves $4d$ records to distribute among three nodes. Each node then has *at least*

$$\left\lfloor \frac{4d}{3} \right\rfloor$$

records. To make this expression easier to compare with that of a B-tree, we remove the floor function to obtain

$$\frac{4d}{3} - \frac{2}{3} \quad \text{records}$$

for the *minimum* number in a node; this expression may be further reduced to

$$\frac{4d - 2}{3} \quad \text{records}$$

Each node has one more pointer than records; so a node would then have *at least*

$$\frac{4d - 2}{3} + 1 \quad \text{pointers}$$

which equals

$$\frac{4d + 1}{3} \quad \text{pointers}$$

Since $(4d - 2) / 3$ (the lower bound for records in a B#-tree) is $> d$ (the lower bound for a B-tree) for all values of $d > 2$, a B#-tree has a greater utilization of storage in the intermediate nodes for all trees of order > 2. If we divide this lower bound for an intermediate node in a B#-tree by its capacity,

$$\frac{(4d - 2)/3}{2d}$$

we obtain

$$\frac{4d - 2}{6d}$$

As the order, d, increases, the -2 term in the numerator will become less significant and the minimum utilization of the intermediate nodes will become $\approx \frac{2}{3}$. This analysis illustrates why a B#-tree has a greater minimum storage utilization than a B-tree.

Other B-Tree Structures

The definition of a B-tree is underconstrained; that is, for a given number of records stored in a B-tree of a given order, *multiple* B-tree structures may exist. For example, if we wish to store five records in a B-tree of order one, *two* B-tree structures exist as shown in Figure 9.4. Both structures meet the definition of a B-tree and both store five records. Structures for storing five records in a B-tree of order one other than these two violate the constraints of the B-tree definition. Can you think of any such trees? Why are they illegal? We will consider such unacceptable structures shortly. Observe that the storage utilizations for the two trees in Figure 9.4 are different. Rosenberg and Snyder [4] discuss the many B-tree representations that are legal for a B-tree of a given order. Figure 9.5 illustrates all the structurally distinct B-trees of order one that can contain 15 records. This group is represented as T_{15}^1 where the general formulation is T_k^M where M is the capacity order and k is the number of records stored. The B-trees in Figure 9.4 would be represented as T_5^1. In Figure 9.5, the ■s represent the nodes and the Λs represent NULL pointers. From the number of pointers emanating from a node, we can tell how many records are stored in it (the number of pointers minus one). In Figure 9.5a, there is one record stored at the root level, two records on the next level, four on the third level, and eight stored on the leaf level (for a total of 15). In Figure 9.5j, there are two records stored at the root level, five at the next level, and eight stored at the leaf level (again for a total of 15). From Figure 9.5, we see that there are *10* legal structures for a B-tree of order one containing 15 records. We have looked then at only two of the many possible algorithms for inserting into a B-tree structure.

Why are we considering the many possible structures for a B-tree? The different structures require different amounts of storage and have different average and worst

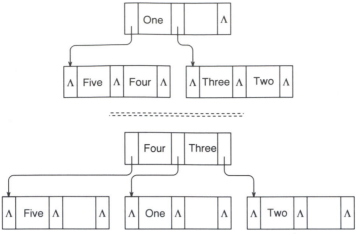

Figure 9.4 Two structurally distinct B-trees of order $T\frac{1}{5}$.

case retrieval performances. Table 9.1 lists the performance and storage requirements for the trees illustrated in Figure 9.5. The conclusion of Rosenberg and Snyder is that

> space optimal trees are nearly time optimal and time optimal trees are nearly space pessimal.

The B-tree structure with the best performance (tree j) requires 12 nodes or 33 percent more storage than the minimum whereas the trees with the best space utilization (trees b and c) require only 8.3 percent more time for probing than the minimum. Attempting to increase the space utilization as we did with the B#-trees was then a reasonable thing to do since the best space utilization gives near optimal performance.

Let us consider a set of structurally distinct B-trees of a different capacity order and number of records. By looking at the valid and invalid members of such a set, we

TABLE 9.1 COMPARISON OF TIME AND SPACE COSTS
OF B-TREE STRUCTURES OF FIGURE 9.5

Tree from Figure 9.5	Time cost (nodes visited)	Space cost (nonterminal nodes)
a	3.27	15
b	2.60	9
c	2.60	9
d	2.53	10
e	2.53	10
f	2.47	11
g	2.47	11
h	2.47	11
i	2.47	11
j	2.40	12

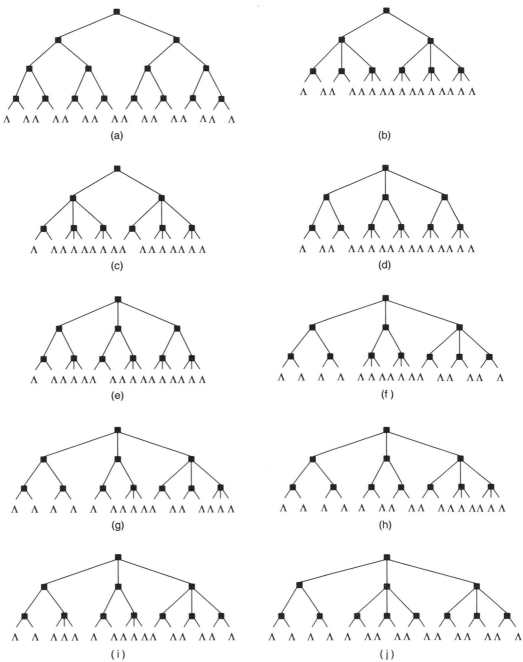

Figure 9.5 Ten structurally distinct B-trees of T_{15}^1.

may acquire a better intuitive feeling of what constitutes a B-tree or one of its variants. How many structurally distinct B-trees are in the set T_3^1? There is just the *one* with the structure

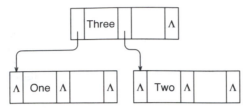

Why is the following tree not in the set?

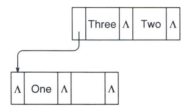

The root node violates the requirement that it have one more pointer than records. It has two records but only one pointer. Such a B-tree could never have been formed from an overflow and splitting. In a splitting, we never reduce the number of nodes on a level as we would have had to do to obtain the previous structure. Similar arguments hold for the structure

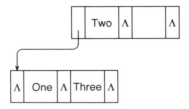

The root node does not have the proper number of pointer values nor would it ever have been formed without a splitting in which at least two offspring resulted.

KEY TERMS

B-tree
    Capacity and utilization
    Illegal structures
    Structurally distinct
    Variant structures
B*-tree

B#-tree
    Capacity order
    Insertion
    Redistributing
    Splitting

EXERCISES

1. Insert the seven words *happy, days, will, soon, be, here,* and *again* into a B#-tree of order one.

2. Repeat Problem 1 but replace *again* with the two words *once* and *more* (for a total of eight words).

3. **(a)** What is the storage utilization of the final B#-tree in Problem 1? **(b)** in Problem 2?

4. Construct a B#-tree of order one for the eight records with the keys inserted in the following order: *Fall, is, the, best, time, of, year,* and *unequivocally.*

5. Construct a B#-tree of order one for the seven distinct records with the keys inserted in the following order: *April, showers, bring, May, flowers, Mayflowers, bring,* and *pilgrims.* Remember that duplicate records are not reinserted. Assume that a blank character appears before the letters in the collating sequence.

6. **(a)** What is the average number of probes for the B-tree of Figure 9.2? **(b)** for the B#-tree of Figure 9.3?

7. Draw the structurally distinct B-trees in the set T_5^2.

8. Draw the structurally distinct B-trees in the set T_6^2.

B+-TREES

The B+-tree is a variant of a B-tree identified as such by Knuth [3]. As the name implies, it has advantages over a plain B-tree. One advantage is that it is usually *shallower* than a B-tree with the same data, which means that fewer probes are needed for direct access. A second advantage of a B+-tree is that it isn't necessary to move up and down the nodes of the tree in inorder to obtain a lexicographic ordering of the data, so sequential processing is faster. We then can improve upon *both* the direct and sequential processing of a B-tree by using a B+-tree.

We obtain a shallower tree with a B+ structure by *increasing* the number of keys we can store per node. Since we usually have a limit (e.g., the physical page size) on the size of a node, we accomplish this increase by storing *only the key fields in the nonterminal nodes* of the tree. Since the key fields are only a subset of the entire record, we can store *more* of them in the fixed amount of space reserved for a node. If we store only the key fields in the nonleaf nodes, where are the actual records stored? We store them in the leaf nodes. The nonterminal nodes then act as an index into the terminal nodes containing the data. In contrast with a B-tree, a B+-tree has the index structure *distinct* from the storage structure. One result of this change in structure is that *each* of the records requires the *same* number of probes to locate it on direct-access retrieval since all the records are stored on the same level.

> A **B+-tree** is a B-tree in which the data records are contained in *terminal* nodes, and the *nonterminal* nodes, consisting only of key values, form an *index* into the data nodes.

We improve the sequential processing by linking the data nodes together. As a result, it is *not* necessary to move about the tree but merely to process a singly linked list of terminal nodes. As a consequence of the procedure for inserting into a B+-tree, the records are ordered alphabetically by node. The B+-tree structure then consists of the nonterminal nodes constituting an **index** or directory and the linked terminal nodes of data called the **sequence set**. The general structure is

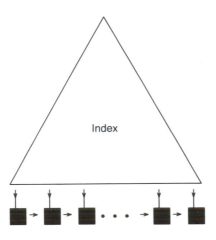

where ■ represents a data node in the sequence set.

Since we are storing only the key fields in the index nodes, their capacity order is normally *much* greater than for nodes in a B-tree. The number of records stored in a data node are many fewer than the number of keys in an index node. The orders of the index and sequence set nodes are *independent*. The depth of a B+-tree is one greater than the depth of the index since the depth of the sequence set is one. As before, the root is at level one.

Note. The index portion of a B+-tree is actually a B-tree; therefore, it *must* conform to the requirements and constraints of a B-tree. In addition, the nodes of the sequence set must be at least half full except, of course, when inserting records into the initial node.

We use the index structure to point us to the correct page for an insertion or a direct-access retrieval. Since the records are no longer stored in the nonterminal nodes, the navigation procedure is modified so that on an equal comparison, we go either left or right to locate the proper terminal node. We arbitrarily choose to go *left on equal*. Other implementations may choose to go right; so you need to be certain that you are aware of the assumptions before you use a particular implementation. The expansion of the index structure for a B+-tree is the same as a B-tree expansion. The major difference between a B+-tree and a B-tree is the manner in which a new record is inserted into a leaf node and what happens when an overflow occurs. In the latter case, the page splits, but only the *key* of the middle record is promoted to the next highest

Algorithm 9.3
B+-TREE INSERTION

I. Navigate the nonterminal nodes of the B+-tree as an index to locate the proper *leaf* node for inserting a new record. If ≤, go left; if > and another record exists to the right, repeat the previous comparison for ≤, else take the branch to the right.

II. If space is available, insert the new record into its lexicographic position in the leaf node, else

    A. Split the overflowing leaf node.

        1. Elevate the *key* of the middle record to the index structure. The middle record is chosen by lexicographic key order from those records in the overflowing node and the record being inserted.

        2. Divide the records of the overflowing node, including the record being inserted, between two nodes such that all the records up to and including the middle record are placed in the left node and the remaining records are placed in the right node.

    B. Process the key that was newly elevated into the index structure as an insertion into a B-tree. Modify the index portion of the structure as necessary (see Algorithm 9.1).

---

level. Since the leaf and index nodes serve different functions in a B+-tree, the insertion of the first node into a B+-tree is a special case. The record is placed in a leaf node and its key is placed in an index node with a pointer to the leaf node, for example,

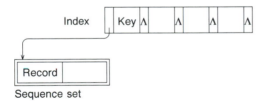

Sequence set

Notice that unlike a B-tree, the root node has a null pointer to the right of the key. The algorithm for insertion into a B+-tree, which appears in Algorithm 9.3, assumes that the first record has already been inserted.

An Example

Let's insert records with the numerical keys that we have inserted in the previous B-tree and B#-tree examples:

80, 50, 100, 90, 60, 65, 70, 75, 55, 64, 51, 76, 77, 78, 200, 300, 150

We use an index capacity order of two and a sequence set of order one. The index nodes may contain up to four keys, and the data nodes may contain up to two records. As with the previous examples, we have chosen such low orders to better demonstrate the mechanisms of the insertion process. In practice, the orders are much higher. We then have two types of node structures: one for the index nodes and one for the sequence set nodes. When we insert the first record with a key of 80, we not only insert it into a terminal data node but we must also *copy the key* into the index so that we have something to compare against. The structure appears as

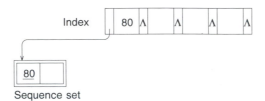

The entire record with 80 as its key is inserted into the data node of the sequence set, and the key is placed in the index. Remember that on an equal comparison, we take a left branch. The underscore below the 80 is to remind us that the *entire* record, and not just the key, is stored in the sequence set. The pointers throughout this example are the actual addresses of the *nodes,* and not fields. These addresses usually refer to secondary storage.

To *retrieve* 80 requires two probes: one of the index and one of the sequence set. In fact all direct-access retrievals will require only *one probe* of the sequence set. (Since we have only one level of terminal nodes.)

Note.   If the index is small enough to fit into primary memory, we need only **one** probe of auxiliary memory to retrieve a record.

Often memory is not large enough to store the entire index, but even storing the root node, which *must* be accessed on every direct-access retrieval, improves performance by eliminating one probe of auxiliary memory for each retrieval. We should store as much of the index as we can, beginning at the top.

Inserting 50 causes no modification of the index since space is available in the terminal data node to which the key maps. The B+-tree structure then becomes

We have rearranged the two records in the data node so that we have a lexicographical ordering. Records with keys of 100 and 90 are added and placed in a new data node to the right of the existing one. The two data nodes in the sequence set are then linked together as shown.

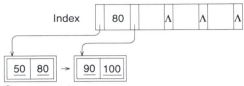

Sequence set

Notice that we can readily process the data sequentially by stepping through the chain of data nodes in the sequence set. When we insert 60, we encounter an overflow condition. There is no room for 60 in the data node that the key maps to; so we need to divide the data in that node between two nodes and elevate another key to the index. The middle key among the records on the overflowing data node and the one being inserted is 60; it is then elevated to the index. We place the records that have keys less than or *equal* to 60 in the left node of the pair formed from the splitting, that is, 50 and 60. 80 is placed in the right node of the pair. The B+-tree structure then becomes

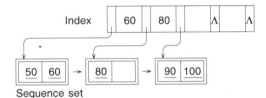

Sequence set

Note.   We process additions to the index as we would insertions into a B-tree.

A difference in the overall processing, though, is that since we have a higher-capacity order in the index nodes of the B+-tree, we require *fewer* splittings.

65 is inserted easily. When 70 is added, we have another overflow condition. 70 is the middle key of the records involved in the splitting process, and a copy of its key is therefore elevated to the index. 65 and 70 are placed in the left node, and 80 goes to the right. The resulting structure is

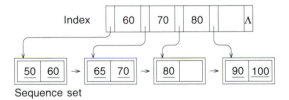

Sequence set

75 is added without causing a split. Inserting 55 causes a split that yields the structure

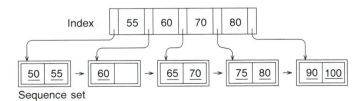

Sequence set

55 is elevated to the index. 50 and 55 are placed in the left node of the pair involved in the split, and 60 is placed in the right node. Inserting 64 causes an overflow condition, with 65 being copied to the index. But the root node in the index is already at maximum capacity. This condition necessitates splitting the root node and creating a new root node and a new level. The middle key of those involved in the splitting is 65, which is then promoted to the new root node. The keys less than 65 are placed in a left index node appearing at a level below the new root. The keys greater than 65 are placed in an index node to the right. The B+-tree is now

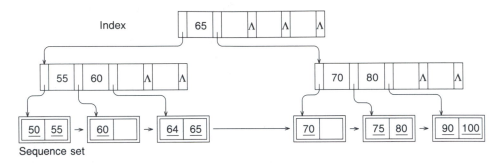

The depth of the B+-tree is three since the insertion caused the depth of the index to increase by one. Both the insertion of 51 and 76 cause data nodes to split. After these insertions the B+-tree becomes

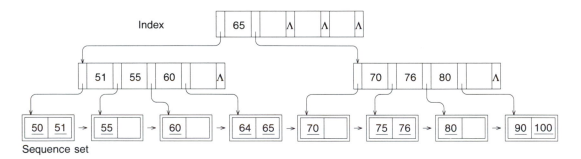

After inserting the remaining records with keys 77, 78, 200, 300, and 150, we obtain the tree in Figure 9.6.

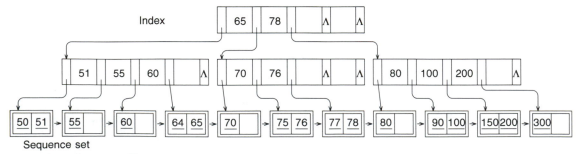

Figure 9.6   B+-tree of index order two with two records per leaf node.

Retrieval

To retrieve the record with key 65, we begin navigating the index structure by comparing 65 with the leftmost key in the root node of the index. Since it is also 65, the comparison is equal and we take a left branch to the index node on the next level. There, we compare 65 against 51. It is greater than, so we continue the navigation process by comparing against the next key in that index node, that is, 55. The comparison is still greater than; so we try the next key in the node, that is, 60. Again the comparison is greater than. However, since this is the last key in the node, we take the right branch, which points to the data node containing the record with key 64 and the one with the desired key, 65.

Deletion

With a B-tree, the nodes are used both to store data and to act as an index for retrieving other records. In a B+-tree, these two functions are separate. The index is used for navigational purposes and the sequence set is used to store the data. As a result of this separation, it is not always necessary when deleting a record to remove the key of the deleted record from the B+-tree index. Only when deleting a record causes the number of records in its sequence set node to fall below the minimum capacity (one-half) is it necessary to reorganize the data in the sequence set. During such a process, if a data node is deleted, one of the pointers in the index and consequently one of the comparison keys are unnecessary and can be removed. Removing a key from the index is equivalent to deleting a record from a B-tree since the index is a B-tree. When deleting a record does not violate the minimum capacity constraint of a node, the existing index, even if it includes the key for the deleted record, will still be effective in providing direction to the proper leaf node during a direct-access retrieval. For example, with the B+-tree of Figure 9.6, when we retrieve the record with key 75, we compare the 75 with the 76 in the index. If we delete the record with key 76 but keep its key in the index, there is no effect on retrieving the record with key 75. We process the index just as before. Deletion from a B+-tree is then straightforward in most cases since it only involves deleting a record from its sequence set node.

Implementation

To process the data in the structure sequentially, we step through the singly linked sequence set. Finding the successor data record requires only a single operation. This is an improvement over a B-tree in which it is often necessary to move up and down the tree to locate the successor data record. The sequence set could be implemented as a *doubly-linked* list to allow processing in both ascending and descending order and to provide greater flexibility in searching subsets of the data. For faster retrieval, the pages of the sequence set should be stored on the same cylinder.

In an implementation, the capacity order of a node is usually determined by the physical properties of the computer system, such as the size of a virtual page and the size of the key field. If a virtual page were 2K bytes, the greatest number of keys that would fit into that size block would determine the capacity order. This order typically ranges between 50 and 200.

As previously mentioned, since every insertion and retrieval on a B+-tree will need to process the root node, performance can be improved if that node is kept in primary memory. A B+-tree insertion algorithm can be simplified if we take advantage of the fact that the nodes that might require additional processing should an overflow situation occur are the nodes on a path from the insertion leaf node to the root. And they need to be processed in that order. What data structure is a natural for remembering where these nodes are and in the proper order needed for processing? Yes, a stack, since it is a last-in first-out data structure. Operating systems frequently use a paging algorithm in which the least recently used (LRU) page is removed from primary memory to make room for an incoming page from auxiliary memory. With such a paging algorithm, the B+-tree pages (nodes) that require splitting are then likely to be in high speed memory.

Indexed Heap

As the number of records in a data node increases, the efficiency of a sequential search decreases. For that reason we may want to use a different searching mechanism such as a binary search to locate a single record within a page. Both a binary search and the efficient sequential processing of the records in the sequence set require that the records appear within a node in lexicographic order. Maintaining a lexicographic order as new records are added to a node may require moving some of the existing records. The need to move *entire* records may slow the insertion process. An alternative data structure to the positioning of records in a node in contiguous locations is the indexed heap. *An indexed heap* is a data block containing *an index* to the individual records within that block. Figure 9.7a illustrates an indexed heap containing three records, *a, c,* and *e.* If we insert a record with relative position *b,* we only need to move the *pointers* to records *c* and *e* and *not* the entire records. See Figure 9.7b. An indexed heap then saves processing time during an insertion.

Variants

To further reduce the depth of the index in a B+-tree, we could limit the key length *in the index* to the fewest number of characters needed to distinguish a key from all other keys in the file. To accomplish this, the user needs to know something about his data in advance, *or* he needs to allow for this limit to change dynamically. Since the fixed limit is simpler, we assume that circumstance. By storing only the fewest number of disambiguating characters, we are storing only the prefix of a key. For that reason

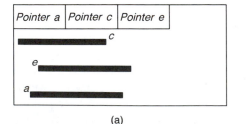

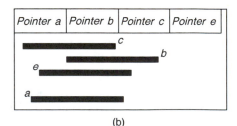

(a)   (b)

Figure 9.7   Indexed heap; pointers are moved, not records.

we call this variant of a B+-tree, a *prefix B+-tree*. We can reduce the depth of a B+-tree since the prefix key lengths are shorter than a full key so we can store more of them in a node of a given length. If the keys consisted of the three-letter animal names in Figure 9.1, what is the minimal length for the prefix to construct a prefix B+-tree? We need to use more than one character since we have two keys that begin with *c: cat* and *cow*. The prefix then needs to be two characters to distinguish the keys.

Could we combine the concept of a B#-tree with a B+-tree to obtain a B+#-tree? (or would it be a B#+-tree)? The answer is *yes*, since the index of a B+-tree is basically a B-tree. We could apply the B#-tree insertion algorithm to improve the storage utilization of the B+-tree index. Is such an improvement in storage utilization as important though with a B+-tree as with a B-tree? No, since the capacity order of a B+-tree index is much larger than that of a B-tree and that circumstance lessens the amount of splitting required. The reduction in splitting indicates that the B+-tree will be at a minimal utilization less often.

Discussion

The B+-tree is currently one of the most widely used structures for organizing data, and the most commonly implemented variation of the B-tree concept. It is especially applicable in situations in which the data is processed *both* sequentially and directly. The foundation data structure for most database management systems is a B+-tree for it is usually not known a priori how the database user will access the data—whether it will be sequentially or directly.

IBM's general-purpose access method VSAM (for <u>v</u>irtual <u>s</u>torage <u>a</u>ccess <u>m</u>ethod and pronounced v-sam) is based upon a B+-tree [5,6]. For all practical purposes, the B+-tree has supplanted the indexed sequential structure that was presented in Chapter 4. There are several reasons for this. The B+-tree is always balanced, so there are never the situations with the long chains of overflow records that we observed with the indexed sequential structure. As a result, direct-access retrieval on a B+-tree requires fewer average probes. Also, since these long chains do *not* occur in a B+-tree file, it is *not* necessary to periodically *reorganize* it to improve performance. Not needing to reorganize a file can result in a considerable savings of time. In addition, processing the data sequentially is faster with a B+-tree since it is only necessary to step through a singly linked list of data nodes rather than processing several levels of an index.

A VSAM file consists of a B+-tree with the data records located in an added leaf level as illustrated in Figure 9.8. Since a leaf, called a **control interval,** is the basic unit in which data is transferred in a single I/O operation, its size is limited by the hardware. A control interval contains one or more data records plus system information about the data and the control interval. A sequence set node is called a **control area.** To improve performance, a control area and all its associated control intervals are placed on the same cylinder. Then once the control area has been located, the disk access arm does not need to move to find the associated control intervals. The system provides a mechanism for the user to specify how much free space is to be included when the file is initially loaded. A 100 percent loading is fitting if the file is to be relatively stable, but it makes subsequent insertions costly. Conversely, a 50 percent loading would waste storage if few insertions were anticipated.

Other advantages of a B+-tree are that neither the insertion nor the retrieval

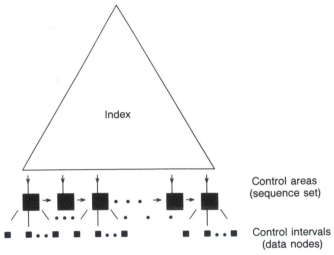

Figure 9.8   VSAM file.

performance is affected by either the insertion order of the records or the distribution of their keys. These two qualities are important in practice for we often cannot modify the insertion order of the data and we usually cannot change the distribution of the keys. For most applications, the keys for the data records do not form a uniform distribution, for example, the names in a telephone book, the words in a dictionary, or the names of products in an inventory. The disadvantages of a B+-tree are the storage needed for the index and the number of probes that may be needed for a direct-access retrieval if the index cannot be stored in primary memory.

KEY TERMS

B+-tree   Doubly-linked list
    Database management   Indexed heap
    Depth   Prefix B+-tree
    Directory   Reorganize
    Index   B+-tree
B#-tree   Sequence set
Control area   VSAM
Control interval

EXERCISES

1. Insert the numbers individually into a B+-tree of index order one with two records per leaf node: 50, 85, 90, 35, 55, 20, 25, 80. Identify the sequence set.

2. (a) If the keys for records to be inserted into a prefix B+-tree were *Yankee, Doodle, went, to, lunch, riding, on, a,* and *pony,* what would the prefix length need to be to distinguish the

keys? **(b)** What would the length of the prefix need to be for the keys *Peter, Piper, picked, a, peck, of, pickled, peppers?*

3. Construct a B+-tree of order two with four records per leaf node. For data, use the nine records with the keys inserted in the following order: *when, everything, seems, to, be, going, well, then,* and *worry.*

4. What length should the prefix be in a prefix B+-tree for the data in Problem 3?

5. What is the **(a)** *maximum* and **(b)** *minimum* number of probes needed to retrieve a record in a B+-tree with leaf nodes at level three? Assume that the index and the leaf nodes reside in auxiliary storage.

6. (a) What is the minimum depth for a B+-tree containing 100 records with an index of order seven and data nodes holding up to six records? **(b)** What is the maximum depth?

7. (a) What is the minimum depth for a B+-tree to store 400 records with an index of order seven and data nodes holding up to six records? **(b)** What is the maximum depth?

8. For a given number of records, does a B+-tree require more, less, or the same amount of storage compared with a B-tree? Explain.

9. List two advantages of a B+-tree over indexed sequential file organization.

10. Discuss the tradeoffs associated with the following alternative to deletion from a B+-tree. Instead of reorganizing the index for a B+-tree when a sequence set node falls below half full, do nothing. If all the records from such a node are deleted, return the empty node to the available space list, set the pointer field in the index that had been pointing to it to Λ, and relink the sequence set.

11. Discuss the tradeoffs associated with a VSAM file storing the data records in an added leaf level rather than in the sequence set.

REFERENCES

1. Comer, D., "The Ubiquitous B-Tree," *Computing Surveys,* Vol. 11, No. 2 (June 1979), pp. 121–137.
2. Bayer, R., and C. McCreight, "Organization and Maintenance of Large Ordered Indexes," *Acta Informatica,* Vol. 1, No. 3 (1972), pp. 173–189.
3. Knuth, D.E., *Sorting and Searching,* Addison-Wesley, Reading, MA, 1973.
4. Rosenberg, A.L., and L. Snyder, "Time- and Space-Optimality in B-Trees," *ACM TODS,* Vol. 6, No. 1 (1981), pp. 174–183.
5. Wagner, R., "Indexing Design Considerations," *IBM Systems Journal,* Vol. 4 (1973), pp. 351–367.
6. Keehn, D., and J. Lacy, "VSAM Data and Design Parameters," *IBM Systems Journal,* Vol. 5 (1974), pp. 186–212.

Chapter 10

Hashing Techniques for Expandable Files

BACKGROUND

When we considered direct file organization in Chapter 3, we discussed several methods for inserting data into such files using hashing and collision resolution. Those techniques provide a means to retrieve data almost directly (we still need on the average about $1\frac{1}{2}$ probes to retrieve a record in a direct-access file with an 85 percent packing factor) and are more efficient for retrieving a single record than previous methods such as a binary or sequential search. Although those hashing techniques are quite important, they suffer from a major deficiency in that they require a *static* table size. The hashing functions used for mapping the key space into the address space use the table size. If we need to add records beyond the capacity of the table, the only way we can increase the table size is to reorganize the table. That is, we must unload *all* the data and completely reenter them plus the new data into a table with a different size and associated hashing function. We need a different hashing function so that the keys can be mapped into a larger address space. We can't merely add records to the bottom of the current table. The reason for reorganizing a table and changing hashing functions is illustrated in the subsequent diagram.

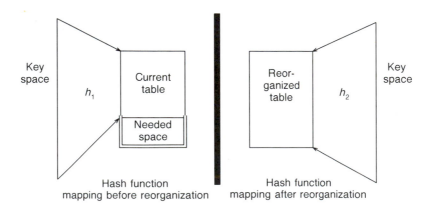

| Key space | h_1 | Current table | | Reor-ganized table | h_2 | Key space |

Hash function
mapping before reorganization

Hash function
mapping after reorganization

Since the reorganization requires considerable time and inconvenience, we would prefer a method that eliminated this drawback of a fixed table size.

Recently, several hashing techniques have been introduced that allow the table size to be changed *without* reinserting the stored records. Since we discussed hashing techniques at some length in Chapter 3, you may be wondering why we didn't consider these hashing techniques for expandable files in that chapter. We are considering them in this part since they use a splitting operation resembling that of a B-tree (and perhaps were influenced by the B-tree splitting). We look at four methods. Initially we describe extendible hashing, dynamic hashing, and linear hashing; then in Chapter 11, after we discuss a different type of tree structure, we look at another type of hashing for expandable files.

EXTENDIBLE HASHING

Extendible hashing [1] is a technique that allows a direct-access table to grow and shrink as records are added and deleted. The technique has many similarities with B+-trees. First, it uses an index into the data storage area just as with B+-trees. Figure 10.1 is essentially the B+-tree index and storage structure *rotated* 90 degrees. Second, all the records are stored in terminal nodes of the structure, and therefore, they all require the *same number of probes on retrieval*. Third, when a page overflows, it is split into two pages, and the index is modified as with a B+-tree although the specific mechanisms are different. If extendible hashing is so similar to a B+-tree, is there a reason that it is preferable? Yes, the index associated with an extendible hashed file is usually of depth one, which means fewer probes than a B+-tree for direct-access retrieval. It is important to note, however, that the data pages of an extendible hashed file are *not* ordered for sequential access.

> **Extendible hashing** is a technique for creating expandable, direct-access files. An index is used as a level of indirection between the key of a record and the data page where it is stored. Using the index allows the file to expand without a complete reorganization of the records.

With extendible hashing, the index is shallow because we use a *hashing* function to determine the data page in which the record of interest is stored. We do not use the hashing function to generate a page number directly as with the previous hashing schemes. Instead we hash the key to obtain a *pseudokey* that determines an entry in the index. The index entry in turn contains a pointer to the proper data page. We therefore have a level of indirection for going between the key and the data page. This indirection allows the table to expand and contract without a reorganization of the data items stored in the table. It provides a flexibility similar to that we previously observed with the indexed heap. The hashing function provides a *pseudokey* of a predetermined number of *bits*.

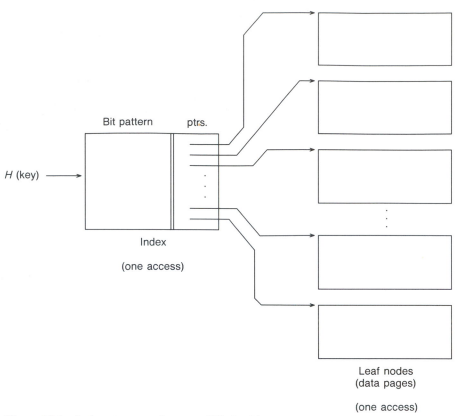

Figure 10.1   Index structure for extendible hashing.

$$H(\text{key}) \rightarrow \text{pseudokey}$$

for example, if $H(\text{key})$ is key **mod** 11 and the number of predetermined bits is four then

$$H(27) \rightarrow 0101 \qquad (\text{right justified})$$

In scanning the index, we are performing a radix search. If the index contains entries of n bits, we compare the n most significant (leftmost) bits of the pseudokey with the index entries to find a pointer into the appropriate data page. The index entries are ordered by their bit patterns; so when we search the index we are performing a search based upon whether the next binary digit is a zero or one. The first n bits of a pseudokey may also be viewed as an index value or subscript of an array or index entry. When a data page overflows and is split into two pages, the records are distributed based upon the first k bits of their pseudokeys. k, which is referred to as the *page depth,* is the number of bits needed to distinguish the pseudokeys on a page from those on other pages. The *index depth, n,* is the *maximum* of all the page depths. If it is not possible to distinguish the current and new records based upon the current value of k, we look another position to the right in the pseudokey values and increase

Algorithm 10.1
EXTENDIBLE HASHING INSERTION

  I. Obtain a pseudokey for the record to be inserted by hashing the actual key for the record.

 II. Use the n (index depth) most significant bits of the pseudokey as the address of an entry in the pseudokey index.

III. Follow the pointer in that index entry to locate the proper page for inserting the record.

IV. If space is available in that page, insert the record, else

    A. Split the overflowing page. Divide those records on the overflowing page plus the incoming record into two groups based upon pseudokey values. Compare the bit values of the pseudokeys from left to right until a division can be made based upon differences in the pseudokey values.

    B. Place each group of records into a separate page: one into the original page and the other into a new page.

    C. Determine the page depths of these two pages.

    D. If these page depths < the index depth, then

        /*   The index does not require expansion.   */

        1. Update those index pointers from the one pointing to the split page through the one pointing to its successor,

    else,

        /*   The index requires expansion.   */

        2. Set the index depth to the maximum (k), expand the index size to 2^n, and adjust the index pointers.

---

the value of k by one. We continue looking to the right and increasing k until we find a point at which we can divide the records into distinct data pages. If the new value of k resulting from the page split is greater than that of n, n is changed to this new maximum value for a k. When we add a bit to the index entries, we *double* the size of the index since its size is determined by 2^n. An index with a length of 1 bit has two entries. A 2-bit index contains four entries, and a 3-bit index contains eight.

    Note. We need to make only *one* access of the index and one access to a data page. If the index can be stored in primary memory, we need only *one* access of auxiliary storage.

    The algorithm for inserting into a file using extendible hashing appears in Algorithm 10.1; it may be helpful to consider it in conjunction with the following example.

An Example

For the example of extendible hashing, we assume that a data page can contain up to three records (just so we have ample splitting). To add continuity, we use the familiar hashing function,

$$H(\text{key}) = \text{key } \textbf{mod } 11$$

to produce the 4-bit pseudokeys (right justified), and we use data similar to that of previous hashing examples. The data and the resulting pseudokeys are

$$f(27) = 5 = 0101$$
$$f(18) = 7 = 0111$$
$$f(29) = 7 = 0111$$
$$f(28) = 6 = 0110$$
$$f(42) = 9 = 1001 \quad \text{instead of } f(39) = 6 = 0110$$
$$f(13) = 2 = 0010$$
$$f(16) = 5 = 0101$$

(Note why the data is not exactly the same as in the previous examples. If we use 39, we would have had four records that contained the identical leftmost 3 bits. This circumstance would have required an additional doubling of the index, which may have complicated the example unnecessarily. Since, in actual practice, we cannot change the data that we are working with, you may consider this a dubious practice. In a nonintroductory situation, however, you would normally have a hashing function that would map the key into a much wider range of values that would mean fewer pseudokeys with the same bit patterns. Typically, the hashing function produces pseudokey values in the range 0 to $2^{\text{word length}} - 1$. But the hashing function of key **mod** 11 is fine for an introductory example, and the fewer expansions of the index may simplify the example.)

Initially, the index is of depth one. In the diagram, the number to the upper right of the index gives its depth. The first three records with keys of 27, 18, and 29 can be inserted without complication into the first data page giving a structure of

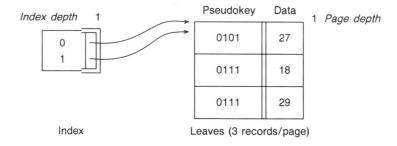

The placement order of the records within a data page is insignificant since the data is not readily processible in a sequential order as it was with a B+-tree. At this stage, both pointers of the index point to this initial page (and *not* to individual records). We consider an alternative implementation later in which a data page has only a single pointer to it. The number to the upper right of a data page is its depth (which must be ≥ 1). Since only the key field of a record is germane to its processing, it is the only one displayed in a leaf node entry. In practice, we would *not* need to store the pseudokey in the data page since it can be easily computed; we include it here only for purposes of explanation and comparison.

When we attempt to insert 28, we have an overflow situation and must split the page. When we try to classify the three previously inserted records plus the record to be inserted, we first look at the most significant bit. Since all the pseudokeys have a zero in this position, we cannot distinguish the records with that bit. If we had been able to distinguish them, we could have added another data page without increasing the size of the index since its two current entries do not point to distinct data pages. We then consider the next bit to the right. Here we have all ones. Finally, we can separate the records based upon the third bit from the left in the pseudokey. The page depth will then be three, which means that the index depth will also need to be increased to three. The index size will then double twice to increase from 2^1 (two) to 2^3 (eight) entries. An insertion that requires an index expansion produces the poorest performance because of the many pointers in the index that need to be set. The structure after inserting 28 is

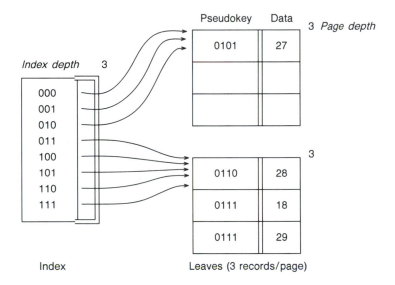

We now have eight pointers to the two data pages. All pointers up to and including the one that corresponds to the first n bits of the pseudokeys on a page point to that page, etc. Any remaining pointers reference the last data page. In this example, the pointers up through and including 010 point to the top data page since the first three bits of the pseudokey of the record stored on that page are 010. The pointers up to and including 011 point to the second page since the first three bits of its pseudokeys begin with 011. The remaining pointers point to the last data page (the second page).

An overflow occurs when we insert 42; it maps to the page with 28, 18, and 29. We can differentiate based upon the *first* bit. So only 42 goes onto a new page, and the other records remain on their existing pages. The result of this insertion is

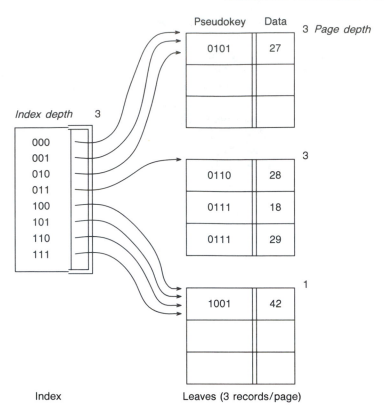

Index   Leaves (3 records/page)

We continue and insert the records with keys 13 and 16 into the top page without difficulty. The final file appears in Figure 10.2.

Instead of processing the most significant bits of the pseudokey in a left to right fashion, we could have used the least significant bits in a right to left manner.

An alternative implementation of extendible hashing is to have *either* a pointer to a data page or a null pointer for each entry in the index. A data page would then have only a single pointer into it and all the records on a data page would have the same n most significant bits in their pseudokeys. Figure 10.3 shows the results of this alternative technique using the example data. The advantages of this alternative are that we do not need to search a data page for a record if we encounter a null pointer, and the page splitting operation is simpler. The disadvantage is that we will have more data pages, especially soon after a splitting operation. A hashing function that has an even distribution will tend to fill all the pages with a few records. The larger the blocking factor of records per page, the greater will be the wasted space. We then have a time-space tradeoff—performance of unsuccessful retrievals vs. storage.

Deletion

Just as with B-trees, deletion is the inverse of the insertion process. Instead of splitting a page when a record is added, we have an opportunity to coalesce data pages when a record is removed. Not only will the coalescing of data pages reduce the amount of storage required to keep the records of the file, but it may also reduce the *depth* of the

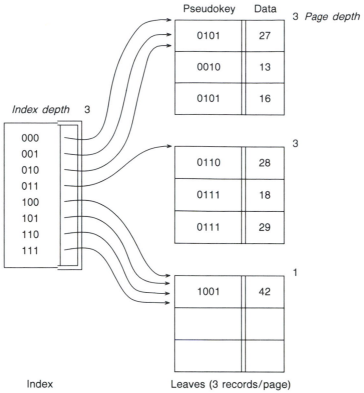

Pseudokey   Data
                                                          3 *Page depth*

0101	27
0010	13
0101	16

Index depth   3

| 000 |
| 001 |
| 010 |
| 011 |
| 100 |
| 101 |
| 110 |
| 111 |

                                                          3

0110	28
0111	18
0111	29

                                                          1

1001	42

Index   Leaves (3 records/page)

Figure 10.2   Extendible hashing example.

index and therefore its storage requirements. Since deletion is simpler with the alternative implementation of extendible hashing, let's first consider it through the example of Figure 10.3. When we delete a record, we check to determine if the *remaining* records on the page of the deleted item can be combined with the records on its *buddy page*. In extendible hashing, a buddy page is the page with the same $k - 1$ most significant (leftmost) bits in its pseudokey index entry but with a 1 in the kth bit position if the pseudokey of the deleted record had a 0 in its kth bit position, or a 0 if the original pseudokey had a 1 in its kth bit position. Remember that k is the depth of the page. Buddy pages are formed during the insertion process when an overflowing page is split into two pages; the records of the overflowing page are redistributed between the buddy pages based upon their pseudokey values. In deletion then, we are merely trying to join together pages that were previously put asunder. What is the buddy page for the page containing the record with key 16? Since the three (k) most significant bits of the pseudokey for 16 are "010", its buddy page has records with "011" in the three most significant bit positions. That is the page with the records with keys 28, 18, and 29.

If we delete the record with key 16, can we combine the remaining records on the page with 16 with the records of its buddy page? No, since its buddy page is already at its maximum capacity with the records with the keys 28, 18, and 29. If, however, we next delete the record with key 29, the remaining records on that page *can* be combined with those of its buddy page since its buddy page now contains only the one record with key 27. When we combine the records on the buddy pages, the depth of

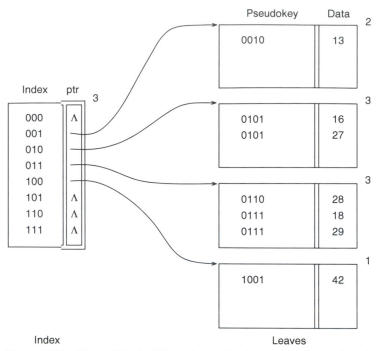

Figure 10.3 Extendible hashing, alternative implementation using Λ link fields.

the index can also be reduced since the depth of the resulting page becomes two. Whenever the records on buddy pages are coalesced, the depth of the resulting page is one less than the depth of the original pages. The maximum depth of any page is now two; so the size of the index can be reduced from eight entries to four entries. The file structure then appears as

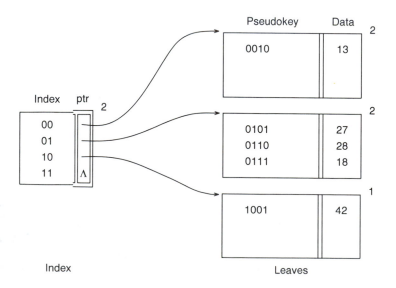

If we remove the record with key 13, since it is the last record on that data page, we return the page to the available space list (as we do with any page no longer needed). We also reduce the depth of its buddy page by one and then decide if the depth of the index could be reduced. In this case, it can; the file structure becomes

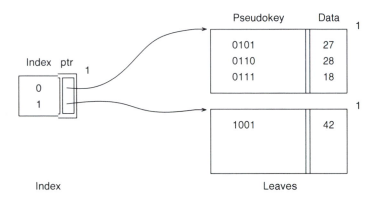

Since checking for buddy pages and determining if they can be combined, plus the determination of the page and index depths, can be time-consuming, rather than perform them on every deletion, make the processing variable by setting a threshold. Only when the number of remaining records in a data page is below this threshold would the coalescing process be performed. As you can observe from this example, an advantage of extendible hashing is that the file can grow and shrink as records are added and removed from it.

Deletion is somewhat more complicated with the original implementation of extendible hashing. Do you know why? It has to do with the processing of the buddy pages. In the original implementation, all the records on a page *do not* necessarily have the same most significant k bits. That is because we conserved storage by filling a page with records with similar but not necessarily the same pseudokeys. Therefore, the buddy page entry in the index may point to the same page as the one containing the record being deleted, for example, the buddy page of the record with key 13 in Figure 10.2. What do we do then to locate the page for possible coalescing with the page of the deleted record? If the buddy page pseudokey is less than that of the deleted key, check the page in the file that precedes the current page. That page is the one with the nearest but prior index entry that does *not* point to the current page. If the pseudokey for the buddy page is greater than that of the deleted record, check in the opposite direction. Then proceed as before in combining adjacent pages and reducing the depth of the pages and the index.

Discussion

To *retrieve* a record using extendible hashing, we first hash the key to obtain the pseudokey. We then note the depth of the index and match the first n bits of the pseudokey with an entry in the index. The index entry points to the data page that the record must be on if it is in the file. We could use any of a number of search techniques to then scan the data page: sequential, binary search, hashing, ... The technique used for searching a page dictates whether the records need to be inserted into a page in a meaningful order.

In conjunction with the insertion example for extendible hashing, you may have asked what happens if two or more keys generate identical pseudokeys? Fortunately, this situation is not as prevalent as the example might suggest. First, the hashing function generates a *much wider range of values* than in the example. In fact, one advantage of extendible hashing over static hashing is that the range of values produced by the hashing function does *not* need to coincide with the address space. Since we are accessing the data pages through the index, we can have a much *larger range of values* output from the hashing function. That means that there is a much smaller chance for identical values. Second, we may have as many identical pseudokeys as there are storage places for records in a data page; only if there are more than that number of identical pseudokeys is special action needed. While we are discussing the hashing function, it should also be noted that it no longer requires a prime number modulo.

What are the pros and cons of extendible hashing? Its primary advantages are (1) that the file can expand and contract *without* reorganizing it and (2) only one probe of secondary storage is necessary to access a record if the index can be stored in primary memory. Its primary disadvantage is the extra storage needed for the index. Another disadvantage, one that is true of the other hashing methods that we have considered, is that we cannot readily process the file sequentially as with B+-trees.

Variants

Extendible hashing has a few other potential disadvantages. One is that the method has oscillatory behavior for insertions; that is, insertions proceed without any splitting for a while, and the time for processing the insertions is relatively low. But then, a number of splittings occur at approximately the same time, so the processing time increases, which is inconsistent with the previous performance. Such disparity may not be acceptable in some situations. Why do the splittings occur at approximately the same time? Recall the two characteristics of a good hashing function: a uniform distribution and simple computation. A hashing function that provides a uniform distribution will map records into the pages evenly. Normally this is good. But that means that all the pages fill up *at approximately the same time.* So the insertion performance looks something like this.

Insertion Costs

Time from initialization

The spikes in the performance graph indicate those periods when most of the pages are splitting.

A second drawback that may occur as the table grows is that the index may eventually become too large to fit into main memory. If we move it to auxiliary storage, that means two accesses of auxiliary storage to retrieve a record, one access for the

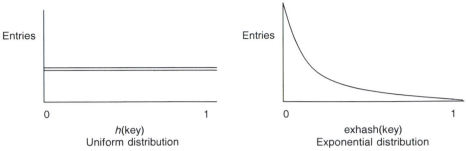

Figure 10.4 Hashing function distributions.

index and one for the data. Such a modification would greatly affect the performance.

A variation of extendible hashing called bounded index exponential hashing was introduced by Lomet [2] to correct these two potential deficiencies. The "exponential hashing" in the title modifies the hashing function to provide an exponential distribution rather than a uniform one. The modified hash function is

$$\text{exhash(key)} = 2^{h(\text{key})} - 1$$

where h(key) is the original hashing function.

Figure 10.4 graphs the distributions of the two hash functions. Exhash distributes its results over the same range as h but has twice as many entries near 0 as near 1 (assuming that we consider a pseudokey value to be a fraction between 0 and 1).

As a result of the nonuniform distribution of exhash, the data pages do *not* fill up at the same rate; so the splitting operations occur regularly over the course of the insertions rather than during a few concentrated periods. These regular splittings will smooth the peaks and valleys yielding a more consistent insertion performance as illustrated by the dashed lines in

The "bounded index" in the title addresses the second deficiency. The index is split in the same manner as it is normally in extendible hashing *until* the index reaches the limit for keeping it in main memory. Then instead of doubling the size of the index when a page overflows, the size of the data page is doubled. A zero in the bit position of the pseudokey beyond the depth of the index means that only the first half of the data page has to be searched while a one indicates that only the second half of the data page needs to be. The data page could continually be doubled in this manner, with the associated bits in the pseudokey used to determine what segment of the page needed to be searched. Figure 10.5 displays how the pseudokey bits are associated with the page segments after doubling. When the page size limit for the computer system is reached, a secondary index needs to be built for the records on auxiliary storage. This additional

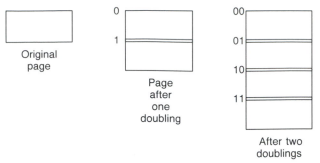

Figure 10.5   Pseudokey bits and page doubling.

index then increases the number of auxiliary storage accesses to two. What the bounded index procedure accomplishes is *postponing* the time when we have to make two accesses of secondary storage.

KEY TERMS

Bounded index exponential hashing	Index
Buddy page	Index depth
Extendible hashing	Page depth
Deletion	Pseudokey
Performance	Radix search

EXERCISES

1. Insert the records with the keys

$$85, 73, 16, 79, 13, \text{ and } 38$$

into a table using extendible hashing. Use

$$f(\text{key}) = \text{key } \textbf{mod } 13 \qquad (4 \text{ bits, right justified})$$

Assume that each data page may hold up to three records. Note the index and page depths.

2. (a) Redo the extendible hashing insertion example of the seven records with the keys

$$27, 18, 29, 28, 42, 13, \text{ and } 16$$

using the least significant bits of a pseudokey in a right to left fashion. **(b)** How do the results of *a* compare with those of Figure 10.2? Consider the size of the directories and the number of data pages in your comparison.

3. (a) Redo the extendible hashing insertion example with the seven records but insert them in the key order

27, 42, 18, 29, 28, 13, and 16

(b) How do the results of *a* compare with those of Figure 10.2? Consider the size of the directories and the number of data pages in your comparison.

4. Use the hashing function key **mod** 17 to insert the keys

27, 18, 29, 28, 39, 13, 16

into an extendible hashing table in which each leaf node holds up to three records. Assume that the pseudokeys are 5 bits, right justified.

5. Show the intermediate file structures leading up to Figure 10.3 when using the alternative implementation of extendible hashing.

6. Redo Problem 1 with the alternative implementation.

7. **(a)** Give a better choice for a hashing function than

$$H(\text{key}) = \text{key } \textbf{mod } 11$$

for setting the 4 bits of the pseudokey in the extendible hashing example. **(b)** Reinsert the seven records of the example using your improved hashing function. **(c)** How do the results of b compare with those of Figure 10.2? Consider the size of the directories and the number of data pages in your comparison.

8. Does the modulo of the hashing function used to generate the pseudokeys in extendible hashing need to be a prime number?

9. **(a)** What is the storage utilization of the data pages in Figure 10.2? **(b)** in Figure 10.3?

10. What determines if the depth of a page may be reduced after a deletion using the original implementation of extendible hashing? Explain.

11. What is the primary advantage of the linear quotient method of constructing a direct-access file compared with extendible hashing?

12. How is the pseudokey determination different in bounded index exponential hashing vs. extendible hashing? (Be specific.)

13. What are the two shortcomings of extendible hashing that are corrected by bounded index exponential hashing?

DYNAMIC HASHING

Dynamic hashing, which originated with Larson [3], is another technique for creating an expandable direct-access file. We are considering it because it uses a different strategy than extendible hashing. The reason for looking at several ways of forming expandable tables is to observe different ways to solve a problem. One thing that is true about data structure selection that we have mentioned several times before is that there is usually not a single way of solving a problem that stands above the others in *all* respects. Each method has its advantages and disadvantages. By looking at a variety of techniques, you should be better able to discover effective solutions to future and unforeseen problems. A concept that may not appear particularly useful today may be just the ticket tomorrow.

> **Dynamic hashing** is a technique for creating expandable, direct-access files. Multiple subdirectories formed from pseudorandom sequences of bits associated with the keys are used to locate the records.

We noted that a disadvantage of extendible hashing is that the splitting operations that expand the index require significant time, which makes the performance uneven. Bounded index exponential hashing was one way to overcome that shortcoming. Dynamic hashing is another. In dynamic hashing, the index grows continually and gradually rather than infrequently and abruptly. A binary tree index is used to point to the appropriate page for storing a record. Rather than generate a fixed-length pseudokey as in extendible hashing, we use a pseudo-random-number generator that can generate a sequence of 1s and 0s of *any* length. The generated binary sequence is uniquely determined by the seed for the function that is produced by applying a hashing function to the key,

$$H_1(\text{Key}_i)$$

As the binary tree index grows, the number of bits returned from the hashing function can be increased accordingly. The hashing function, B, for Key_i then generates a sequence of binary digits such that

$$B(H_1(\text{Key}_i)) = (b_{i0}, b_{i1}, b_{i2}, \ldots) \qquad b_{ij} \in \{0,1\} \text{ for all } j$$

where a 0 is returned one-half of the time and a 1 the other half. We use the output from this generator to navigate the binary tree index. If the first bit is a zero, we go left; if it is a one, we go right. The tree is divided into internal nodes (○) and external ones (■). The external nodes contain pointers to the actual data pages (▦). The internal nodes are used to direct a search to the proper external node. If the level one branch leads to a terminal node, we are finished traversing the index. If not, we apply the next bit from the pseudo-random-number generator hashing function, going left if it is zero, and right if it is a one. Since the hashing function produces ones and zeros with the same frequency, the binary tree should be stochastically balanced.

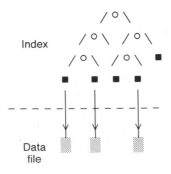

For flexibility in expanding the index, it is implemented as a linked structure. To simplify storage management, both the internal and external nodes are the same physical size. The fields of the nodes are

Tag	Parent
Left	Right

Internal node
tag = 0

Tag	Parent
Record count	Bucket pointer

External node
tag = 1

The *parent* link is for moving up the tree. In an internal node, the *left* and *right* links are for moving down the tree. In an external node, the *record count* keeps a count of the records in a data page or bucket. The existence of this field can save retrieval and insertion time. On retrieval, if this field is zero, we do *not* need to make an access of secondary storage to determine that a record is not in the file. On insertion, this field can inform us when we have an overflow situation (assuming that we do not need to check for duplicate keys). The *bucket pointer* is a pointer to the data page or bucket holding the associated records.

To keep the index from becoming deep (which would require much searching), dynamic hashing uses a *forest* of binary trees rather than a single tree. A hashing function

$$H_0(\text{Key}_i) \rightarrow \{0, 1, \ldots, n\}$$

determines which subtree to search. H_0 should be independent of H_1. Having multiple directories to search is a *divide and conquer* strategy similar to having multiple pseudolink chains to search in collision resolution. The basic index and storage structures of dynamic hashing are depicted in Figure 10.6. The numbers above the subtrees, which identify them, are generated by H_0. Notice that all the external nodes do *not*

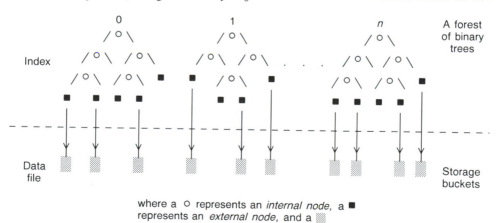

where a ○ represents an *internal node*, a ■ represents an *external node*, and a ▨ represents a storage bucket.

Figure 10.6 Forest of dynamic hashing directories.

Algorithm 10.2
DYNAMIC HASHING INSERTION

I. Apply H_0 to the key of the record to determine the proper subdirectory for inserting it.

II. While the current node is not an external node, navigate the subdirectory using the next bit of the pseudorandom sequence obtained from the generating function B with H_1 (key) as a seed. Go left on a zero bit and right on a one bit.

III. If the external node does not yet have a data page associated with it, that is, if the number of records stored in that bucket is zero, get a new page, insert the data record into it, and set the pointer from the external node to that page, else if the associated data page is not full, insert the new record, else repeat until an overflow situation no longer exits.

    A. Convert the one external node that originally pointed to the data page into an internal node and create two offspring external nodes.

    B. Reinsert the records of the overflowing page and the record to be inserted into a left or right data page using the next bit of the pseudorandom sequence to determine the proper placement. If no records map to one of the pages, it does not need to be allocated, but an overflow condition still exists for the other of the twin pages.

have to be on the same level or point to a data page. Figure 10.6 is the basic structure of an index and a data file. We search the index by first selecting the proper subdirectory by using H_0 and then use H_1 and B to navigate that subdirectory.

What happens on insertion when we map to a page that is full? The page splits as with extendible hashing, but the modifications to the index are smoother. On overflow, an external node converts to an internal node and two external nodes are added as offspring. These two external nodes may add another level to the index. Records are placed in the left bucket if the next bit in the binary sequence is a zero and are placed in the right bucket if it is a one. An overflow condition never causes a doubling of the index—only gradual increases; the modification to the index caused by an overflow condition is local. The algorithm for inserting into a file using dynamic hashing appears in Algorithm 10.2.

An Example

For the example of dynamic hashing, we again use the familiar records with the keys of

$$27, 18, 29, 28, 39, 13, 16, \text{ and } 36$$

We use

$$H_0 = \text{key } \textbf{mod } 3 \quad \text{and} \quad H_1 = \text{key } \textbf{mod } 11$$

TABLE 10.1	PSEUDORANDOM BIT SEQUENCES FOR SEEDS 0–10	
$B(0) = 1011$		$B(5) = 0101$
$B(1) = 0000$		$B(6) = 0001$
$B(2) = 0100$		$B(7) = 1110$
$B(3) = 0110$		$B(8) = 0011$
$B(4) = 1111$		$B(9) = 0111$
	$B(10) = 1001$	

Since it is difficult to have a real time pseudo-random-number generator available, we have provided Table 10.1, which contains sequences of random bits for the 11 possible seeds. These sequences use only 4 bits, but that will be sufficient for the example. In practice, you would generate only as many bits as you need to navigate the tree. These sequences of random bits are *not* the binary representations of the seeds. The data pages hold up to two records each.

27 is inserted first. $H_0(27) = 0$; so the record should be inserted into the zeroth subdirectory. $H_1(27) = 5$; so that is the seed for the binary random sequence generator. Since the subdirectories contain only a root external node initially, we do not need to use the binary sequence in this insertion. The record is inserted as

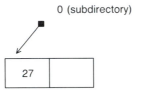

18 maps to the same subdirectory, and since there is room for it in the data page, it can be inserted without difficulty. Maintaining the records within a page in a sorted order simplifies the searching at the page level but does not affect the insertion technique.

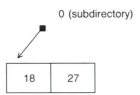

29 is inserted into subdirectory 2 and 28 into subdirectory 1.

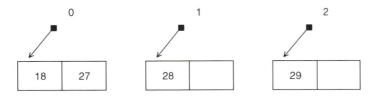

When 39 is inserted into the zeroth subdirectory, an overflow situation arises. The one external root node becomes an internal node, and two offspring external nodes are added. The records in the overflow page are reinserted. Since the leftmost bit in the sequence $B(H_1(18)) = B(7)$ is one, 18 is inserted into the right bucket. The leftmost bit of $B(H_1(27))$ is zero; so 27 is placed into the left bucket. $B(H_1(39))$ has a zero as its leftmost bit; so it is placed into the left bucket with 27.

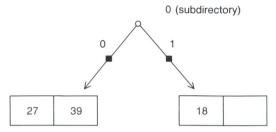

13 is inserted into subdirectory one without causing a splitting operation.

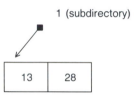

16 also maps to subdirectory one, but now the one data page is full and an overflow occurs. The index structure changes as before. We then attempt to place the three records, 13, 16, and 28 into the two data pages. But note that

- $B(H_1(13))$ has a left bit of zero.
- $B(H_1(16))$ has a left bit of zero.
- $B(H_1(28))$ has a left bit of zero.

All three keys have the same leftmost bit generated by B; so all three records are mapped into the left bucket. Since it can only hold two records, we need to perform another splitting operation on the left external node. Since we were not able to distinguish the records based upon the leftmost bit of the binary sequence, we look at the next bit position to the right. Using that bit position, we are able to separate the records as shown in

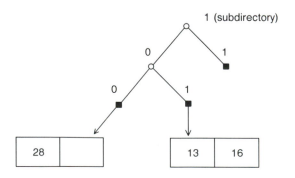

Finally we add 36 to the zeroth subdirectory, giving the result

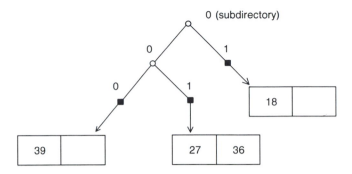

Notice that the leaf nodes are not on the same level. If the index is stored in primary memory, however, each retrieval will require the same number of probes of auxiliary storage.

Discussion

On *retrieval,* we use a similar process. We employ H_0 to determine which subdirectory to search. Then we apply B with the seed given by $H_1(\text{key})$ to navigate the structure if the root node of the index is an internal node. If the root node is external, it points directly to the proper data page. If a record is not in the file and its key maps to an external node that has a zero record count, we do not need to make an auxiliary storage access. All the retrievals require approximately the same amount of time since the binary tree indexes should be stochastically balanced. If we can store the subdirectories in main memory, then only *one* access of auxiliary storage is necessary.

To retrieve the record with key 27, we first apply $H_0(27)$ to determine that we need to search subdirectory zero. We next use $H_1(27) = 5$ to get the seed of B. B then generates the binary sequence 01 . . . , so we follow the path left, then right to locate the external node pointing to the data page containing 27.

Variations

As with extendible hashing, a potential disadvantage of dynamic hashing is that the index could expand to the point where it no longer fits into primary memory. If we did not take special action, this situation would require that we put a portion of the index into auxiliary storage; doing that would double the number of auxiliary storage accesses on retrieval. Scholl [4] has suggested two alternatives to an index too large for main memory. One variant is called dynamic hashing with deferred splitting. Rather than splitting a bucket when it overflows, we *defer* the splitting until *after* some $\beta \times$ bs records have been directed to a bucket, where bs is the bucket size and β is the real number deferral factor > 1. Only when the $\beta \times$ bs limit is *exceeded* is a bucket split. For example, if β is 1.5 and bs is 2, splitting occurs only *after* three [1.5 × 2] records have been added to a bucket, that is, when the fourth record is added to that bucket.

What do we then do with the overflow records? We must create an overflow bucket for the records that cannot fit into the normal bucket. A data page that has overflowed then appears as

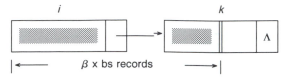

where i is the normal bucket and k is an overflow bucket. ▓ represents the portions of the buckets that contain data. Having the overflow records stored on a separate page means that two probes of auxiliary storage are needed to retrieve the records on that page. Not all records require two probes of secondary storage, only those on the overflow page. Deferring the splitting reduces the index by a factor of β. By choosing β to be the smallest value that will allow the index to fit into primary memory, we minimize the average number of retrieval probes. The number of secondary storage accesses is at most $\lceil \beta \rceil$, the ceiling of β, that is, the smallest integer greater than or equal to β. The deferred splitting solution is better than putting the index on secondary storage because that would require two probes for *all* retrievals. Notice that β does not have an upper limit, so we can have more than the one overflow page.

A second suggestion by Scholl to reduce the size of the index is to store *only* the terminal nodes of the index. We navigate to a terminal node by using a convention of how the nodes are split. The convention is *not* based upon which page overflows. Instead, we split the *least recently split page* after a fixed number of records, $\tau \times$ bs, have been added to the subdirectory, where τ is a multiplier ≥ 1. In that way each index tree is a complete binary tree such as

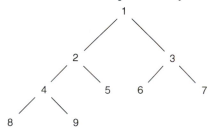

You will recall the relationships among the nodes of a complete binary tree

- lchild(i) = 2 * i
- rchild(i) = 2 * i + 1
- parent(i) = $\lfloor i / 2 \rfloor$

that can be used to move about it. If we rewrite the offspring relationship as

- offspring(i) = 2 * i + b$_i$

    where b_i is the next bit in the binary sequence, and the offspring is a lchild if the bit is a zero and an rchild if it is a one,

we can then navigate the tree in the same manner as a regular dynamic hashing index!
However, we do need to recognize when we reach a terminal node so that we can
terminate the traversal of the subdirectory. We can *compute* this information by
knowing how many records are in the tree and the fact that we split it regularly *after*
each $\tau \times$ bs insertions. Each time we split a node, an external node becomes an internal
one and two new external nodes are formed. The number of splittings then is equivalent
to the number of internal nodes, and that provides a division point between the internal
nodes and the external ones.

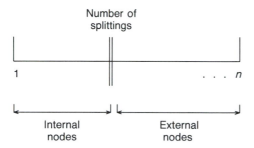

As we move through the nodes of the virtual binary tree subdirectory, we check each
new node position to determine if it is a terminal node (by comparing it against the
number of splittings that have occurred). If it is a terminal node, we terminate the
traversal and process the terminal node pointing to a data page.

Since we split the least recently split page rather than one that is overflowing, it
is possible that overflow pages may be needed occasionally to store records. The
overflow page (or pages) is usually *only* needed until that page or chain of pages
becomes the least recently split page. With a good pseudo-random-number generator,
the subdirectory should be stochastically balanced so few overflow pages should be
needed. Just as with deferred splitting, a few records may then require more than a
single probe of auxiliary storage. But this condition is better than moving the entire
subdirectory to secondary storage.

This second variation is called dynamic hashing with linear splitting based upon
the splitting occurring at regular intervals. The basis for dynamic hashing with linear
splitting rests upon an important principle: *compute information when there is insufficient room to store it.* We can compute the path to an external node by making use
of the regularity of the splitting. By doing so, we store only the terminal nodes of the
index and thereby reduce its size. See Scholl [4] for additional details on linear splitting.

KEY TERMS

Binary tree	Deferred splitting
Forest	External nodes
Stochastically balanced	Internal nodes
Complete binary tree	Linear splitting
Dynamic hashing	Pseudo-random-number generator

EXERCISES

1. Insert the records with the keys 80, 53, 26, 17, 62, 18, 35, 51 using dynamic hashing with $H_0(\text{key}) = \text{key} \bmod 3$, and $H_1(\text{key}) = \text{key} \bmod 11$. Assume that a page (bucket) can hold up to two records and use the given pseudorandom bit strings.

2. What is the (a) maximum and (b) minimum number of buckets needed to store 10 records using dynamic hashing with a bucket size of 2, assuming that all records hash to the same subdirectory?

3. Use dynamic hashing to insert the records with the keys

$$20, \ 40, \ 60, \ 73, \ 45, \ 85, \ 64$$

into a dynamic file. Let $H_0(\text{key}) = \text{key} \bmod 4$ and $H_1(\text{key}) = \text{key} \bmod 9$ and a bucket size of 2.

4. Repeat Problem 3 using dynamic hashing with *deferred splitting* with $\beta = 1.5$.

5. What is the primary advantage of dynamic hashing with linear splitting vs. deferred splitting?

6. (a) For equivalent data, does extendible hashing or dynamic hashing usually require less storage for its directory? (b) Why?

7. In an extendible hashing directory, the pseudokeys are the same length whereas in a dynamic hashing directory the external nodes may be at different levels, which is equivalent to pseudokeys of different lengths. Does this latter strategy lead to *more* or *fewer* entries in a directory as compared to a strategy requiring equal-length pseudokeys (or the equivalent)? Explain.

8. How many internal and external nodes are in a dynamic hashing tree index with linear splitting in which 217 records have been inserted? (Assume $\tau = 2$ and bs $= 3$.)

9. Devise a formula for the number of splittings that have occurred in a subdirectory for dynamic hashing with linear splitting. Give the formula in terms of τ, bs, and r, the number of records that have been added to the portion of the file indexed by the subdirectory.

10. In dynamic hashing with linear splitting, what is the criterion for determining whether a node in an index is an internal or external node?

11. For a subdirectory in which the number of nodes, n, is large, what is the approximate percentage of storage saved when storing the subdirectory using dynamic hashing with linear splitting with $\tau = 1$ vs. regular dynamic hashing? Prove it.

12. Sketch a procedure for deleting a record from a file that has been created using dynamic hashing. What is the inverse of a page splitting operation and what triggers it?

13. Write an algorithm in either outline or pseudocode form for dynamic hashing with linear splitting.

LINEAR HASHING

The primary advantage of the two expandable file hashing methods that we have previously considered, extendible hashing and dynamic hashing, is that they allow the file to expand without reorganizing it. On the negative side though, both methods require additional space for an index. The third method we want to consider, linear hashing [5], also permits file expansion without reorganization, but it does *not* require space for an index. Why are the fixed table size hashing schemes fixed? It is a consequence of the hashing function needing the table size to map the much larger key space into the smaller address space. Linear hashing allows the table to expand by *changing the hashing function* during the insertion process.

One reason for considering linear hashing is that it applies an important principle for us to consider when designing future data structures. Keep an open mind about what may be changed and what must be fixed; *don't be limited in your thinking.* There is nothing about hashing that *requires* that you use only a single hashing function for a file.

Another feature of linear hashing is that the storage utilization, or packing factor, may be specified. The data records are stored in *chains* of pages linked together. *Remember* that a hashing function in this context maps a key into a chain of data pages rather than the location of a single record *or* a single data page. A chain split occurs when we exceed the upper bound for storage utilization, and a coalescing of buddy chains occurs when we go below the minimum packing factor after a deletion.

> **Linear hashing** is a technique for creating expandable, direct-access files. Rather than use an index, it has two hashing functions active at a time, so the file may expand without reorganizing the records. A hashing function maps a key into a *chain* of data pages.

We split the chains in a file in a regular fashion and use the regularity to allow us to *change* hashing functions as we expand the table. Remember that we also took advantage of regular splitting in dynamic hashing with linear splitting. At any one time with linear hashing, we have *two* hashing functions active: one for the current level of splitting and one for the next level.

We keep a pointer to the NEXT chain to be split. The initial chains, of number N, constitute level zero of the file. The hashing function for a level is given by

$$h_{level}(\text{key}) = \text{key } \textbf{mod } [N (2^{level})]$$

When we create a hashing function for the next level, we *double* the modulo, which allows us to *double* the number of chains that we can map records into. This doubling of the chains allows us to expand the address space (or chain space) which we could not do with a single hashing function. One hashing function allows us to access records stored in the basic portion of the file and the second hashing function allows us to access records in the expanded portion of the file.

On each level, we split the chains in order from chain 0 to the maximum chain on that level, which is calculated by the expression $N (2^{level})$; this expression results

Algorithm 10.3
LINEAR HASHING INSERTION

I.   Determine the chain, m, which the record maps to using $m = h_{level}$ (key).

II. Check whether the chain has split by comparing m with NEXT. If $m <$ NEXT, the chain has split, then set $m = h_{level+1}$ (key).

III. Insert the record into chain m.

IV. Check the upper space utilization bound. While it is exceeded then

    A. Create a new chain with index equal to NEXT $+ N*2^{level}$.

    B. For each record on the chain NEXT, determine whether to move it. If $h_{level+1}$ (key) $\neq$ NEXT then move the record to the new chain.

    C. Update parameters. Set NEXT = NEXT + 1. If NEXT $\geq N*2^{level}$, all the chains on the current level have split, then reset NEXT to 0 and create a new level by incrementing level to level + 1.

---

from the number of chains doubling from one level to the next higher one. After all the chains on the current level have been split, we increment the current level and begin the splitting process over again with chain 0.

How do we know which hashing function to use, h_{level} or $h_{level+1}$? After we determine, m, the chain a record maps to, we then compare it with NEXT. If $m <$ NEXT, we use the hashing function $h_{level+1}$; otherwise we use h_{level}. NEXT acts as a separator or dividing point between the two hashing functions; it also divides the chains that have been split from those that have not been.

The chains to the left of NEXT have been split on the current level, so we use the alternate hashing function. Again because of the regular mechanism for determining which chain to split, it is an easy matter to choose the proper hashing function. This is another instance of the basic principle that we noted with dynamic hashing with linear splitting: if a method has regularity or a pattern to it, it may be possible to *compute information rather than store it.*

The algorithm for linear hashing insertion appears in Algorithm 10.3. To facilitate understanding, we summarize its variables and specify which ones change during the lifetime of a linear hashed file:

    level   = specifies current round of splitting; *changes*
    NEXT   = performs two functions—it specifies which chain to split next and it divides the chains into those that use the hashing function for

the current level and those that use the one for the next level; *changes*

N = number of chains initially; *fixed*

m = chain to search for insertion or retrieval, output from hashing function; *changes*

h_{level} = hashing function for current level; *fixed* for each level

An Example

To briefly explain the splitting and retrieval processes of linear hashing, we consider the example from [6]; later we consider a detailed example with the standard data. The records with the keys

$$10, 20, 3, 7, 13, 14, 17, 21, 25, 16, 22, \text{ and } 30$$

are placed into four chains (consisting of primary and overflow pages) using the hashing function

$$h_0(\text{key}) = \text{key } \textbf{mod } 4$$

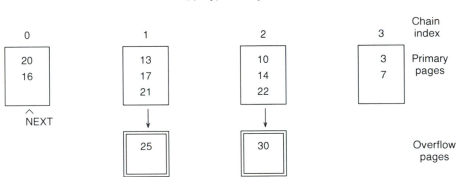

The primary pages hold three records and the overflow pages hold two. The storage utilization bounds are (40 percent, 80 percent). The current storage utilization is $\frac{12}{16}$, or 75 percent which is within the bounds.

When we add a record with a key of 9, it maps to chain one. Notice that the storage utilization is now $\frac{13}{16}$, or 81 percent, which exceeds the upper bound for the packing factor, so we must split a chain. Which one? Because **NEXT** points to chain zero, we split it and apply the hashing function for the next level

$$h_1(\text{key}) = \text{key } \textbf{mod } 8$$

to redistribute its records over chain zero and a new chain four. Since 20 **mod** 8 = 4, that record maps to chain four and since 16 **mod** 8 = 0, that record remains on chain zero. **NEXT** is then incremented to one. The results of the insertion are

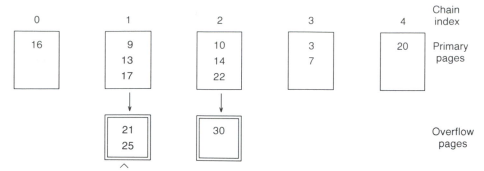

On retrieval, we first apply h_{level} to the key to determine m. If $m \not< \text{NEXT}$, we have located the chain that the record must be on if it is in the file. If $m < \text{NEXT}$, we use $h_{\text{level}+1}$. To retrieve a record with key 20, we first apply h_0. That gives us a value of zero for m. Since NEXT is now one, m is less than it. So we apply h_1 to determine the chain that a record with key 20 must be on. We locate 20 on chain four.

Linear hashing allows a file to expand or contract without a reorganization. Rather than use an index to locate a data record, it uses a hashing function to go directly to the chain of data pages that a record must be in if it exists in the file. The form of the hashing function changes as the table size varies; at any given time, two hashing functions are active. Only the hashing functions, the level number, NEXT, N, and the storage utilization bounds must be stored. Unlike extendible hashing in which the index size doubles periodically, in linear hashing, only the *potential* number of data chains *doubles* whenever we move to a new level.

Another Example

We now go into more detail with a second example. In this example the primary pages hold two records and the overflow pages hold one. We are using these small values to force splitting. Normally we would use much larger values for the page sizes. The space utilization bounds will be the same as in the first example: (40 percent, 80 percent). Let's use the familiar key values from examples of previous methods. The keys that we insert are

$$27,\ 18,\ 29,\ 28,\ 39,\ 13,\ 16,\ 51,\ \text{and}\ 19$$

N is two and

$$h_0(\text{key}) = \text{key } \mathbf{mod}\ 2$$

Level and NEXT are initially set to zero.

The first three keys are inserted into the file without difficulty. $h_0(27) = 1$; $h_0(18) = 0$; $h_0(29) = 1$. At this stage in the insertion, if the key is even, it maps to chain zero. If it is odd, it maps to chain one. The file then looks like

(NEXT)

When we insert 28 into chain zero, the storage utilization becomes 100 percent, exceeding its upper bound. So we need to split a chain; since NEXT is currently zero, we split that chain using

$$h_1(\text{key}) = \text{key } \textbf{mod } 4$$

That gives the result

0	1	2
28	27	18
	29	

with NEXT being incremented to one.

To retrieve 18, we apply

$$h_0(18) = 0 < \text{NEXT} \quad \text{so we use} \quad h_1(18) = 2 \quad \text{for retrieval}$$

We next add 39, which maps to chain one, which has not been split, so we do not need to apply h_1. But notice that the primary page is full, so we then need to add an overflow page. We can have any number of overflow pages. The file now looks like

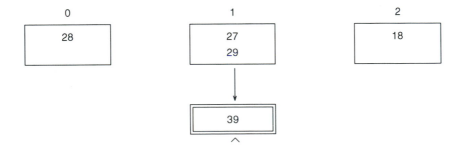

We check the storage utilization. Since $\frac{5}{7} = 71$ percent we are within the bounds.

We next insert 13, which also hashes to chain one. After it is inserted onto a second overflow page, the storage utilization is an acceptable 75 percent. The file now appears as

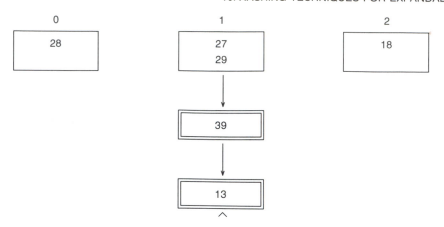

The record with key 16 maps to chain zero using h_0. *Recall* that we must compare this chain value against the value of NEXT to determine if we need to apply h_1. Since 0 is < NEXT, we apply h_1 and, in this case, it also maps 16 to chain zero. Inserting that record causes the space utilization bound to be exceeded at $\frac{7}{8}$, or 87 percent, so we must split. Since NEXT is one, we split that chain and its records are redistributed using $h_1 = $ key **mod** 4. NEXT is incremented but since

$$\text{NEXT} + 1 \geq N * 2^{\text{level}}$$

we have split all the chains on the current level, so we set

$$\text{NEXT} = 0 \quad \text{level} = 1 \quad \text{and} \quad h_2(\text{key}) = \text{key } \textbf{mod } 8$$

The file now looks like

0	1	2	3
28	29	18	27
16	13		39

Notice, however, that the space utilization has *not* changed since we replaced two overflow pages holding one record each with one primary page holding two records; so we must split again. Chain zero is split since NEXT points there. The two records on chain zero, with keys of 28 and 16, are then redistributed between chain zero and a new chain four using h_2. 16 hashes to chain zero and 28 hashes to chain four. NEXT is then updated to one. The file now appears as

0	1	2	3	4
16	29	18	27	28
	13		39	

The 70 percent space utilization is acceptable.

The final entries of 51 and 19 hash to chain three. There is no space for them in the primary page; so they are inserted into overflow pages using h_1. The final file structure is then

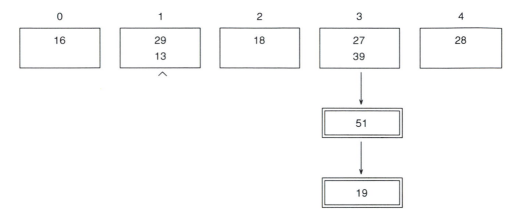

Its 75 percent space utilization is within the specified bounds.

Discussion

Let's retrieve the records for keys 28 and 19. Since the current level is one, we apply $h_1(28) = 0$. But since $0 <$ NEXT, we then apply $h_2(28) = 4$ to locate the chain on which the record is stored. Since 28 is stored in a primary page, we need only one auxiliary storage access to retrieve it. To retrieve 19, we obtain $h_1(19) = 3$. Since 3 $\nless$ NEXT, we do not need to use h_2. We scan the primary and overflow pages for chain three until we locate 19 on the second overflow page. As a result of its location, three auxiliary storage accesses are needed to retrieve it. Remember that linear hashing does *not* use an index. What do we pay for this feature? All the example records except 51 and 19 require only a single access of secondary storage. But 51 and 19 require two and three accesses, respectively. So to retrieve all the records in the file once, we need an average of 1.33 probes. In extendible hashing and dynamic hashing, only a *single* access of auxiliary storage is necessary when the index fits into primary memory. The average probes for successful searches with random data are given in Table 10.2. The results for a primary page size of two records are from [7], and the others are from [5].

Notice that the page sizes in Table 10.2 are larger than the single record per

TABLE 10.2   AVERAGE RETRIEVAL PROBES FOR LINEAR HASHING

Primary page size	2 records			10 records			20 records			50 records		
Overflow page size	1	1	1	4	3	3	7	6	5	15	14	12
Packing factor	75%	85%	95%	75%	85%	90%	75%	85%	90%	75%	85%	90%
Average probes	1.43	1.68	2.22	1.14	1.33	1.57	1.08	1.24	1.44	1.05	1.20	1.35

bucket that we have been considering with the other hashing methods. With a bucket size of one, the performance is poorer than for any of the results given in the table.

Since a primary deficiency of linear hashing is the relatively large number of retrieval probes needed to access a record, several suggestions have been made for improving its performance. Larson suggested doubling the file size in a series of partial expansions [8, 9, 10]. He has also suggested placing the overflow records into primary pages, thereby eliminating the need for overflow pages [9] and using multiple overflow chains [10]. Mullen [11] has also proposed a method that eliminates the overflow pages. Ramamohanarao and Sacks-Davis [12] suggest handling overflow records by applying linear hashing to them recursively.

If we do *not* have space for an index and the table will be expanding and contracting, then linear hashing is an appropriate method.

KEY TERMS

Linear hashing   Performance
    Chains   Primary page
    Deletion   Storage utilization
    Overflow page

EXERCISES

1. Insert records with the keys

$$80, 53, 26, 17, 62, 18, 35, \text{ and } 51$$

into a file using linear hashing with the primary pages having a capacity of two records and overflow pages having a capacity of one record. Let $N = 2$ initially with space utilization bounds of (40 percent, 80 percent) and $H_0 = $ key **mod** 2 and $H_1 = $ key **mod** 4, etc.

2. Repeat Problem 1 but with the data

$$80, 53, 35, 29, 44, 76, 43, \text{ and } 32$$

3. Explain why the retrieval performance of linear hashing improves as the primary page size increases.

4. Does an overflow page size of 50 records or 10 records give better performance for linear hashing when the primary page size is 50? Explain.

5. Explain why the retrieval performance of linear hashing degrades as the upper bound for storage utilization increases.

6. What is the primary advantage of linear hashing over computed chaining?

7. What is the primary advantage of computed chaining over linear hashing?

8. Can the order of insertion of records using linear hashing have an impact upon the retrieval efficiency? Explain.

9. Write an algorithm in either outline or pseudocode form for the deletion of a record from a linear hashed file.

REFERENCES

1. Fagin, R., J. Nievergelt, N. Pippenger, and H.R. Strong, "Extendible Hashing—A Fast Access Method for Dynamic Files," ACM *TODS,* Vol. 4, No. 3 (September 1979), pp. 315–344.

2. Lomet, D.B., "Bounded Index Exponential Hashing," ACM *TODS,* Vol. 8, No. 1 (March 1983), pp. 136–165.

3. Larson, P., "Dynamic Hashing," *BIT,* Vol. 18 (1978), pp. 184–201.

4. Scholl, M., "New File Organizations Based on Dynamic Hashing," ACM *TODS,* Vol. 6, No. 1 (1981), pp. 194–211.

5. Litwin, W., "Linear Hashing: A New Tool for File and Table Addressing," *Proceedings of the Sixth International Conference on Very Large Data Bases* (October 1980), pp. 212–213.

6. Burkhard, W.A., "Interpolation-Based Index Maintenance," *BIT,* Vol. 23 (1983), pp. 274–294.

7. Ruchte, W.D., "A Method for Improving the Retrieval Performance of Linear Hashing," Internal Report, North Carolina State University, 1986.

8. Larson, P.A., "Linear Hashing with Partial Expansions," *Proceedings of the Sixth International Conference on Very Large Data Bases* (October 1980), pp. 224–232.

9. Larson, P.A., "Linear Hashing with Overflow-Handling by Linear Probing," ACM *TODS,* Vol, 10, No. 1 (March 1985), pp. 75–89.

10. Larson, P.A., "Performance Analysis of a Single-File Version of Linear Hashing," *Computer Journal,* Vol. 28, No. 3 (July 1985), pp. 319–329.

11. Mullen, J.K., "Tightly Controlled Linear Hashing without Separate Overflow Storage," *BIT,* Vol. 21, No. 4 (1981), pp. 390–400.

12. Ramamohanarao, K., and R. Sacks-Davis, "Recursive Linear Hashing," ACM *TODS,* Vol. 9, No. 3 (September 1984), pp. 369–391.

Chapter 11

Other Tree Structures

TRIES

Yes, you read the title of this section correctly, that is, if it made it through the editing processing intact. To many people, the term *trie* looks like a misspelling for "tree," which it is not, although it is a type of tree structure. The term originated in the sixties when computer science was still in its infancy, and many of the terms were given unusual or interesting names (e.g., do you recall the term "garbage collection" from a previous course?). The term *trie* comes from the word re*trie*val since the structure is used in the retrieval of data [1]. The problem with that pronunciation is that it is the same as the word "tree." To distinguish the two terms, some individuals spell *trie* letter by letter, *t, r, i, e,* but that seems to be a nuisance. The more common solution is to pronounce the term *not* as it appears in retrieval but rather to pronounce it as if it were spelled T-R-Y.

> A **trie** is a character by character representation of a key.

Until now, we have considered the key field of a record as a single entity, as an entire unit. When we hashed a key, we applied the hashing function to the entire key value. But treating the key as a single entity makes it difficult to manage variable-length keys, such as people's last names. If we reserve a fixed amount of space for each field, we waste considerable space, plus the blanks in the key value may have a detrimental effect on the evenness of the distribution of the hashing function.

Rather than apply a numeric function to the key value, we build a *trie* of the characters in the keys for the file and use the *trie* to map the keys to addresses by following the appropriate path through the *trie*. We will give the algorithms for traversing a *trie* later. Another reason for using a *trie* structure to map a key to an address is that many keys in actual practice are alphabetic rather than numeric. So rather than going through a transformation process to convert an alphabetic key to a numeric quantity, we can use the alphabetic key directly. A *trie* acts like an index in ascertaining the proper data page for storing a record.

A *trie* is a data structure that should receive more attention. It is often a natural representation for commercial applications, as we will see later. Its lack of newness and unusual name should not prevent it from being considered seriously. We have talked about a trie, but you still may not have a clear idea of what one is. To remedy that situation, let's look at two examples.

Sequential Tries

A trie may be represented as either a sequential or a linked structure. Let's first consider the *sequential* trie. Figure 11.1 contains an example of a sequential trie formed from the names of animals used in the B-tree example: *cat, ant, dog, cow, pig, rat,* and *gnu*. The first word *cat* is entered into column one, row *c,* because the first letter of the key is a *c.* The next key is *ant,* which also goes into column one but row *a* instead of *c.* The third key of *dog* goes into column one, row *d.* Figure 11.1a shows the trie at this stage. We are storing the actual key values in the trie entries for illustrative purposes, whereas in practice an entry would more likely contain a pointer to where the entire record for that key was stored. Next we add *cow.* But here we have a difficulty because

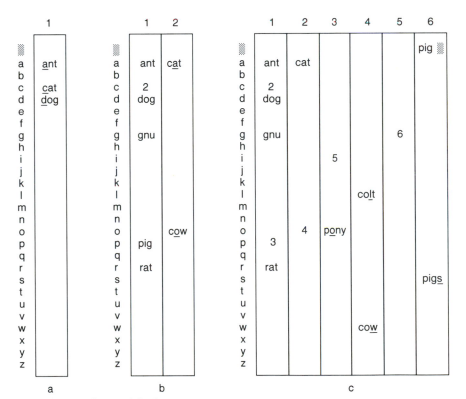

Figure 11.1 Sequential trie.

row *c* of the sequential trie is already occupied with *cat*. What this means is that we cannot distinguish these two keys on the first character, so we must move to the second character. We remove *cat* from column one, row *c,* and replace it with a pointer to a column number in which we can distinguish on the second character. We then add a column two to the sequential trie. *Cat* is placed in column two, row *a,* since the second character is an *a*. Likewise, *cow* is placed in column two, row *o*. The next three keys from the earlier example, *pig, rat,* and *gnu* are added to column one without difficulty. The trie at this stage appears in Figure 11.1b.

Let's add two other keys, *pony* and *colt*. When we try to add *pony* to column one, we have a collision with *pig,* so we must set a pointer to another column. Can we use column two? *No,* absolutely not, because everything in column two *must* have a *c* as the first character. Since *pig* does not, we must add a column three into which we insert *pony* into row *o* and *pig* into row *i*. Let's then add *colt*. Here we have a collision with *cow* in column two, row *o*. We must then use the third character to distinguish these two keys. We must also add another column, four. Everything in column four will then have a *co* in the first two characters of the key. *Cow* is inserted at row *w* and *colt* is placed at row *l*. Let's conclude with one more key, *pigs*. This key collides with *pig* at column three, row *i*. We then look at the third character position to see if we can differentiate them at that point. We can't, but we still must *add* to the trie structure. We need to add a column five that would contain all entries for words beginning with *pi* such as *pike*. We place a pointer in row *g* to the next column to examine. We finally can separate the two keys in the fourth position. We need to add a sixth column that will contain all keys that have *pig* in the first three positions, for example, pigeon. Into that column, we place *pig* into row "▓", where the "▓" represents a blank character, and *pigs* into row *s*. The final trie appears in Figure 11.1c.

As you can see, this sequential trie wastes considerable space. Why? When we add a column to a sequential trie to distinguish two keys that have a common initial substring, it contains slots for *all* possible symbols that may occur in the distinguishing position, not just the two slots needed to differentiate the two current keys. If we have a large proportion of *dense* keys, that is, those which share common characters with other keys, we need many columns in the sequential trie to differentiate those keys, and most of the slots in those columns will remain empty. On the other hand, if we have *sparse* keys, those which have few characters in common with other keys, the storage utilization will be greater. For example, if we have 25 keys and each one begins with a different letter, a single column is sufficient to represent each key.

Linked Tries

Sequential storage is one of the two basic representations for storing information; the other is linked. Let's next consider *linked* tries. Instead of having a two-dimensional table with slots for all possible combinations of characters for keys, a linked trie stores *only* those sequences of characters that actually form keys. That has the attraction of possibly saving much storage. We store only what we have and not what we might have. We will insert the keys again into a linked trie. The characters of the key will be linked together, *character by character,* with each succeeding character on a new level. We begin the construction process with *cat.* Notice that each character of the key appears on a different level to produce the linked trie

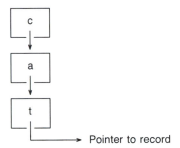

Pointer to record

If we had only the one key, we could map it to the record by storing only the first character. In fact we need to store only the fewest number of disambiguating characters to retrieve a record.

Next we add *ant.* Since it begins with a different letter than *cat,* we need to create another root node. The trie then becomes

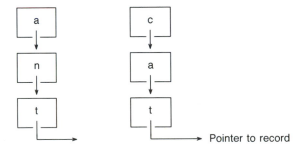

Pointer to record

When we add *cow,* since it has the same first character as *cat,* we can use the *c* node that already exists in the linked trie. So, unlike sequential tries where common characters are a detriment, they are an asset in linked tries. The linked trie then appears as

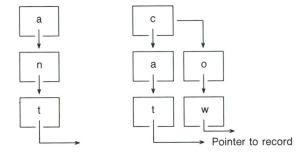

Pointer to record

Inserting *pig* adds another subtree, with *p* in its root. The trie is then

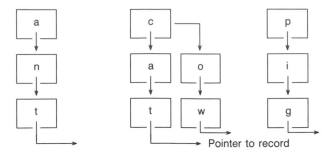

We omit *rat* and *gnu* in this example since they would merely add additional subtrees. We then add *pony,* which can share a *p* with *pig.* (Unusual, a sharing pig.) The linked trie grows to

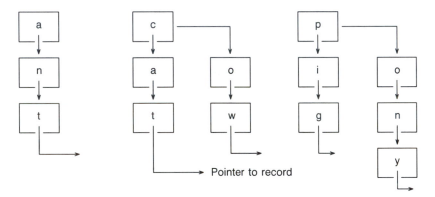

When we add *colt,* it can share both the *c* and the *o* with the *cow,* which illustrates that the *denser* the keys are in a linked structure, the *more efficient* is the storage utilization. To conclude this example, we add *pigs* and the generic term *pet.* In adding *pigs,* the first three characters are the same as in *pig,* so they can be shared. However, we must also add a fourth character to *pig,* a blank (▓), to distinguish it from its plural form. A blank character appears explicitly only when it is used to distinguish a word from a prefix. The final linked trie appears in Figure 11.2.

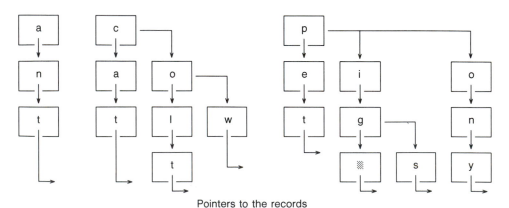

Figure 11.2 Linked trie.

Algorithm 11.1
CONVERTING A TREE INTO A BINARY TREE

I. For each node in the tree,
    A. Join it to its right adjacent sibling by a *right link.*
    B. Connect it to its leftmost offspring by a *left link.*

It then provides an *index* to the eight records associated with the inserted keys. The denseness of the keys produces a trie that uses storage efficiently. When the keys are sparse, a linked trie is not as efficient as a sequential trie since the pointers require additional storage.

Binary Linked Tries

The linked trie of Figure 11.2 has a complication with its representation. Notice that the p node has three branches emanating from it. In fact that node could have had as many branches as there are symbols from which the keys are composed, as many as 27 in this example. What is a common way to represent a general tree structure that may have multiple branches? Yes, convert it into a binary tree. We can accomplish this with the straightforward procedure described in Algorithm 11.1.

The linked trie of Figure 11.2 may then be transformed into the binary linked trie of Figure 11.3 in which the right links are horizontal and the left links are vertical.

An advantage of the resulting binary structure is that the algorithm for searching for a record or inserting the key of a new record becomes rudimentary. The search algorithm may be summarized as "search via right links until a match, then go left, and repeat on the next level." The complete algorithm appears in Algorithm 11.2. The algorithm for a nonbinary linked trie is more complicated.

The binary representation of a linked trie is preferred to that of a general trie because of the efficiency of its representation (each node needs only two pointers) and the simplicity of its search algorithm. It also has an advantage over a sequential trie in the case of *dense* keys.

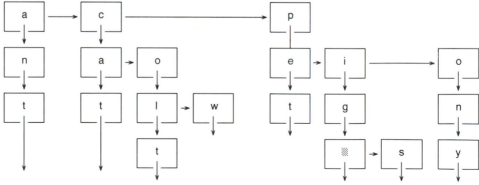

Pointers to the records

Figure 11.3 Binary linked trie.

Algorithm 11.2
SEARCHING A BINARY LINKED TRIE

I. Begin the search by comparing the leftmost character of the key against the nodes on the root level of the linked binary trie.

II. While the current node of the trie is not a terminal node

    A. Search for the current character of the key against the current level of the trie via *right links* until either (1) it is located or (2) a null right link or a character that is lexicographically greater than the character sought is encountered.

    B. If 2, terminate, else upon finding a match of characters, follow a *left link* to the next lower level.

III. The current node contains the desired record.

Length-Segmented Lists

An important principle for improving the retrieval performance of an information system is to preprocess the information by dividing it into mutually exclusive search segments. On *retrieval* then, we need to search only the relevant segments *instead* of the entire search space. We do the task faster by doing less. We have seen this principle applied before, for example, with multiple probe chains in collision resolution and with multiple subdirectories in dynamic hashing. We observe that principle being applied again to *extend* the trie concept to length-segmented lists [2]. Rather than having one trie index, we have several based upon the *length* of the keys stored in that subdirectory. Instead of using a hashing function to determine which segment of the information to search, we simply have to count the characters in the key. A length-segmented list then appears as

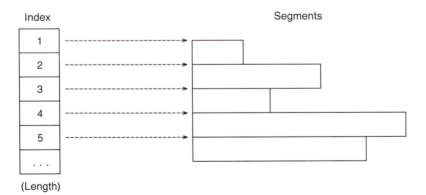

where each *segment contains a subdirectory trie* consisting of all keys of a length specified in the index. The first segment contains an index to all keys of length one. The next segment contains an index to all keys of length two, and so on. Instead of searching one large index, we search only *one* smaller index. A length-segmented list for the keys in the linked binary trie of Figure 11.3 is

The length-segmented list is a refinement of a trie structure to improve retrieval performance.

Now that you are familiar with the trie structure, we are ready to consider applications. As we mentioned earlier, the trie is perhaps an underutilized data structure. With the examples of applying a trie described in the next few sections, your imagination may be sparked to discover many other applications.

APPROXIMATE STRING MATCHING

Until now, when we have searched for a record with a given key, we have demanded an *exact match* between the search key and the key of the stored record. That is certainly an important requirement to ensure that we retrieve the record that we intend. But what do we do in the frequent situations in which we *do not know the exact key of the item that we want to retrieve?* What are examples of such situations? Have you ever needed to look up the spelling of a word in a dictionary? In such a circumstance, you do *not* know the exact key, for if you did, you would not be making the search. Have you ever gone to the library to look for a book by a certain author when you weren't sure of the spelling of her last name? The search was more difficult and time-consuming, wasn't it? Here's an example involving computers. Many of you have first or last names, or both, which may have multiple spellings (or should we say misspellings?). What if information about you is then entered into a computer and you or someone else wants to retrieve it? Often when you have a problem with a bill at a department store, a utility company, a hospital, or the like, when the clerk goes to retrieve your record, his response is that he has no record of you in the computer. At this point, his expression is something like "Sorry, Charlie, but I can't help you without the information from the computer." This is when it helps to know a little about how information is stored within a computer, to know that most retrieval systems require an exact match with the key field. For then you may ask the clerk to try various misspellings of your name until he finally locates your record. Wouldn't it be pleasant if computers possessed approximate string matching in such a predicament?

> **Approximate string matching** is being able to locate a record when you know only *approximately* the key of the record sought.

Here's an example to illustrate the importance of approximate string matching. Some years ago in a course on document retrieval, the students in the class had an opportunity to use a commercial, computerized, document retrieval system. Each student had an opportunity to retrieve records on a topic of his interest. One student was interested in attending graduate school at MIT and wanted to know of articles that dealt with that institution. So he typed in *Massachusets Institute of Technology* as one of his keyword descriptors. Since this was a rather long descriptor and the system's response time was slow that day, my initial reaction was to inform him of his spelling error prior to his pressing the return key on the terminal. But on second thought, I decided that it would be instructive if I allowed him to enter his descriptor with the misspelling. When he didn't obtain any documents back from the system, I would then ask him the reason, since MIT is certainly a well-known institution. He would then look back at the descriptor that he had entered and notice the misspelling. He or I would then comment on how the computer needed an exact match. But you know what happens to the best-laid plans of mice and men. Lo and behold, the system retrieved a document. In a way, though, this was even more instructive because it illustrated the disadvantage of requiring exact matches in a document retrieval search. For all practical purposes, this retrieved document was *lost* in the system. Unless the user misspelled the descriptor in exactly the same way, the document would *never* be retrieved. And that could be serious.

Hall and Dowling [3] discuss the importance of approximate string matching and techniques for handling it. One technique is the trie. The reason that the trie is an appropriate data structure is that it records *environment*. At any position in the trie, it is possible to determine other keys that are similar to the desired key. This is a result of the character by character nature of a trie. An example of approximate string matching that applies a trie appears in the next section.

Spelling Correction

Either because of the lack of typing skill or the ignorance of how to spell a word, much information and many requests to a computer are entered incorrectly. Because of the exacting nature of most computer software, the response to an incorrect command is usually negative. Or if a document contains misspelled words, the reaction of the person reading it may be negative. Correct spelling *is* important. Can the computer help? As you are aware, many computer software systems today contain spelling correctors. Be clear on the difference, though, between a spelling checker and a spelling corrector. A checker merely searches a table for an exact match between an entered word and one in the table and then notes if it does not exist in the table. A corrector has the more formidable task of finding appropriate alternatives in the event the word cannot be located in the table.

One algorithm for correcting spelling [4] uses a linked binary trie. It uses a linked binary trie as the basic data structure for spelling correction because (1) it provides *context* that is so important in deciding alternatives and (2) the resulting algorithm is straightforward in processing the types of spelling errors that are likely to occur.

If we do not assume ignorance, all spelling errors are the result of mistyping. Shaffer and Hardwich [5] studied the kinds of errors that typists are likely to make. It could be argued that individuals using a computer keyboard would make similar ones. They categorized the errors in the order of likelihood of occurrence as

Algorithm 11.3
SPELLING CORRECTION USING A TRIE

I.   For a substitution error, skip one character in the input string and skip one level in the trie.

II. For a deletion error, skip one level in the trie.

III. For an insertion error, skip one character in the input string.

IV. For a transposition error, reverse the order in which the levels of a trie are searched. Search the level after the incorrect character, then the level with the incorrect character, and finally the remaining levels to match the remaining characters of the string.

- Substitution—an incorrect character appears.
- Deletion—a character is missing.
- Insertion—an extra character appears.
- Transposition—two adjacent characters appear in the reverse order.

A substitution error could be classed as an insertion error coupled with a deletion error, and a transposition error could be categorized as two adjacent substitution errors. But having the four distinct classes of errors gives the correction algorithm greater flexibility.

If we store the dictionary as a linked binary trie, we can then navigate the structure *letter by letter* to locate a word. The point at which we no longer find a match provides the context for the correction processes and signals special action. The algorithm merely varies the order in which characters of the key are compared with the characters in the trie. Each of the four types of spelling errors can be managed as described in Algorithm 11.3.

We demonstrate the spelling correction algorithm with the linked binary trie in Figure 11.3. If we enter the string *ciw,* it appears to contain a *substitution* error of the letter *i* for *o* in *cow.* In searching the trie for *ciw,* no match is found on level two after matching a *c* on level one. The algorithm says to skip one character in the input string and one level in the trie. We would then search for a *w* in level three of the subtree containing the matched *c* at level one. We find a match, and to have arrived at that node, the second character must have been an *o* to give us the correct spelling for *cow.*

If *clt* is entered, we would again not find a match after the *c* at level one. In checking for a substitution error, we would locate the word *cat* as a possible correct spelling. The algorithm can be made to stop either after the first possible correction or after all possibilities. In this case, let's continue beyond the first possibility and search for a *deletion* error. The algorithm skips one level in the trie and continues searching at level three in the subtree with the matched *c* at level one. An *l* is found at level three and a *t* at level four. We then have a match if the second character is an *o* for a correct spelling of *colt.* No other corrections exist with the linked binary trie of Figure 11.3 for the input of *clt.*

For an *insertion* error, let's assume that *pket* appears. We have a match of a *p* at level one, but no match at level two. For an insertion error, the algorithm skips one character in the input string and continues searching. Since an *e* matches on level two and a *t* on level three, the proper spelling is *pet*.

Finally, let's consider a *transposition* error. Given the input string of *pnoy,* a match occurs with the letter *p* at level one but no match is found at level two. We then search for an *n* at level three under the subtrie with the matched *p* at its root. We find an *n*; we next search the parent node in the subtrie to examine whether it matches the next character in the input string, an *o,* which it does. We then continue matching the remaining character, a *y,* at level four. The correct spelling is *pony*.

A goal of this text is to encourage the reader to examine and analyze the given data structures and processing algorithms and to search continually for improvements. Can you think of an alternative method for correcting a *transposition* error? One option is to vary the order of checking within the *input string* rather than within the *trie*. If a failure occurs at position i in the input string, then search for character $i + 1$ at the current level of the trie. If successful, then look for character i at the next level in the trie. If found, the search continues for character $i + 2$ at the subsequent level of the trie and so forth. If we use the example input string *pnoy* as before, a match occurs with the letter *p* in level one of the trie. The *n* does not match at level two of the trie; so we search for the next letter in the input string, that is, *o,* on the current level of the trie. We have a match, so we proceed by searching for the *previous* character in the input string on the *next* level of the trie. The *n* in the input string is matched, so we then continue the search with the fourth character in the input string on the fourth level of the trie. Upon matching the *y,* we again recognize the word *pony*.

Which method is preferable? It is considerably easier to change the order of searching within the input string than within the trie. In changing the search order of the input string, only the *subscripts* to a sequential array containing the input need to be varied whereas in modifying the search order of the trie, a different series of *pointers* needs to be followed. In addition fewer nodes in the trie need to be checked with the alternative implementation. In the original one, a search at level $i + 1$ could involve searching several subtries.

Since the example trie contains only a few data items, we did not have many possible corrections for a misspelling. An advantage of this algorithm is its flexibility. As previously mentioned, we can terminate with the first alternative for a misspelling or we can find all alternatives. If we are looking for only the single most likely alternative, we can easily vary the order in which the four types of errors are processed. To provide individualized results, we can store the kinds of errors that each user is prone to making and then process the trie in an order appropriate for the person entering the data.

The algorithm does not locate *all* spelling or typographical errors. It does not handle phonetic errors (such as "ph" for "f" in *Filipinos*). If multiple error types appear in a string, at least one correct character must separate them. Nevertheless, when combined with human intelligence, the algorithm is a useful application of tries. A person can complement the algorithm by selecting the proper alternative, for example, *cat* or *colt* in the previous example, or by making a modification of a string when the algorithm cannot locate a possible correction. The algorithm complements the person by locating all the words in a document that are not in the dictionary and by providing possible corrections if they are misspelled.

As an aside, it is interesting to note the reaction of many to this algorithm when it was introduced. A common reaction was that the algorithm required too much processing to be practical. But in the intervening years, computers became faster and memory became cheaper. So just because something does not appear achievable today doesn't mean that it will always be such. *Don't restrict your thinking.*

If the basic data structure for a spelling corrector were a length-segmented list, it would be necessary to check not only the index of the length of the misspelled word but also the index of one size larger for an insertion error and the index of one size smaller for a deletion error.

Another application of tries that is beginning to appear is in telephones that store telephone numbers. On some of the advanced models that let you key in the name of the individual or organization you want to call, you only need to key in the fewest disambiguating characters and the phone then begins dialing. The system is searching an abbreviated trie with entries for only the disambiguating characters of the keys and when it encounters a terminal node (a pointer to a number) during the search, it then executes.

Many computerized file systems now use a trie index for locating a file. In searching for a file, the user only needs to enter a single letter and the system will immediately point to the first file beginning with that letter. If more than one such file exists, the user then needs to continue entering letters until the fewest disambiguating characters are entered. As the letters are input, the system continually updates its pointer to the first file with the current prefix of letters.

A trie may be used to implement *completion* in a text or program editor. With *completion,* after typing in the initial disambiguating characters of a lengthy, but previously used term, the user merely presses a special key instead of typing the remainder of the term. The system completes the term for the user. If more than one *completion* is possible, the system allows the user to choose the proper term. *Completion* aids the conscientious programmer who uses mnemonic variable names and the writer who uses long words. The *completion* for "po" when using the trie in Figure 11.3 is "pony".

KEY TERMS

Approximate string matching
Error
    Deletion
    Insertion
    Spelling
    Substitution
    Transposition
    Typographical
Key
    Dense

Sparse
    Variable-length
Length-segmented lists
Spelling correction
Tree
    Converting to binary
Tries
    Binary linked
    Linked
    Sequential

EXERCISES

1. **(a)** Into which row and column should the key *pigeon* be inserted in the sequential trie of Figure 11.1c? **(b)** the key *pike?*

2. Construct a binary linked trie for the six words, *fe, fi, fo, fum, fumed, Fred.*

3. Construct a length-segmented list for the data in Problem 2.

4. Compare a length-segmented list with a single trie index in terms of **(a)** storage requirements and **(b)** performance.

5. Give two applications (uses) of approximate string matching in a document retrieval system.

6. Insert the words *gnat, ram,* and *cobra* into the sequential trie in Figure 11.1c.

7. Construct a sequential trie for the following keys: *hear, here, night, knight, nite, feet, feat, heal, heel, hill.*

8. Construct a binary linked trie for the following keys: *Frequently, freckled, Freddy, a, fresh-man, freeloaded, freebies, from, the, freightyard.*

9. Using the spelling correction algorithm of Algorithm 11.3, which of the following possible corrections for English words would be returned *first* if the string *caot* were entered? *clot, cabot, coat,* or *cart?*

TRIE HASHING

We want to consider one last hashing mechanism, trie hashing [6], for several reasons. One is that it is an application of a *trie,* which is, as previously mentioned, an *underused* data structure. This application of tries gives you more exposure to them and should help in making you think of them when you are designing future data structures. Second, trie hashing has a property that we have not seen previously in the hashing methods; the resulting file is *ordered* so that you can process it *sequentially* or *directly* without taking uncommon efforts. It also has features that we have seen in some of the previous hashing schemes in this part:

- An index to map into the data pages.
- A single access to a data page if the index can fit into primary memory.
- The file can expand and contract without reorganization.

> **Trie hashing** employs a *trie* index to *hash* (transform) a key into a storage address such that the records of the expandable file are sequentially ordered.

In this method, we do not apply a function to a key as we have in previous hashing schemes. Instead, on insertion or retrieval, we sift the key through the trie index. At each node, we make decisions based upon both a character and its position in the key when determining whether to continue left or right. For *less than or equal, go left; otherwise, go right.* In forming the index, we use the leftmost distinguishing character of the **middle** key of the affected records to separate the keys into two groups. We record both the character and its position in the key in the node. The index has two kinds of nodes: internal nodes that point to other nodes in the index and external nodes that point to the data pages. The basic node structure is

Left link	Tag	Character value	Position	Tag	Right link

The **tag** tells us whether the corresponding *link* field points to another index node or to a data page. The **character value** is the character that we are comparing against, and **position** gives the position of the *character value* in the key.

As we insert more data records, the index continues to grow. Just as with dynamic hashing, when a data page overflows we divide it into two pages; an external data page becomes an internal node in the index with pointers to two external nodes. One external node is the existing one and the other is new.

An Example

To understand the method, let's go through an example similar to one in [6]. Since we will not be applying a mathematical function to the keys, we do not need them to be in numeric form. Therefore, we use alphabetical keys consisting of the most frequently used words in English inserted in decreasing order of occurrence [7]. The frequency of occurrence, however, has no bearing on the functioning of the method. For this example, the bucket size is four. We begin with a pointer to the initial page. The first four records with the keys *the, of, and,* and *to* are inserted into the file giving

```
and
of
the
to
```

The records are placed in the node alphabetically to facilitate sequential processing and node splitting. When we attempt to add *a,* we encounter an overflow condition. The alphabetical *middle* key of those stored in the page plus the key of the record being inserted is *of.* We can then divide the records into two groups using the leftmost character of the middle key. The letter *o* in position 1 is the information we need to store in the newly created internal node. The file now appears as

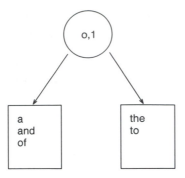

The internal nodes are represented by circles and the external nodes by boxes. The only information that we need to make explicit in an internal node is the *character value* and the *position*. *In* is inserted into the left node, and *that* goes into the right node.

When we try to insert *is,* since it maps to the left node, we have an overflow condition in that node. We again order the five keys associated with that node: *a, and, in, is,* and *of. In* is the middle key. We can separate the keys into two groups with the leftmost letter of *in.* (The division of the records into the two groups is not always balanced.) The file now becomes

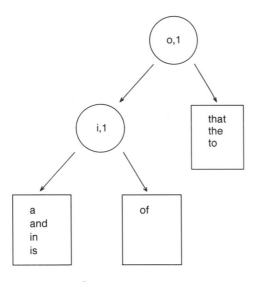

If we wish to retrieve *and,* we filter it through the trie index. We first compare the *a* with the character value stored in the root node of the index; it is less than, so we go left. We encounter another internal node. This time we are comparing the *a* with the *i* stored in the internal node. Again a less than comparison, so we proceed to the left. We have reached an external node, so we search it for the occurrence of the desired record.

Let's continue with the insertions. When we add *i*, it maps to the leftmost leaf page. Since this page is already full, we have an overflow. We order the five records affected by the split. Notice that there is something different in this insertion. *i* is the middle key. But if we compare against an *i* in position 1, we cannot separate the five records into two groups because *in* and *is* would map to the left in addition to *i*, *a*, and *and*. We then need to go to the next character position in *i*. Since *i* has just one character, we use a blank as the delimiter for the word and consider it the next character. We divide the five words into two categories by comparing against an *i* ▯. The bottommost internal node represents a comparison against the two-character pair *i* ▯.

Note: As the position of the character increases, we are comparing against a *concatenation* of the characters appearing in positions to the left of the current character. We obtain these values from the parent nodes in the index. If less than or equal, we go left; otherwise, we move to the right external node. In the example *a, and,* and *i* map to the left node, *in* and *is* to the right. The structure is then

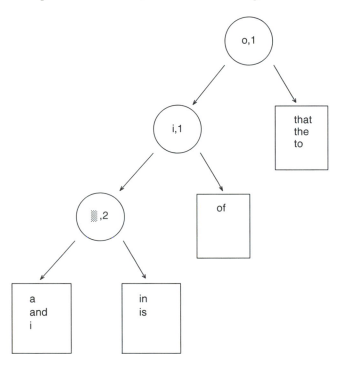

At this point, we deviate from using keys in the order of frequency of appearance in English so that we may readily illustrate another situation that may arise during an insertion with trie hashing. If we add the records with the keys *by* and *be*, we have another overflow situation in the leftmost leaf node. The middle key of the affected five records is *be*. All the keys mapping to this page were compared against *i* ▯. Can we concatenate onto this comparison sequence and produce a longer sequence to separate the five records? No, because *be* is not a string with the prefix *i* ▯. In this case, we revert to comparing on the first character of the middle key, which is *b* in this situation. In traversing the trie, we know not to concatenate at (*b*, 1) since the character position that we are comparing against is less than that in its parent node. The final structure is then

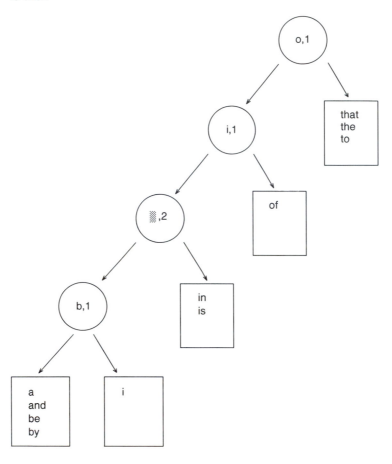

Let's look at one other situation that may arise during insertion. If we have an external node with the keys *Peter, Piper, picked,* and *pickled* as

```
Peter
picked
pickled
Piper
```

and we add the word *peppers,* we have an overflow situation. But the first character cannot be used as a distinguisher since it is the same for all the keys. What we then do is proceed to the next character position and attempt a distinction there. Since the middle word is *picked,* we cannot divide the group based on the second character position since we would be comparing against *pi* and all the keys would map to the same left node. We then proceed to the third position. There we can differentiate on the *c* (with a comparison against *pic*) so that the tree becomes

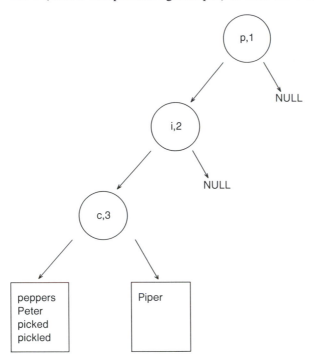

The NULL pointers reserve space for future insertions. In this example, we expanded the index to the point where we could distinguish the five keys.

Discussion

We can process all the records in a trie hashed file in *sequential* order by traversing the *terminal* nodes in a left to right order. We can process a single record *directly* by traversing the index to the proper data page. A true advantage of this method is that we can process the data both ways. In addition, if the index fits into main memory, we need only one access of auxiliary storage for direct access. We can also expand and contract the file without reorganizing it. On the somewhat negative side is the fact that this index requires more space than the directories in extendible hashing and dynamic hashing since we are storing considerably more information. Also the space utilization is usually around 70 percent, which is not as good as in some other methods but is comparable with B#-trees.

Since both trie hashing and B+-trees allow one to process data both directly and sequentially, what are the advantages of each when compared? A B+-tree stores all the records on the same level, so its performance is consistent. Processing the sequence set of a B+-tree is also easier than processing the external nodes of a trie hashed structure since the sequence set nodes are linked together directly. On the other hand, trie hashing permits variable-length keys, not handled easily with B+-trees, which set a maximum key length.

Deletion in a trie hashed structure is handled in a manner similar to that with dynamic hashing. When a record is deleted from an external node, a check is made to determine if the remaining records in that node and the records stored in its **twin** node (the one on the same level with the same parent) can be stored in a single node (data page). If they can, the records of the two pages are combined and their parent node is eliminated from the index.

KEY TERMS

Trie hashing
    Deletion
    Direct access
    External node

Internal node
Sequential access
Space utilization

EXERCISES

1. Use trie hashing to insert records with the following keys into a dynamic file with a bucket size of four.

    Hollies, with, red, berries, make, handsome, holiday, decorations, remarked, Holmes

2. Use trie hashing to insert records with keys that are the months of the year given in calendar order; use a bucket size of four.

3. What is the primary advantage of trie hashing over the other methods for constructing dynamic files?

4. Which, if any, of these standard orders for traversing a binary tree suffices for processing the external nodes in a trie hashed structure in a sorted order? *preorder, inorder,* or *postorder.*

5. Does the order of insertion of the records affect the depth of a trie hashing index? Explain.

PATRICIA TREES

The PATRICIA tree [8] is another structure originating in the sixties with an unusual name. PATRICIA is an acronym for <u>P</u>ractical <u>A</u>lgorithm <u>T</u>o <u>R</u>etrieve <u>I</u>nformation <u>C</u>oded <u>In</u> <u>A</u>lphanumeric. Yes, it does seem that the acronym may have existed before the meaning. Nevertheless, it is a useful variant of a trie for situations in which the keys are quite similar. Consider how you would represent the following two keys in a trie dictionary

antidisestablishmentarianism

and

antidisestablishmentarianist

Would you really want to have to search 28 levels before you could distinguish these two keys? Obviously, that would not be efficient retrieval. What a PATRICIA tree allows us to do is to store the number of positions we should move forward before making the next comparison. In this way, we can eliminate *all* the nonessential compares and improve performance considerably. A node in a PATRICIA tree contains both the number of positions to *move forward* and the character to compare against at that position. A $\leq$ comparison signals a left branch and a $>$ comparison indicates a right branch.

> A **PATRICIA tree** is a binary trie index with skip ahead capabilities for eliminating comparisons.

A PATRICIA tree for the previous two keys is

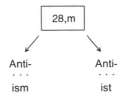

If we then add a record with a key of *ant,* we need to modify the structure to distinguish the new key from the two already in it. At what position can we make a distinction? Since the two existing keys both begin with *a,n,t,* we need to assume that a blank is appended to the end of the key *ant.* We could then differentiate in the fourth position from the left. The PATRICIA tree would then become

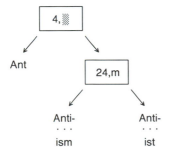

Since the position values are move ahead values, the 28 in the original root node becomes 24 in the corresponding node in the revised structure since we are already at position 4 when we enter that node. Let's add another record to the structure, one with a key of *antidote.* In order to distinguish that key from the two long keys, we need to compare at position 6. The PATRICIA tree then becomes

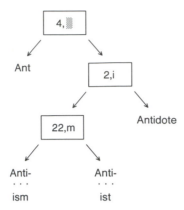

On *retrieval,* we navigate the PATRICIA tree by comparing characters in the specified positions. To retrieve *antidote,* we first compare in position 4. Since the *i* of the sought key is > the "▓" in the root node, we follow a right branch. The next comparison position is two more than the current position, which is six. We then compare an *o* from the key with an *i* in the decision node. Again we have a > comparison, so we take a right branch to the terminal node pointing to the information about *antidote.*

If more than two keys are distinguishable in a single position, it is necessary to add another level and node (with a zero move ahead) to the tree since a PATRICIA tree is binary. For example, if we add a record with the key of *ants* to the previous tree structure, it also requires a test in the fourth position to separate it from the other keys. The PATRICIA tree becomes

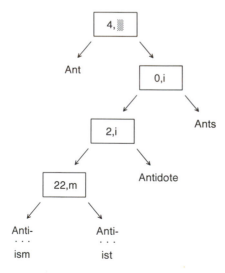

Discussion

A PATRICIA tree is applicable primarily in those situations in which multiple keys have several contiguous, leftmost characters in common. As previously noted, a disadvantage of the PATRICIA tree is that since it is a binary structure, it only provides a two-way branch. If more than two keys are distinguishable in the same character position, extra nodes need to be added to the index to separate them. In the previous example, two nodes were needed to separate *anti . . . ism, anti . . . ist,* and *antidote.* If *n* records are distinguishable in the same character position, *n* − 1 nodes are needed in the index to separate them. If many such cases existed, a plain trie structure might be preferable to a PATRICIA tree, depending upon the number of nodes saved by a PATRICIA tree vs. the number of nodes added to process multiway branches.

DIGITAL SEARCH TREES

Earlier in this part when we considered binary search trees, we assumed that all the records in a tree were accessed *equally often.* That assumption was critical to the efforts to improve retrieval performance. When we considered AVL and IPR trees, the rotations applied to improve performance moved some nodes higher in the tree but concurrently moved a smaller number of nodes lower in the tree. In this way, the *overall* number of accesses on the tree was reduced even though an individual record might have more. Although the equal accesses assumption is appropriate in many applications, it is *not* suitable for all applications. You can recall situations from your own experience in which you access some information much more frequently than others. One example is telephone numbers. You have some numbers that you reference regularly and others that you use rarely if at all. In such a circumstance, an equally likely assumption does not give the best retrieval performance.

We next consider a tree in which the records have different *frequencies of occurrence.* In such a case, we want the *most frequently* accessed records stored near the *root* of the tree. A digital search tree is a binary tree in which the records are stored based upon their frequencies of occurrence [9]. To accomplish this, the records are inserted into the tree in a descending order of occurrence frequency. Obviously, then, we need some information a priori, historical or predicted, on the relative likelihood of retrieval on each of the records. Rather than use the characters of the keys for comparison purposes, the insertion process uses the bit patterns of the characters of the keys. By using the bit patterns instead of the actual characters, the resulting tree tends to be more balanced since the 1s and 0s occur equally often. We have two symbols to compare on instead of 2^i, where i is the bit length of a character. In traversing a digital search tree, we go left on a zero bit and right on a one bit. Each bit searched will correspond to a distinct level in the digital search tree.

> A **digital search tree** is a binary tree that is formed on the basis of the frequency of occurrence of the records and the bit patterns of the associated key fields.

An Example

As data in this example, we again use the two-digit numerical keys that we used with the hashing examples, that is,

$$27 \equiv 011011$$
$$18 \equiv 010010$$
$$29 \equiv 011101$$
$$28 \equiv 011100$$
$$39 \equiv 100111$$
$$13 \equiv 001101$$
$$16 \equiv 010000$$

For their binary representation, we use a 6-bit field, right justified; the number of bits depends upon the range of the key values. We insert the records with the given order assumed to be in decreasing frequency. We insert the record with key 27 into the root node. The tree is then

```
┌──────┐
│  27  │
└──────┘
  ⊥      ⊥
```

where ⊥ represents a null pointer

We insert the remaining records into leaf positions of the tree using the bit pattern of the key to determine the path. Once a record is inserted into a position, it remains in that position. We next insert 18, which follows a path to the left of 27 because its leftmost bit is a zero. The digital search tree becomes

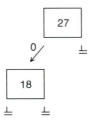

29 is inserted to the right of 18 since its left most two bits are a zero and a one; the tree then appears as

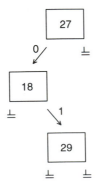

28 is then placed to the right of 29 since its third bit from the left is a one. The tree now looks like

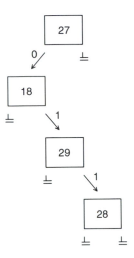

At this point, the tree does not appear particularly balanced since the quantity of example data is not sufficiently large to provide a nearly random distribution of the keys.

On insertion, we are assured that the process will eventually locate a terminal node since at some point the bit value needs to be different from those in the tree for otherwise it would be a duplicate record. If a possibility of duplicates exists, we need to check all the keys on the path to a terminal node.

We next insert the record with a key of 39 to the right of the root node since its leftmost bit position is a one. The tree is now

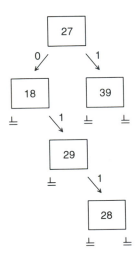

After inserting the final two records with keys of 13 and 16, the tree appears as

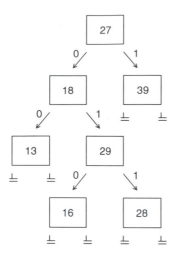

The tree has become more balanced with the addition of more records. As the tree continues to grow, the most frequently accessed records will appear near the top, and the least frequently accessed records will appear near the bottom.

Discussion

For *retrieval,* we follow the same procedure as with insertion; that is, we use a bit representation of the keys to guide us through the tree. We check the key value of each node touched along the way. The procedure terminates upon locating either the desired node or a null pointer. The latter condition signifies that the record is *not* in the file.

Note that an important difference between a digital search tree and a binary search tree is that an inorder traversal of the nodes in a digital search tree *does not give a lexicographic ordering* of the data records.

The primary advantage of the digital search tree is that the overall direct-access retrieval performance is *improved* by placing the most frequently accessed records near the root of the tree.

Couldn't we have used a binary search tree to achieve the same results? Wouldn't it have been simpler merely to insert the keys into a binary search tree in the order of descending frequency of occurrence? The difficulty with that is what happens when the frequency order coincides with the lexicographic order. Let's consider the previous example data in sorted order, that is,

$$13 \equiv 001101$$
$$16 \equiv 010000$$
$$18 \equiv 010010$$
$$27 \equiv 011011$$
$$28 \equiv 011100$$

$$29 \equiv 011101$$
$$39 \equiv 100111$$

If we place these keys into a binary search tree in the order shown above, the resulting tree is skewed to the right as

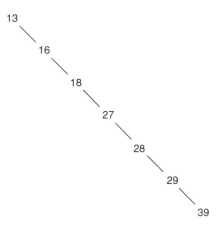

which leads to poor retrieval performance. The corresponding digital search tree is

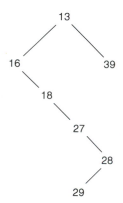

Although the depth of the digital search tree is reduced only by one with this data, if the keys had been more evenly distributed, the digital search tree would have been even more balanced than a binary search tree. For example, if 27 is replaced by

$$23 \equiv 010111,$$

and 29 is replaced by

$$34 \equiv 100010,$$

the resulting digital search tree is

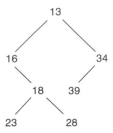

which is three levels *less than* the equivalent binary search tree with the frequency of occurrence order identical to lexicographic order. If the binary representations of the original data were not evenly distributed, as with the example data, we could apply a hash function to the key field to generate a more evenly distributed set of binary digits. The hashing function could be similar to that used with dynamic hashing; for example, one to generate a binary sequence of length l could be specified as

$$H_1(\text{Key}_i) = (b_{i0}, b_{i1}, b_{i2}, \ldots), b_{ij} \in \{0,1\} \qquad \text{for all } j$$

The advantage then of using the bit patterns rather than the letters in constructing the digital search tree is that they tend to balance the tree more, which means fewer retrieval probes and therefore better performance.

The digital search tree is an alternative to the binary search tree in situations in which the frequency of access of the records varies widely. In such situations, the average number of probes needed to retrieve each record once is less for a digital search tree than for a binary search tree since the internal path lengths of each node are weighted by the frequency of access. In achieving this better direct-access performance, though, we lose the ordered sequential processing of the data achieved by an inorder traversal of the tree. Just as with a binary search tree, the number of probes needed to retrieve the records near the bottom of the tree may become prohibitively large as the number of records stored becomes quite large. The advantage of a digital search tree, however, when compared with a binary search tree, is that those records are retrieved relatively infrequently.

KEY TERMS

Digital search tree   Retrieval
PATRICIA tree   Frequency of occurrence

EXERCISES

1. Insert the records with the keys: *shortbread, shortchanging, shortcakes, shortchanged, and shortchanges,* into a **PATRICIA** tree.

2. Comment on the similarities and differences between a **PATRICIA** tree and a trie hash index.

3. Insert the records with the keys 73, 15, 44, 37, 30, 59, 49, and 99 (using a 7-bit, right justified

representation) into a digital search tree. Assume that the given order is the order of decreasing frequency of occurrence.

4. Will the inorder traversal of a digital search tree visit the nodes in ascending order?

5. Will the inorder traversal of the terminal nodes in a PATRICIA tree visit the nodes in ascending order?

6. Draw a PATRICIA tree for the five records with the keys: *Parading, paradigms, for, problem, solving.*

7. (a) Insert the records with the *decimal* keys

$$04 \equiv 00100$$
$$09 \equiv 01001$$
$$11 \equiv 01101$$
$$18 \equiv 10010$$
$$19 \equiv 10011$$
$$29 \equiv 11101$$

into a binary search tree in the order given. (b) Repeat using the *binary* representations of the keys and insert into a digital search tree.

8. (a) Compare the average number of accesses needed for an unsuccessful search vs. a successful one in a PATRICIA tree. (b) Repeat for a digital search tree.

REFERENCES

1. Fredkin, E., "Trie Memory," *CACM,* Vol. 3, No. 9 (September 1960), pp. 490–500.
2. Turba, Thomas, N., "Length-Segmented Lists," *CACM,* Vol. 25, No. 8 (August 1982), pp. 522–526.
3. Hall, Patrick, and Geoff R. Dowling, "Approximate String Matching," *Computing Surveys,* Vol. 12, No. 4 (December 1980), pp. 381–402.
4. Muth, Frank E., and A. L. Tharp, "Correcting Human Error in Alphanumeric Terminal Input," *Information Processing and Management,* Vol. 11 (1977), pp. 329–337.
5. Shaffer, L.H., and J. Hardwich, "Typing Performance as a Function of Text," *Quart. J. Exper. Psychol,* Vol. 20, No. 4 (1968), pp. 360–369.
6. Litwin, Witold, "Trie Hashing," *ACM SIGMOD Conference Proceedings,* 1981, pp. 19–29.
7. Knuth, D.E., *Sorting and Searching,* Addison-Wesley, Reading, MA, 1973.
8. Morrison, Donald R., "PATRICIA—Practical Algorithm to Retrieve Information Coded in Alphanumeric," *JACM,* Vol. 15, No. 4 (October 1968), pp. 514–534.
9. Coffman, E.G., and J. Eve, "File Structures Using Hashing Functions," *CACM,* Vol. 13, No. 7 (July 1970), pp. 427–432, 436.

Chapter 12

Secondary Key Retrieval Revisited

K-d TREES

When we considered secondary key retrieval in Chapter 6, we studied range searching and schemes for retrieving records with multiple attribute values. *K-d trees* (which is short for *K*-dimensional trees) combine both primary and secondary key retrieval into a single structure [1]. A *K-d* tree is applicable in those situations in which we want to retrieve all records with secondary key values over a range, for example, all employees who earn more than $45,000, or ranges, for example, all employees who earn more than $45,000 and have more than two dependents. Instead of performing primary key retrieval as we have done with the other tree structures of this part, the *K-d* tree retrieves all records with values within designated ranges over *d* dimensions. Primary and secondary key retrieval are then the same. To simplify this discussion, we limit *d* to two dimensions. A 2-*d* tree is similar to a binary search tree except that instead of comparing along *one* dimension at each and every level of the binary tree, we alternate the comparisons between two dimensions at succeeding levels of the binary tree. If we had a 3-*d* tree, we would search on a different dimension for each of three successive levels and then repeat the comparisons over the three dimensions. A 4-*d* tree would be processed similarly.

The 2-*d* tree acts as an *index* into where the records are stored and *not* as a place to store them. The index is constructed from the field values of representative data records. It is then used to guide the insertion of the data records into the data pages. Since the index may not evenly distribute the records among the data pages, we need a page overflow mechanism.

Another difference between a 2-*d* tree and a binary search tree is that since we are performing range searching, we may have more than one search path active. At a decision node, we may be interested in records that compare both ≤ and > ; so we need to follow both alternatives. As with several of the previous tree structures, a ≤ comparison implies a left branch and a > compare means a right branch.

> A **K-d tree** is a <u>K</u>-dimensional tree index used for secondary key retrieval with range searching.

An Example

The example 2-*d* tree organizes personal data using the dimensions of height (in inches) and weight (in pounds). Only the representative records, given in Table 12.1, and not all the data, are needed to build the index.

In building and processing the 2-*d* tree index, we alternate between the attributes of height and weight when deciding how to branch on successive levels of the tree.

We insert the two comparison fields for Sleepy at the root node, giving the initial 2-*d* tree of

Sleepy (36,48)

The __ indicates which field is used to compare against in each node. When inserting Happy's comparison fields, we compare his *height* of 34 inches with the height in the root node (that of Sleepy). Since the comparison is <, we follow a left branch. The current left branch is null, so that tells us to insert Happy's data to the left of the root. The 2-*d* tree is then

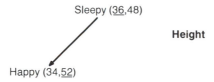

When we insert the data for Doc, we compare his *height* of 38 inches with that in the root. In this case, we have a > comparison, so Doc's data is inserted to the right of the root, yielding the tree

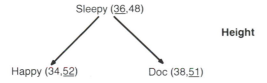

Since the 2-*d* tree is now full through level two, we know that the next insertion will require a level two comparison on weight. We then insert Dopey. His *height* of 37 inches gives a > comparison at the root node. We then compare his *weight* (the

TABLE 12.1 REPRESENTATIVE DATA RECORDS

Name	Height	Weight	Other data
Sleepy	36	48	
Happy	34	52	
Doc	38	51	
Dopey	37	54	
Grumpy	32	55	
Sneezy	35	46	
Bashful	33	50	
Ms. White	62	98	

second dimension) of 54 pounds with that of Doc's. We again have a $>$ comparison, so the data is inserted to the right. The tree is then

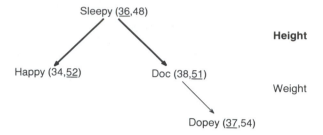

When inserting Grumpy's data, we compare his *height* of 32 inches with that in the root node. It is $<$, so we go left. At the next level, we are comparing on weight. His *weight* of 55 pounds is greater than that in the comparison node (that of Happy), so we insert to the right. The tree becomes

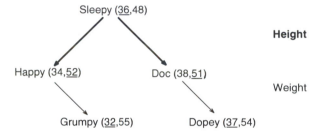

To insert Sneezy's comparison fields, we again alternate the field values we compare against. His *height* compares $<$ at level one; his *weight* is also $<$ at level two, so his data is inserted to the left of Happy's. Bashful's data is inserted to the left of Sneezy's data since his *height* of 33 inches compares $<$, his *weight* of 50 pounds compares $<$, and his *height* (used for a second comparison) is also $<$. The 2-*d* tree at this time is

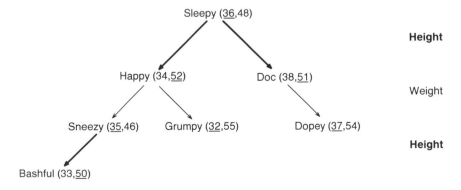

Ms. White's data is inserted at the extreme rightmost position since her height and weight are superior to those of the other individuals. The final tree appears in Figure 12.1.

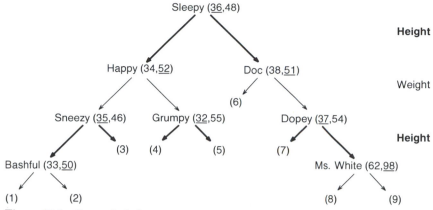

Figure 12.1   Example 2-*d* tree.

Interpretation

The 2-*d* tree in Figure 12.1 acts as an *index* into nine subdivisions or buckets for storing records using the dimensions of height and weight.

Note.   We do *not* store the entire records in the index nodes, only the secondary key values that we are comparing against.

The complete records are stored in the nine divisions referenced by the index. The divisions are labeled 1 through 9 in the 2-*d* tree diagram and appear where null links normally occur in nodes of degree less than two in a binary tree. Why are there nine divisions? You may have seen a theorem and/or a proof in your data structures course that stated that a binary tree of *n* nodes has $n + 1$ null links. Since we have eight nodes in this binary tree, there are nine null links or, in this case, nine divisions. Each node in the 2-*d* tree is actually dividing the two-dimensional search space into two parts: $\leq$ and $>$. As we move down the tree, the search space continually narrows. For instance, in the example, we have the dimensions of height and weight. The first node divides the search space *horizontally* along the height dimension at 36. We could represent that as

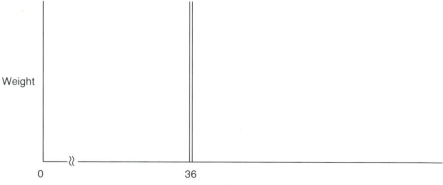

If we further subdivide the search space to the left of 36, we use a weight of 52 to divide it *vertically* into two subdivisions to give us the representation of

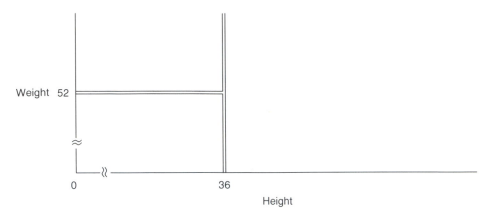

When we continue down the tree through Sneezy and Bashful, we add another *horizontal* and another *vertical* subdivision to obtain

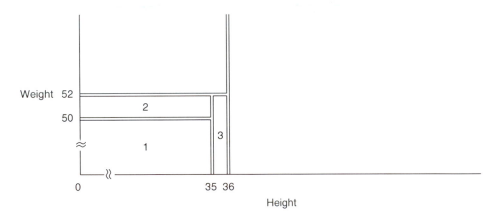

The pointers to the data pages in the example 2-*d* tree define an area corresponding to one of the subdivisions in the search space. Since the boundaries of each subdivision are contiguous, the records within the corresponding data page have similar characteristics. The three numbered and completely enclosed areas of the diagram have these associations and meanings:

- 1, left link of Bashful— $\quad 0 < \text{height} \leq 35$
  $\quad\quad\quad\quad\quad\quad\quad\quad\quad 0 < \text{weight} \leq 50$
- 2, right link of Bashful— $\quad 0 < \text{height} \leq 35$
  $\quad\quad\quad\quad\quad\quad\quad\quad 50 < \text{weight} \leq 52$
- 3, right link of Sneezy— $\quad 35 < \text{height} \leq 36$
  $\quad\quad\quad\quad\quad\quad\quad\quad\quad 0 < \text{weight} \leq 52$

Implementation

In choosing the records for constructing the index, the root record should have a near median value for the dimension being compared against. By doing that, approximately half of the records are sorted to the left and half to the right. A similar strategy should be followed on the second dimension at the next level.

By knowing the file size and the bucket size in advance, one can estimate how many representative records are needed for constructing the index. The number of representative records, which equals the number of nodes in the 2-*d* tree, is one less than the number of data buckets desired. Since only one field is compared against in each node, it is sufficient to store only that one field, for example, the one with the ___ in the example.

Storage

Each record is passed through the index to determine its storage location. For simplicity, we assume that only the representative records have been stored. Which of them are mapped into each of the nine divisions? The mappings are

Sleepy	3
Happy	2
Doc	6
Dopey	7
Grumpy	4
Sneezy	1
Bashful	1
Ms. White	8

Retrieval

If we want to locate all true dwarfs, where a true dwarf is defined to be someone less than 37 inches in height and less than 50 pounds in weight, we can search the 2-*d* tree in Figure 12.1 to locate the desired records. As we step through the 2-*d* tree making the appropriate decisions at each node, we narrow the final search space to divisions one and three. At Sleepy (the root node) we take a left branch since the desired height is the same as Sleepy's; at the Happy node since the search weight of 50 is < 52 we again go left. At Sneezy, the search value of 37 inches is not entirely in the left or right branch, so we follow *both* branches. The right branch leads to division three, so we are done along that path. We continue along the left path and take a left branch at Bashful since the search weight of 50 pounds is equal to that stored in the node. The left branch leads to division one. We have limited the search to those records in divisions one and three. If we limit ourselves to the representative records, of the two records in division one, we find that only Sneezy is a true dwarf according to the definition. The record for Bashful in division one may be viewed as something akin to a false drop. Since the search bounds did not coincide with those of division one, there was a possibility that some of the records in that division did not meet the search requirements, so each record had to be checked to eliminate those which did not. But only those records in the data pages identified by passing the search values through the index need to be searched exhaustively, rather than the alternative of *all* the records in the file. In division three, we note that Sleepy is also a true dwarf.

Discussion

The *K-d* tree is another filtering mechanism for narrowing a retrieval search. We are again able to *improve* the search performance by *reducing* the amount of data that needs to be searched exhaustively. A *K-d* tree is appropriate for those situations in which there is not a primary key field that predominates the retrievals, for example, when searching for houses with certain attributes. It is also applicable where range searches are prominent.

In addition to a *K-d* tree being used as the sole data organization for a file, it could be used as an alternative to partial match retrieval considered in Chapter 6. When using it with a separate primary key organization, the buckets pointed to by the *K-d* index would be pointer pages similar to those in Figure 6.7.

A potential disadvantage of *K-d* trees is that all retrieval fields may not be evenly distributed, so the resulting index structure may degenerate as the value for *K* increases, which would in turn decrease its performance effectiveness. If the number of overflow pages in a file created with a *K-d* tree index becomes excessive, it would be necessary to reorganize the index with a different set of representative records and reinsert the data.

KEY TERMS

2-*d* tree	Index
False drop	Range searching
Filtering	Secondary key retrieval

EXERCISES

1. **(a)** Which of the nine divisions of the 2-*d* tree in Figure 12.1 does the record for Prince Charming map into? He is 78 inches high and weighs 240 pounds. **(b)** Repeat the question for Clever, who is 36 inches high and weighs 53 pounds.

2. Using the 2-*d* tree in Figure 12.1, which buckets (data pages) need to be searched exhaustively to find all average people where average is defined to be a height < 60 inches but > 36 inches and a weight < 100 pounds and > 51 pounds.

3. Construct a 2-*d* tree of depth three using the following data with gpa as the first dimension (compare against a record with a 3.0 gpa) and age as the second dimension, etc. In choosing age values for comparison, attempt to equalize the number of records in a terminal bucket; likewise for the second use of gpa. Note which records are inserted into each bucket.

Name	Age	gpa
Joe	17	3.5
Mary	17	2.0
Bill	18	2.2
John	18	3.2
Sam	19	2.8
Denise	19	3.8
Jeff	19	3.1
Don	20	3.0
Pat	20	2.0
Joan	27	2.6
Jenny	20	2.85
Roland	20	3.7
Sharon	21	1.6

4. **(a)** How many representative records are needed to build a 3-*d* tree index into a file of 1000 records? Assume that the index provides an even distribution and that each data page holds 10 records. **(b)** How deep will the resulting index be?

5. Can one representative record be used to set the comparison value in more than one node in a *K-d* tree? Explain.

6. What are the advantages of a *K-d* tree for primary key retrieval as compared with a hashing technique? What are the disadvantages?

7. How does the searching of a *K-d* tree index compare for an unsuccessful search vs. a successful one?

8. What modifications to a 2-*d* tree file would be needed if one of the attribute values of a representative data item changed, for example, what if Sneezy grew by 2 inches?

GRID FILES

The grid file [2] is another data structure for providing efficient retrieval of records with multiple attributes. Like the *K-d* tree, it accomplishes this goal by limiting the search space that must be considered for retrieval. It is another example of the principle of *doing it faster by doing less.* We postponed consideration of the grid file until now, rather than in Chapter 6 with other multikey retrieval techniques, because of its relationship to *K-d* trees, extendible hashing, and dynamic hashing. In the example with the 2-*d* tree, the search space was represented by a rectangle that continually shrank as we correspondingly moved down the tree index. As the area of the rectangle became smaller, we had fewer data items to consider as likely solutions to the query. In the general case, the search space of a grid file may be considered as a multidimensional box in which each dimension represents an attribute for possible retrieval. A two-dimensional instance is a rectangle just as with the 2-*d* tree. A point in the space represents the records with the set of values that intersect to form that point. Figure 12.2 represents this concept with three dimensions; the point shown represents all people with a certain height, weight, and age. What we need to describe then in what follows is how we can answer a point query, that is, access such a point (or points for a range query). In other words, we need to characterize what form the index or directory structure should take.

> A **grid file** organizes data into buckets or pages corresponding to the intersection points of a multidimensional box. An index provides access to the individual buckets.

The multidimensional search space such as that in Figure 12.2 may be considered a bitmap structure in which a "1" indicates that at least one record exists that contains the values forming the coordinates of that location and a "0" indicates the absence of such records. Let's use the three dimensions of height, weight, and age from Figure 12.2 with the assumptions that the height axis has values from "0" to "90" inches, that the weight axis has values from "0" to "500" pounds, and that the age axis has values from "0" to "100" years. How dense would you expect the search space to be if the data was students at your university in your major? Not very dense at all, for there are over $4\frac{1}{2}$ million points in this search space. Most of the points would be zero because both the amount of data is much less and the data are not evenly distributed, for example, you are unlikely to find a 90-year-old computer science major who is 6 feet 8 inches and weighs 350 pounds. We have the same situation as we encountered with sparse matrices in data structures. It is obviously inefficient to represent all possible points in the search space. We need a compression technique. One way to limit the storage structure to the nonnull points is to have an index into them.

Let's next consider the storage and index structures. The records are stored in buckets or pages; the number of such buckets is allocated only as needed. Typically these buckets contain between 10 and 1000 records. Until we go through an example, though, we do not need to specify a bucket size. For simplicity, we treat the search space as having two dimensions, those of height and weight. As records are added and storage buckets overflow, the search space is partitioned into smaller units. Initially assume that

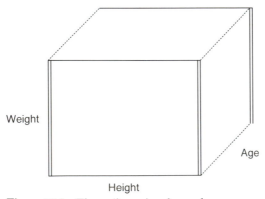

Figure 12.2 Three-dimensional search space.

there exists one large partition of the search space. We refer to the segments of a partitioning as *grid blocks*. Initially, then, there is one grid block containing a pointer to a bucket, for example,

Search space

All the records in the file, even though they have quite different attributes, are stored in this single bucket. Notice similarities between this partitioning process and the splitting process in extendible hashing. When additional records are added to the extent that the bucket overflows, the grid block needs to be partitioned along one of its dimensions. Each grid block contains a pointer to a bucket containing data corresponding to that section of the search space. Each block points to only one bucket, but a bucket may have more than one pointer to it. When we divide the grid block, which occurs at the midpoint of one of the dimensions, in this case, "height," we obtain

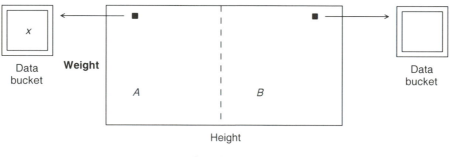

Search space

The dimension selected for partitioning is usually chosen in a cyclical manner. With only two dimensions, we alternate between "height" and "weight," just as we did with the *K-d* tree example. Since there are two buckets and two grid blocks, when a bucket overflows, another partitioning is needed. This time we partition on "weight." The search space now has four partitions, but storage is not needed to create two new buckets, so at this stage, we have two grid blocks pointing to the same bucket. We partition two grid blocks but split only one bucket. Let's assume that bucket x overflowed. We need more space for records in what had been partition A but not for those in partition B. Therefore, a single bucket may be used for the new partitions B and D. The structure after the partitioning appears as

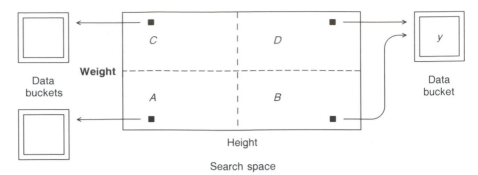

If data bucket *y* subsequently overflows, we might be able to add a new bucket without another partitioning of the search space (it would depend upon how the data values mapped into the search space). After the addition of a fourth bucket, the structure might appear as

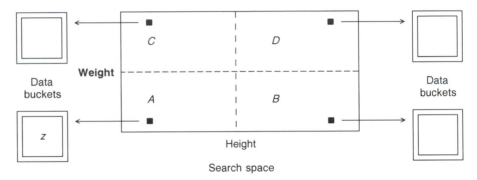

Let's assume that bucket *z* overflows next; we then need to partition the search space along the "height" dimension. The resulting structure appears in Figure 12.3.

Again we have fewer buckets than grid blocks.

What we need is an index to map the grid blocks of the search space to the appropriate buckets. A two-dimensional *grid array* that provides a level of indirection similar to that of the index of extendible hashing is used to perform this function. It is one of the two components of the *grid file index.* The other component is a set of *linear scales,* one for each dimension, for converting an attribute value to the correct element in the grid array. For Figure 12.3, we need a linear scale for "height" with three intervals and a linear scale for "weight" with two intervals. Using the previous ranges for attribute values and the partitioning of Figure 12.3, the height linear scale is [22, 1; 45, 2; 90, 3] and the weight linear scale is [250, 1; 500, 2]. The ranges for the linear scales should be wide enough to accommodate all possible data records. Each element pair in a linear scale represents the {*maximum value for the interval, the interval number*}. Figure 12.4 represents the grid file index.

The pointers in the grid array point to the data buckets, and the underlined numbers adjacent to the linear scales are the interval numbers. In general, the grid index is large enough that it needs to be kept on auxiliary storage but the linear scales could

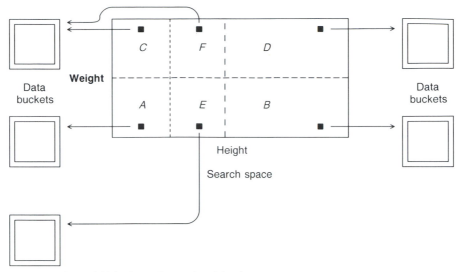

Figure 12.3 Grid blocks and associated buckets.

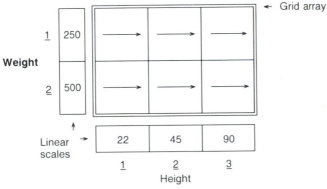

Figure 12.4 Example grid file index.

probably be kept in primary memory. Then only two disk accesses are necessary to make a point query: one to the grid array and one to the data bucket. To locate individuals who are 6 feet tall and weigh 300 pounds, the height linear scale sends us to interval 3 and the weight linear scale sends us to interval 2. These coordinates then direct us to the element of the grid array in the lower right-hand corner. From the grid array, we can go immediately to the bucket containing the desired records. A bucket may be organized in any manner; if the number of records in a bucket is relatively small, a sequential organization with a subsequent linear search may suffice. As more records are added to the file, the grid file index would expand until it might appear as in Figure 12.5. A linear scale for "weight" might be [125,1;156,2;187,3; . . . ; 375.9; 500,10].

The basic structure of a grid file is then quite simple, as illustrated in Figure 12.6. Like so many structures that we have considered previously, it consists of an index into data. The data is organized into buckets based upon the attributes of the records; a

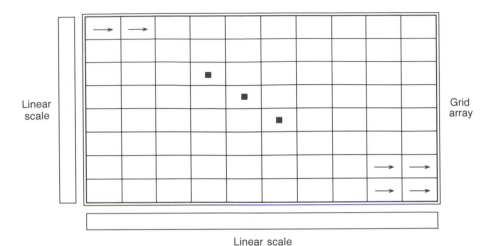

Figure 12.5 General grid file index for two dimensions.

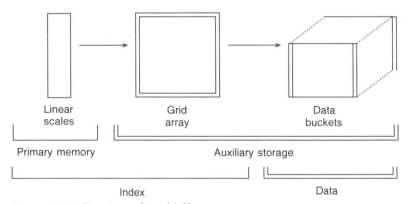

Figure 12.6 Structure of a grid file.

two-stage index allows access to an answer bucket via a path from the desired attribute values to an entry in a multidimensional array which in turn points to the data bucket.

An Example

To better understand grid files, let's work through an example using the data with the attributes of height and weight from Table 12.1. In order to illustrate the partitioning and splitting mechanisms, we use an unrealistically small bucket size of two. Since we know that the data has a limited range of values, we set the height scale to range from 0 to 70 inches and the weight scale to range from 0 to 100 pounds. We insert the records in the order that they appear in the table. We initially insert the records for Sleepy and Happy without partitioning. The grid file is then

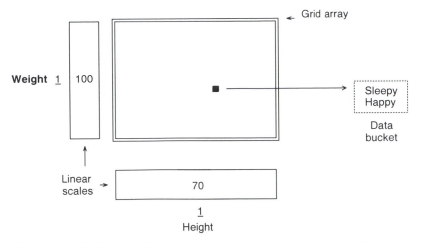

When we add the record for Doc, an overflow situation occurs in the one data bucket. We need to split it and partition the search space. We arbitrarily choose to split on height first and then we alternate dimensions for subsequent partitionings. Remember that we partition the search space independent of the data values. If we choose the midpoint of the height dimension, the linear scale is [35,1; 70, 2]. The record for Happy then maps to the first interval and those for Sleepy and Doc go to the second one. The grid file is then modified to

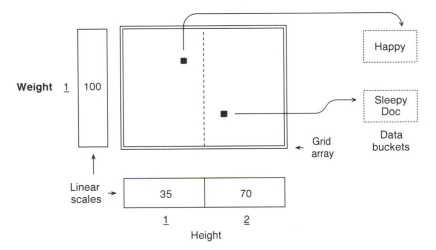

The record for Dopey is routed to the bucket already containing Sleepy and Doc. We then need to split that bucket, and since it is pointed to by a single grid block, we also need to repartition the search space. This time we use the weight dimension to yield a linear scale of [50,1; 100,2]. Rather than create *two* new buckets such that there would then be one corresponding to each grid block, we create only a single bucket, since that is all the additional space that we need. Two grid blocks will point to the bucket with the record for Happy. The grid file then becomes

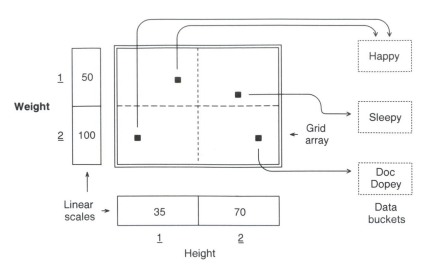

The record for Grumpy maps to the bucket with Happy, which has space for another record so a splitting is not necessary. Sneezy maps to the same bucket which now contains Happy *and* Grumpy so we need to split. We need to redistribute the records for Sneezy, Happy, and Grumpy; again note the similarity with extendible hashing. Since this bucket has multiple pointers to it, it may be possible to perform a splitting without a partitioning of the grid array. Two of the records occur in one grid block and the other in a different grid block, so the splitting may occur *without* a partitioning. The resulting grid file is

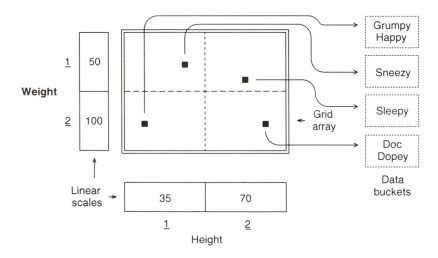

The record for Bashful maps to the bucket with Sneezy which has space available so a splitting is not required. Finally, the record for Ms. White is added, causing an overflow in the bucket containing Doc and Dopey. Since each bucket is now pointed to by just one grid block, both a splitting operation and a partitioning are needed. We partition the height interval that is causing the overflow, that is, the interval between

36 and 70. The revised linear scale is then [35,1; 52,2; 70,3]. This partitioning converts two grid blocks into four. But as in a previous partitioning, we do not need buckets corresponding to all four grid blocks. The final structure for the grid file appears in Figure 12.7.

Discussion

If it were possible for a bucket to have more records with an identical set of associated attributes than its capacity, then a special overflow procedure would be needed. This overflow situation is not to be confused with the one basic to grid files in which records with unlike attribute values cause an overflow. A special overflow procedure would dictate more than two accesses per point query since an overflow bucket would need to be accessed to answer a query. For the remaining discussion, we assume that a bucket's capacity is large enough to hold all records with identical attribute values.

To answer the point query of finding all people who are 50 inches high and weigh 50 pounds, we need to consider only one grid block. With the example data, the interval values for that grid block are height two and weight one. The grid array directs us to the bucket containing Sleepy. After retrieving the record for Sleepy, we find that his attributes do *not* match those sought in the query. What we have here then is an *unsuccessful search.* For an unsuccessful search of a point query, as with a successful search, we need to make two accesses of auxiliary storage: one to the grid array and the other to the bucket. We observed the characteristic of requiring the same number of accesses for both successful and unsuccessful searches as a feature of the direct organization methods that we considered for expandable files.

To retrieve all true dwarfs from the example grid file in Figure 12.7, as we did with the 2-*d* tree example previously, we need to use the linear scales to locate all height intervals less than 37 inches and all weight intervals less than 50 pounds. Those are

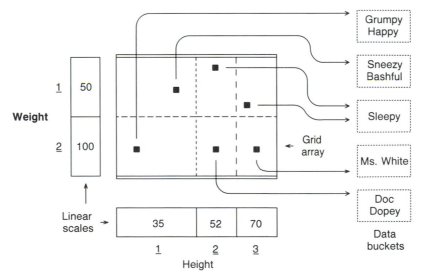

Figure 12.7 Example grid file.

height intervals one and two and weight interval one. These linear scale interval values direct us to two elements in the grid array that point to the bucket with the records Sneezy and Bashful and the bucket with the record for Sleepy. Since the boundaries of the grid blocks do not correspond with the boundary values of the range query, it is necessary to check each record to determine if it does indeed meet the query requirements. This is something like checking for false drops. Since this is a range query rather than a point query, it is necessary for us to examine more than one bucket; we have to make three auxiliary storage accesses to answer this query: one to the grid array and two to the data buckets. After performing the checking, we find that Sneezy and Sleepy are the only true dwarfs in this example database. Notice that we obtained the same answer with grid files as we did with 2-*d* trees. We now have two new methods for performing multikey retrieval; these two methods have similarities but also significant differences.

All searching techniques may be considered either as ones that organize the data or ones that organize the embedding space into which the data are placed. The *K-d* tree index organizes the data based upon a portion of their values. In contrast, the grid file partitions the embedding space without regard to a specific set of data and in so doing overcomes two potential disadvantages of *K-d* trees. Depending upon the distribution of the data, the *K-d* index may become unbalanced, which may necessitate more nodes in a retrieval path than with a balanced index; this longer path in turn will degrade performance. Second, once the *K-d* index becomes established, it does not change as new records are added. If many additions map to one terminal node in the index, an overflow handling procedure is needed. If overflow pages are added, the number of accesses needed to retrieve an item will increase. A grid file eliminates the possibility that retrieval on one attribute will take much longer than on any other; it also suggests that retrieval performance will not be degraded as a result of records being added to the file. An argument in favor of the *K-d* tree is its simplicity. A grid file is characterized as being *symmetric* and *adaptable*. Symmetric means that each key is treated as the primary key, and adaptable suggests that as records are added, the partitions will change and conform to the current set of data.

The grid file has a data storage utilization of $\approx$ 70 percent.

Grid files have an advantage over B+-trees for concurrent processing applications, for example, an airline reservation system. In a grid file, the access paths for different queries are disjoint; whereas, with a B+-tree, all inquiries must go through the same root node. With a B+-tree in a concurrent system, when an insertion needs to modify the root node, all other processes need to be locked out until the root modification is completed. The alternative is a complicated locking protocol.

Deletion

How do we handle deletions? What do we need to do other than removing the deleted record from its data bucket? With deletion being the converse of insertion, we have to consider how to handle the reverse of the partitioning and splitting operations. Partitioning involves the index. We usually ignore modifying the index because of the effort needed to do so, and the fact that a file that is not decreasing in size will require an index of the current proportions. An exception is a shrinking file. We made a similar decision with the index in B+-trees for essentially the same reasons. We will, however, be interested in the reverse of the bucket splitting operation: a bucket merging process. This merging process is quite similar to that in extendible hashing. When a block splits to form two blocks, those two blocks are referred to as "buddies." When the capacity of a block decreases below some threshold, a check is made of its buddy block to see whether its contents and those of the current block can be combined. This type of merging requires that two blocks can be combined only if their corresponding grid blocks were formed from the same parent grid block. Other procedures exist for combining buckets from other "neighboring" grid blocks, but they are more complicated. See [2] for details.

Implementation

Regnier [3] observed that depending upon how the grid file index is implemented, the grid file may be considered as a generalization of either dynamic hashing or extendible hashing. As a result, Regnier refers to the grid file algorithms as multikey dynamic hashing or multikey extendible hashing. Let's compare these two implementations. In this discussion, we have used the mechanism that is an extension of dynamic hashing. Figures 12.8a,b,c, and d illustrate the expansion of the grid array leading up to Figure 12.3. The numbers represent pointers to the data buckets; duplicate numbers represent multiple pointers to a bucket. Figure 12.8e shows the form of the grid array if an additional partitioning of the lowest interval on the weight dimension were to occur. The expansion of the grid array is continual and gradual just as is the index for dynamic hashing. This contrasts with the expansion of the index for extendible hashing, which is periodic and abrupt. For one dimension, the expansion of the grid file index as we have described it would be like that of a dynamic hashing index. Figure 12.9 represents the multikey extendible hashing method for implementing the grid file index in which the size of the grid array doubles each time the shortest interval is split. During the doubling process, a large number of pointers must be set, just as with extendible hashing. The grid array in Figure 12.8e has nine entries, whereas the one in Figure 12.9e has 16. With the multikey extendible hashing method of index expansion, considerable activity occurs during a doubling; then no partitioning is required again until another doubling. For one dimension, the index expansion for multikey extendible hashing is equivalent to extendible hashing. You can observe that with either dimension in Figure 12.9. A disadvantage of this scheme is that when the data is not uniformly distributed, the index may become excessively large. In studies on performance [3], Regnier con-

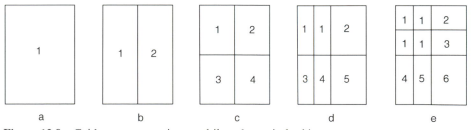

         a   b   c   d   e

Figure 12.8 Grid array expansion; multikey dynamic hashing.

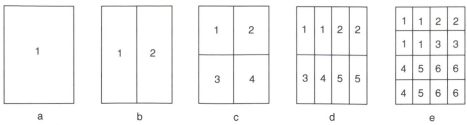

Figure 12.9   Grid array expansion; multikey extendible hashing.

cludes that multikey dynamic hashing should be used when the bucket size is less than 20 or when the distribution is not uniform. The advantage of multikey extendible hashing is that it is readily implemented.

Other implementation considerations are given in [2] and [4].

KEY TERMS

Bitmap structure	Grid array
Bucket splitting	Grid block
Grid file	Grid file
Buddy system	Index
Concurrent processing	Grid file
Grid file	Linear scales
Dynamic hashing	Multikey retrieval
Multikey	Partitioning
Expandable file	Grid file
Extendible hashing	Point query
Multikey	Range query

EXERCISES

1. When inserting into a grid file, does a splitting operation always require a partitioning operation?

2. How many points are in the search space for the example with the records from Table 12.1 (all attribute values are integers)?

3. Give two similarities between grid files and B+-trees.

4. Give a major difference between grid files and partial match retrieval using page descriptors.

5. Using the example grid file of Figure 12.7, how many buckets (data pages) would need to be searched exhaustively to find all average people where average is defined to be a height < 60 inches but > 36 inches and a weight < 100 pounds and > 51 pounds.

6. What effect does a nonuniform distribution of field values have on the size of a grid file index (as compared with a uniform distribution)? Explain.

7. Construct a grid file using the following data with gpa as the first dimension to partition on and age as the second dimension, etc. Let the gpa dimension range from 0 to 4.00 and the

age dimension range from 10 to 30. Assume that each bucket has a capacity of two records. Note which records are inserted into each bucket of the grid file.

Name	Age	gpa
Joe	17	3.5
Mary	17	2.0
Bill	18	2.2
John	18	3.2
Sam	19	2.8
Denise	19	3.8
Jeff	19	3.1
Don	20	3.0
Pat	20	2.0
Joan	27	2.6
Jenny	20	2.85
Roland	20	3.7
Sharon	21	1.6

8. Insert the example data, that from Table 12.1, into a grid file using the multikey extendible hashing method for implementing the index. Show the contents of the grid index and the buckets after each splitting or partitioning.

REFERENCES

1. Bentley, Jon L., and J. H. Friedman, "Data Structures for Range Searching," *Computing Surveys,* Vol. 11, No. 4 (December 1979), pp. 397–409.
2. Nievergelt, J., H. Hinterberger, and K. C. Sevcik, "The Grid File: An Adaptable, Symmetric Multikey File Structure," *ACM TODS,* Vol. 9, No. 1 (March 1984), pp. 38–71.
3. Regnier, Mireille, "Analysis of Grid File Algorithms," *BIT,* 1985, pp. 335–357.
4. Hinrichs, Klaus, "Implementation of the Grid File: Design Concepts and Experience," *BIT,* 1985, pp. 569–592.

PART SUMMARY

In this part we explored a variety of tree structures for numerous purposes. The tree structures provided us with *flexibility, search efficiency, naturalness of representation,* and *power of representation.* In summary, we want to review the many techniques and principles studied independent of the particular structure in which they were introduced. The individual data structures may come and go but the general techniques and principles will have more lasting value.

TECHNIQUES

- Subdivide a search space into many parts so that only a portion of them need to be considered when searching (e.g., dynamic hashing, length-segmented lists, grid files).
- Compute information when there are storage limitations. (e.g., dynamic hashing with linear splitting).
- Spend additional time structuring the information during insertion to reduce the retrieval time; store once, retrieve often (e.g., AVL trees).
- Realize that an improvement in space utilization does not always degrade performance (e.g., B#-trees).
- Use a level of indirection to gain flexibility in accessing information (e.g., extendible hashing, grid files).
- Eliminate unnecessary decisions (e.g., PATRICIA trees).
- Consider storing information according to different characteristics (e.g., frequency order vs. alphabetical order—digital search trees).

PRINCIPLES

- Be flexible when deciding what is fixed or variable in a procedure (e.g., the use of multiple hashing functions in linear hashing).
- Expand your thinking beyond what is feasible with today's technology (e.g., the spelling corrector using tries).

- Continue searching for improvements to a technique even though it has been widely used for some time (e.g., IPR trees vs. AVL trees).
- Apply known techniques in new environments (e.g., B-tree page splitting in hashing dynamic files).
- Identify deficiencies in a procedure and eliminate them (e.g., bounded index exponential hashing).

PART EXERCISES

1. Which of the following collision resolution methods uses the least amount of storage for the index? Extendible hashing; bounded index exponential hashing; dynamic hashing; dynamic hashing with linear splitting; linear hashing; trie hashing. Explain.

2. Which of the collision resolution methods mentioned in Problem 1 requires the fewest average number of probes of secondary storage (assuming that the index is kept in primary memory)?

3. For each data structure given in the column on the left, select its distinguishing feature(s) from the characteristics given in the column on the right. (An answer from the right column may be used more than once.)

Data structure	Distinguishing feature
_____ a. B+-tree	i. Any number of elements may be added to the file without reorganizing it.
_____ b. B-tree	
_____ c. Computed chaining	ii. Page splitting occurs evenly rather than oscillatorily.
_____ d. Extendible hashing	
_____ e. Bounded index exponential hashing	iii. The right and the left siblings are checked for space before a page splitting occurs.
_____ f. Trie hashing	iv. Does not require an index.
_____ g. Dynamic hashing with linear splitting	v. Retrieval performance is independent of the key distribution.
	vi. The data may be readily accessed directly or sequentially.
	vii. Only a portion of the index needs to be stored.
	viii. The entire record is stored in the index.

PART FOUR

FILE SORTING

PREVIEW

A common way in which to organize a file is to order it, or *sort* it, based upon the values in a field of the records in the file. This type of ordering is immensely helpful to us humans when we want to search for the data associated with a particular key. Can you imagine a dictionary in which the words were *not* ordered alphabetically? It would be next to impossible to locate a word. With a dictionary being ordered, we know approximately where to begin the search: at the front, back, or middle. After our first approximation, because of the ordering, we know whether to move in a forward or a backward direction to locate the sought word. We can usually locate the word without looking at many pages. This process is similar to the interpolation search that we considered in Chapter 2. Most computer output for human consumption is ordered in some fashion. Sorting, or the ordering of data, by a computer is a common operation. Knuth[1] noted that computer manufacturers estimate that 25 percent of the running time of computers is spent on sorting.

An ordering of data is a basic component of many data structures. Many of the file organizations that we have examined required an ordering of the data for them to function properly, for example, the binary search. As mentioned during the discussion of Bloom filters, ordering data is also a technique that can be used to eliminate duplicates in a list; if duplicates exist, they must then be in adjacent positions in a sorted list. Another application of sorting occurs when we wish to check if one file of records contains another. An ordering of the two files simplifies the task since we then need to examine the data records only once. A variation of this task occurs when we wish to determine which records from one file are contained in another, for example, if we wish to know how many students in course *A* are also taking course *B*.

To some extent, because of the great popularity and importance of sorting, we consider sorting algorithms *less* than some of the other file organization techniques. That doesn't seem logical. If sorting is so important and popular, shouldn't we discuss it in proportion to its usage? One reason is that some files organizations maintain the data in sorted order so explicit sorting is not needed. However, the primary reason that

[1]Knuth, D. E., *Sorting and Searching,* Addison-Wesley, Reading, MA, 1973.

we study it less is that every computer system and many software packages have *built-in* sorting procedures. So when a user needs to order data, he merely invokes a system sort, rather than writing his own. On the other hand, though, most of the system sorting procedures are general-purpose and may require considerable system overhead to access and use them. So there are a few occasions, for example, for a relatively small number of records, when it is useful for a computer user to be able to develop his own sorting procedure.

The reasons then for considering sorting algorithms are to

- Structure data for use with another file organization technique.
- Structure data in situations in which it is too costly to invoke a system sort.

We have already considered one technique for ordering data; that was the binary search tree. With it, we can obtain an ordering of data by traversing its nodes in inorder. This technique, however, is only suitable if the binary search tree existed or were to be used for another purpose. Building it only for sorting requires considerable storage. In the remainder of this part, we look at more efficient techniques that are intended solely or primarily for sorting. Most of these techniques will sort data internally, that is, perform the operation entirely in primary memory. For large files, however, an external sort, or one that uses auxiliary storage, may be necessary. External sorting will be discussed near the end of this part.

Chapter 13

Sorting

INSERTION SORT

One of the simplest sorting procedures is the *insertion sort*. You may have seen or programmed it in an assignment in a previous course, and it is one that you have most likely used in your ordinary activities.

> An **insertion sort** orders records by placing each one individually into its proper position in an ordered list of previously inserted records.

This is the procedure that we are likely to follow in processing a hand of playing cards. Or it may be how we organize our records or cassettes. Each time we receive a new one we insert it into its proper position, so that our collection remains ordered.

If we have a file of records to be ordered, a simple strategy to follow is to process each of the records of the file as if it were being reinserted into an ordered file of previously reinserted or processed records. We first locate the record with the lowest key, assuming that we will sort the records into ascending order. That record is then swapped with the first record in the file. Then we process each of the remaining records, from position two forward, as if it were being reinserted into an existing ordered file. The portion of the records that have already been processed constitutes the sorted file. The key for the next record processed or reinserted is compared with that of its left adjacent neighbor, and if it is less than, the two records are swapped. The key of the record that has been swapped to a lower position is in turn compared with *its* left adjacent neighbor. This comparing and swapping continues until the left member of the most recently swapped pair of records is greater than its left adjacent neighbor (the proper position of the reinserted record has been found) or until the beginning of the file is encountered. Algorithm 13.1 defines the basic procedure.

The record with the lowest key is initially swapped into the first position of the array to ensure that the records preceding (to the left of) *position* are always in sorted order. In other words, it is as if we are always inserting into an ordered list. This initialization also ensures that the algorithm will execute properly in most programming languages (see Problem 3 at the end of this section).

**Algorithm 13.1
INSERTION SORT**

proc insertion_sort

       /* The *n* elements of the file are located in positions 1 through *n*. */

1   swap the record in position one of the file with the record with the lowest key.

       /* Move the next record into its ordered position */

```
2       for position = 2 to n do
3           current := position
4           while current > 1 and
5               key[current−1] > key[current] do
6                   swap(record[current],record[current−1])
7                   current := current − 1
8           end
9       end
end insertion_sort
```

An Example

We use our standard data that we have processed so many times previously, that is, records with the keys

27, 18, 29, 28, 39, 13, 16, 38, and 53

The first step is to locate the record with the lowest key and move it into position one. The data then appears as

13, 18, 29, 28, 39, 27, 16, 38, and 53
∧

where ∧ indicates the location of *position* (the record being "reinserted" into the file) and ♦ the location of *current* (the left member of a swapped pair). The data is then sorted as

13, 18, 29, 28, 39, 27, 16, 38, and 53.
∧
13, 18, 29, 28, 39, 27, 16, 38, and 53.
∧
13, 18, 28, 29, 39, 27, 16, 38, and 53.
♦ ∧
13, 18, 28, 29, 39, 27, 16, 38, and 53.
∧
13, 18, 28, 29, 39, 27, 16, 38, and 53.
∧
13, 18, 28, 29, 27, 39, 16, 38, and 53.
♦ ∧

13, 18, 28, 27, 29, 39, 16, 38, and 53.

13, 18, 27, 28, 29, 39, 16, 38, and 53.

13, 18, 27, 28, 29, 39, 16, 38, and 53.

13, 18, 27, 28, 29, 16, 39, 38, and 53.

13, 18, 27, 28, 16, 29, 39, 38, and 53.

13, 18, 27, 16, 28, 29, 39, 38, and 53.

13, 18, 16, 27, 28, 29, 39, 38, and 53.

13, 16, 18, 27, 28, 29, 39, 38, and 53.

13, 16, 18, 27, 28, 29, 39, 38, and 53.

13, 16, 18, 27, 28, 29, 38, 39, and 53.

13, 16, 18, 27, 28, 29, 38, 39, and 53.

The file is now sorted. Observe that upon completion of the **while** loop in the algorithm, the records to the left of the *position* pointer are in sorted order.

Discussion

The insertion sort requires less effort to sort partially ordered files because fewer swaps and comparisons are necessary. In fact, a file of n records that is already in sorted order requires only $n - 1$ compares and no swaps in processing line 2 of the algorithm.

In general though, the algorithm is not that efficient. In fact, the worst case computational complexity of the algorithm is $O(n^2)$, which is inefficient for practical applications when n becomes large. In this case, it is easy to grasp how the computational complexity is obtained. The **for** loop in the algorithm (line 2) is executed $n - 1$ times, and each execution of the **while** loop (line 4) contained therein may cause (*position* $- 1$) comparisons. Therefore, the maximum number of comparisons is

$$\sum_{position=2}^{n} (position - 1) = \frac{n^2}{2} - \frac{n}{2}$$

Removing constants, the time complexity is then $O(n^2 - n)$, which becomes $O(n^2)$. Unfortunately, the average performance is also $O(n^2)$. But for small n, because of its simplicity, it is a useful algorithm. It can also be used as a component of more complex algorithms; for example, when a larger file is continually subdivided to the point where n is small, then the insertion sort can be invoked. Knuth [1] discusses several variations of the insertion sort, and Bentley [2] considers techniques for implementing it efficiently.

KEY TERMS

Binary search tree   Insertion sort
      Sorting

EXERCISES

1. Sort the records with the keys

    73, 15, 44, 37, 30, 59, 49, and 99

    using the insertion sort algorithm. Show the contents of the file and the location of *position* and *current* after each change.

2. (a) How many comparisons were required to sort the data in Problem 1? (b) How many swapping operations were required?

3. (a) What happens in many programming languages if Algorithm 13.1 is implemented such that the record with the lowest key is *not* moved into the first position prior to executing the **for** loop in line 2? (b) Why can't we just sort all n records beginning with line 2 of the algorithm?

4. Given the months of the year in calendar order, sort them using an insertion sort. How many swapping operations are required?

5. (a) What is the minimum number of comparisons (excluding those needed to locate the record with the smallest key) needed to perform an insertion sort on n records? (b) the minimum number of swaps? (c) the maximum number of compares? (d) the maximum number of swaps? (e) How would you characterize a file that requires the maximum number of swaps?

6. What is the relationship between the number of compares and the number of swaps in the **while** loop of line 4 in Algorithm 13.1 when sorting a file of n records?

QUICKSORT

One strategy for processing information that we have seen repeatedly in this text is one of *divide and conquer*. Rather than dealing with the entire set of information as a single unit, we divide it into smaller groups that we can process more efficiently. That strategy carries over to sorting when performed with the quicksort [3].

> **Quicksort** orders the records of a file by partitioning them into subgroups based upon a comparison with a selected member of the file. It then recursively sorts each subgroup.

When a subgroup contains only a single item, the sorting operation terminates on that subgroup. When the sorting process has terminated on all the subgroups, the entire file is sorted. The groups of data are subdivided based upon a comparison with the first

Algorithm 13.2
SWAPPING RECORDS IN QUICKSORT

proc   quicksort__swap

1   DIVIDER := location[first record of file or subfile]
2   **for** position = 2 to *n*
3   **if** key[position] < key[first] **then**
4   DIVIDER := DIVIDER + 1
5   swap(record[DIVIDER],record[position])
6   **end**
7   swap(record[first],record[DIVIDER])
end quicksort__swap

record in each subgroup. All the records with key values less than the key of the first record are placed in one group (the lower group) and all the records with key values greater than the first record are placed into another group (the upper group), that is,

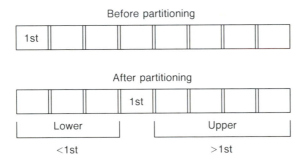

This partitioning of the records requires that some records be exchanged or swapped. Perhaps for that reason, this sorting method is also called the partition-exchange sort. How should the exchange operation be implemented? Bentley [2] argues for a technique suggested by Nico Lomuto of Alsys, Inc., which, although not the fastest procedure, is simple. *Simplicity* is certainly an important design criterion when developing software, especially if it is to be maintained or modified by others. Lomuto's procedure keeps a pointer, DIVIDER, to separate those records that have keys that are less than the key of the first record, that is, the separator key, and those that are greater. As it scans the data (excluding the first record) sequentially comparing each key with the separator key, when the comparison is greater than or equal, it does nothing. When the key is less than, however, that means that the associated record needs to be moved into the lower group. To accomplish this, the DIVIDER pointer is incremented by one so that it points to the first record of the upper group. That record of the upper group and the record that compared less than are then swapped. DIVIDER then continues to separate records in the lower group from those in the upper group. This swapping procedure is described in Algorithm 13.2.

In the following diagram, ∧ represents DIVIDER, < represents those records

Algorithm 13.3
QUICKSORT ORDERING

I. If multiple data records are to be sorted, partition them into three groups using the "quicksort__swap" procedure: [those records with keys that are lexicographically less than the first record, the first record, and those records with keys that are lexicographically greater than the first record],

    A. Quicksort the data records in the lower group into three parts.

    B. Quicksort the data records in the upper group into three parts.

II. Else terminate.

---

that have keys lower than the first, $>$ represents those records that have keys greater than the first and ? represents those records that have not yet been scanned; | points to the current record scanned (the variable *position*).

If the current record has a key that is less than the first, then it should be in the lower group. We increment DIVIDER by one

and then swap the current record with that now pointed to by DIVIDER, $>_x$. As a result, the records in the lower group are kept together, giving the arrangement

Upon completion of the scan, the record in the first position is swapped with that of the DIVIDER record, $<_y$, yielding

The lower and upper groups are sorted in turn using the same procedure. A recursive definition of the quicksort algorithm appears in Algorithm 13.3.

An Example

Let us demonstrate how quicksort works using the familiar data that we have processed previously, records with the keys

27, 18, 29, 28, 39, 13, 16, 38, and 53

The first step is to partition the records into those that are less than the first record and those that are greater than it. DIVIDER is set to the first record, that is, 27. When we compare 18 with 27, we have a less than compare that signals a swap. DIVIDER is incremented to point to the next record, the one with key 18. And the current record, 18, is swapped with the record pointed to by DIVIDER, also 18. Such a condition would not occur frequently, so it is actually preferable to perform the unnecessary swap than to check for equality of these pointers in all cases. After that partitioning, the data are then

16 18, 13, 27, 39, 29, 28, 38, and 53.

such that all the records with keys less than 27 appear to its left and those greater are to its right. We then partition the lower and upper subgroups in turn, obtaining

13, 16, 18, 27, 38, 29, 28, 39, and 53.

Three of the remaining subgroups now contain only a single item, so processing on them is complete. We then continue with the partitioning of the remaining subgroup with multiple records. The data then becomes

13, 16, 18, 27, 28, 29, 38, 39, and 53.

The remaining subgroup is sorted. Now all the subgroups contain only a single member, so the sorting is complete.

13, 16, 18, 27, 28, 29, 38, 39, and 53

Discussion

An unanticipated characteristic of the quicksort algorithm is that the closer the original records are to a sorted order, the *poorer* is the algorithm's performance. In fact, the worst case is when the records are already in a sorted order! In that circumstance, each application of the algorithm reduces the number of records that need sorting by only one. You can imagine how inefficient this is for a large number of records. For the example set of data, we needed to invoke the quicksort—swap algorithm five times. If the data had appeared in a sorted order, we would have needed eight such calls to the procedure. In general, for n records in sorted order, $n - 1$ calls to quicksort—swap are needed. An improvement to the algorithm to avoid this poor performance when the data are nearly sorted is to use a random record instead of the first one for comparison purposes. Use a pseudo-random-number generator to choose one record from among those to be sorted. An alternative comparison value suggested by Singleton [4] is the median of the three values: the first, the middle, and the last record from the group of records to be sorted. Choosing the comparison value from a sampling of the data reduces the chance of a poor choice, but it takes more time to determine than selecting the first or a random record.

When the number of records to be sorted is small, the quicksort algorithm requires a disproportionately high percentage of time for overhead. To lessen the effects of this situation, another improvement to the algorithm is to use a simpler sorting algorithm, for example, the insertion sort, when the number of records falls below a threshold. Bentley [2] found the number 15 to be the best choice for the threshold for his environment.

Although the quicksort algorithm has a worst case computational complexity of $O(n^2)$, its average complexity is only $O(n \log n)$ where n is the number of records sorted. So for large values of n, the quicksort algorithm is preferable to the insertion sort.

Improvements and modifications to the quicksort algorithm continue to be made, for example, see [5] for a variation to handle those cases when the input data are sorted or nearly sorted.

KEY TERMS

DIVIDER   Quicksort
Partition-exchange sort   Separator key
Pseudo-random-number generator

EXERCISES

1. Sort the records with the keys

   73, 15, 44, 37, 30, 59, 49, and 99

   using the quicksort algorithm. Show the contents of the file after each call to the procedure.

2. How many exchange operations, including those of a record with itself, were needed in sorting the data in Problem 1?

3. Describe how you could modify Lomuto's partitioning scheme to eliminate the swap of the first record at the end of a scan [6]. Hint: Use a sentinel and scan the records from right to left.

4. (a) Given the months of the year in calendar order, sort them using quicksort. (b) How many exchange operations are required?

5. (a) To what record, in general terms, does DIVIDER point? (b) What is its function?

6. (a) What is the minimum number of exchanges needed to quicksort a file of n records? (b) the maximum number? (c) How would you characterize a file that requires the maximum number of exchanges?

HEAPSORT

The heapsort is a fascinating algorithm because it is simple but elegant. In addition, it has characteristics that distinguish it from the insertion sort and the quicksort. Your being aware of a variety of sorting routines will allow you to make a good choice when you need to order information; you will not need to just "make do." The heapsort derives its name from its underlying data structure, the heap.

> A **heap** data structure is a complete binary tree in which the key at each node is at least as great as the key of either of its offspring.

This definition implies that the root will contain the record with the highest key. Figure 13.1 illustrates the relationships of a heap. Do not confuse this use of the term "heap" with its earlier use in the "indexed heap" data structure.

The heapsort takes advantage of the special properties of a complete binary tree that we have used previously with the binary tree collision resolution technique and dynamic hashing with linear splitting. We can store the records in a sequential array such that the following relationships hold for a node i:

- lchild(i) = 2 * i,
- rchild(i) = 2 * i + 1, and
- parent(i) = $\lfloor i / 2 \rfloor$

These relationships allow us to *move* records into a sorted order within a single, sequential storage structure. The subscript on an element in Figure 13.1 corresponds to its index in the sequential array.

> A **heapsort** is a two-stage sorting procedure in which a heap is built in the first stage, and the ordering is accomplished in the second stage by repeatedly removing the root record while maintaining a heap structure for the remaining records.

The heapsort originated with Williams [7]; nevertheless, we will follow the implementation suggested by Bentley [8]. Let's begin with the first stage that builds the heap. We assume that the n records to be sorted are stored in an n-element **sequential**

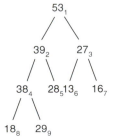

Figure 13.1 A heap.

**Algorithm 13.4
PLACING A RECORD AT ITS
PROPER POSITION IN A HEAP**

proc sift__up(i)

> /* Set the key of the root node's parent to δ, a value larger than any key, so that special checking may be eliminated within the loop */

```
1      key[parent(1)] := δ
2      while key[parent(i)] < key[i] do
3             swap(record[parent(i)],record[i])
4             i := parent(i)
5      end
end sift__up
```

array, record[1:*n*]. We process each record in turn to preserve a heap structure with the previously processed records. We are essentially sifting a new record up the tree until it has a value at least as great as all its offspring, that is, until the conditions of a heap are satisfied. The algorithm for accomplishing this placement appears in Algorithm 13.4.

Figure 13.2 illustrates the effect of applying the sift__up procedure to the record with key 53 on the existing heap of Figure 13.2a.

The second stage of the heapsort algorithm removes the root record from the heap by swapping it with the last record of the current heap; this in turn reduces the size of the heap. In Figure 13.1, we swap 53 with 29 (the record with the highest subscript); we then sift 29 down to its proper position to maintain a heap structure. The resulting heap has eight elements. Figure 13.3 illustrates this sifting down process with the record with key 29.

The record with key 39, now in the root node, has the next highest key. The algorithm for sifting down appears in Algorithm 13.5; the parameter "heap__size" conveys the bounds of the remaining heap. Sifting up is less complicated because each

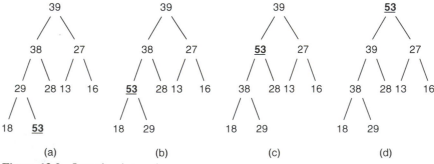

Figure 13.2 Inserting into a heap.

Algorithm 13.5
MAINTAINING A HEAP AFTER THE REMOVAL
OF THE ROOT RECORD

```
proc    sift_down(heap_size)

1       i := 1
2       while rchild(i) ≤ heap_size and
3       (key[i] < key[lchild(i)] or key[i] < key[rchild(i)]) do
4               if key[lchild(i)] > key[rchild(i)] then
5                       do
6                               swap(record[lchild(i)],record[i])
7                               i := lchild(i)
8                       end
9               else
10                      do
11                              swap(record[rchild(i),record[i])
12                              i := rchild(i)
13                      end
14              end
15      if 1child(i) = heap_size and key[i] < key[1child(i)]
16      swap(record[1child(i)],record[i])
end sift_down
```

node has just one parent; in sifting down, a node may have up to two offspring. If both offspring are greater than the parent, we choose the larger one.

With the procedures for sift_up and sift_down, the heapsort algorithm is straightforward, as presented in Algorithm 13.6.

How is that for a simple algorithm? After applying the heapsort algorithm to data in a sequential storage area, that *same* area will contain the records sorted in ascending order.

An Example

As a result of the previous discussion and the simplicity of the method, you probably already have a firm grasp on how the heapsort performs. But for completeness, let's

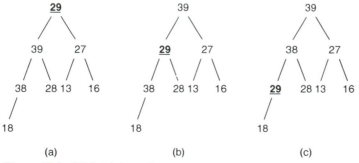

Figure 13.3 Maintaining a heap.

Algorithm 13.6
HEAPSORT

---

proc heapsort

/*   Build the heap   */

1   **for** i = 2 to n **do**
2   sift_up(i)
3   **end**

/*   Order the records and maintain the heap   */

4   **for** i = n to 2 **do**
5   swap(record[1],record[i])
6   sift_down(i − 1)
7   **end**
end heapsort

---

work through an example using the standard data, that is,

27, 18, 29, 28, 39, 13, 16, 38, and 53

The heap of Figure 13.1 is the result of processing the data records in phase one. In phase two, the record with key 53 is the one moved to record[n]. Figure 13.3c shows the remaining heap; from it, we can observe that the record with key 39 is moved to entry record[n − 1]. The complete ordering process appears in Figure 13.4.

Discussion

The heapsort is a fun algorithm that has elegance and simplicity. In Figure 13.4, you can observe the formation of the heap by following the ♦s moving to the right, and you can see the ordering of the records by following the ■s moving to the left. The two phases of the algorithm are also apparent. Figure 13.5 illustrates how the sequential storage area changes during the ordering process.

Notice that a standard traversal of a heap does not yield a sorted order of the records; it is only after each record is processed in turn that the records become ordered.

What is the worst case computational complexity of the heapsort algorithm? The number of operations to be performed in each call to either the sift_up or sift_down procedure is bounded by the depth of the corresponding heap. Since the depth of each binary tree heap is bounded by $\log_2 n$, and we need $\approx 2n$ calls, the worst case computational complexity is $O(n \log_2 n)$. The average computational complexity is also $O(n \log_2 n)$.

Advantages of the heapsort algorithm are that it has a lower average computational complexity than the insertion sort and it requires no additional storage as does the quicksort. Its big advantage, though, is that it has a lower worst case performance than do either of the other two sorts. On the average, however, the quicksort performs better than the heapsort. In the heapsort, the moving of records with large keys to the left in the first phase and then back to the right in the second phase takes its toll.

Record index	1	2	3	4	5	6	7	8	9	
Initial values	27,	18,	29,	28,	39,	13,	16,	38,	53.	
	27,	18♦	29,	28,	39,	13,	16,	38,	53.	
	29,	18,	27♦	28,	39,	13,	16,	38,	53.	
	29,	28,	27,	18♦	39,	13,	16,	38,	53.	Build heap
	39,	29,	27,	18,	28♦	13,	16,	38,	53.	
	39,	29,	27,	18,	28,	13♦	16,	38,	53.	Use sift–up
	39,	29,	27,	18,	28,	13,	16♦	38,	53.	
	39,	38,	27,	29,	28,	13,	16,	18♦	53.	
(Compare with subscripts in → Figure 13.1)	53,	39,	27,	38,	28,	13,	16,	18,	29♦	
	39,	38,	27,	29,	28,	13,	16,	18,	▪53.	
	38,	29,	27,	18,	28,	13,	16,	▪39,	53.	
	29,	28,	27,	18,	16,	13,	▪38,	39,	53.	Order the records;
	28,	18,	27,	13,	16,	▪29,	38,	39,	53.	Maintain the heap
	27,	18,	16,	13,	▪28,	29,	38,	39,	53.	
	18,	13,	16,	▪27,	28,	29,	38,	39,	53.	Use sift–down
	16,	13,	▪18,	27,	28,	29,	38,	39,	53.	
	13,	▪16,	18,	27,	28,	29,	38,	39,	53.	

Figure 13.4 Heapsort example.

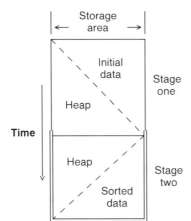

Figure 13.5 How the storage area changes during the heapsort process.

The heap is a useful data structure independent of its sorting capabilities. If you have not encountered it previously, you may want to note that it, together with the sift__up and sift__down algorithms, can be used to implement a priority queue. A **priority queue** processes its elements based upon priority or importance values associated with each element rather than the pure first-in first-out (FIFO) arrangement of regular queues.

KEY TERMS

Complete binary tree	Heapsort
FIFO queue	Priority queue
Heap	

EXERCISES

1. Build a heap for the records with the keys

$$73, 15, 44, 37, 30, 59, 49, \text{ and } 99$$

2. Sort the records in Problem 1 using the heapsort algorithm. Show the contents of the sequential storage array after each call to sift—down.

3. What would happen if the **while** condition "rchild(i) $\leq$ heap—size" were removed from Algorithm 13.5?

4. Bentley [8] suggests that the heapsort algorithm can be improved if the heap is built in phase one in a right to left direction with a modification of the sift—down procedure. Write a procedure incorporating this change and explain why the time required to build the heap is then O(n) instead of O(n $\log_2$ n).

5. (a) Build a heap for the months of the year given in calendar order. (b) How many exchange operations are required to build the heap?

6. (a) Sort the months of the year given in calendar order using the heapsort. (b) How many exchange operations are required to sort the data (after the initial heap is built)?

7. What is the computational complexity of inserting an item into a FIFO queue vs. a priority queue?

8. What changes do you need to make to the given heapsort algorithm to place data into descending order?

EXTERNAL SORTING

Internal vs. External Sorting

The three sorting algorithms that we have considered thus far in this chapter have been intended primarily for *internal sorting,* that is, they process the records entirely in primary memory. But how does one sort data that are too large to fit into primary memory? The best and most obvious solution is to use a built-in or system sort. Considerable effort has gone into developing these packages and many factors were considered for optimizing their performance. It would be a difficult task and an unwise use of his time for each programmer to develop his own *external sorting* routine, that is, one in which the data are not ordered entirely in primary memory. But there are a few cases, such as when it is not possible to invoke a system sort from within a program, when it is necessary for the programmer to develop an external sorting routine. Since external sorting routines access auxiliary storage during the ordering process, they are *much* slower than internal sorting routines, so the latter are obviously the procedures of choice where possible.

Although many methods exist for performing external sorts [1], the most common methods use a merging process in which previously ordered subfiles are combined. These ordered subfiles are called *runs.* Before considering such external sorting procedures, we look at the merge sort, which although an internal sorting procedure, will provide the proper background for understanding the external ones that merge data.

SORTING BY MERGING

The Internal Merge Sort

Merging records from two or more previously ordered files is a simple way to create a larger ordered file. For ascending order, one repeatedly moves the record with the smallest key to the output file. Since the input files are ordered, only the first record in each file needs to be considered. When a record is removed from a file, the next record then becomes the record for comparison. Algorithm 13.7 describes this process for two files.

But what if instead of processing previously *ordered* files, we encounter a file of *n unordered* records that we wish to order? In that case, this file of *n* records could be considered to be *n* sorted files of one record each. Adjacent file pairs could then be merged using Algorithm 13.7 to obtain approximately half as many files but with two records each. This merging process would be continually repeated on the resulting adjacent files until a single ordered file remained.

> A **merge sort** orders the records of a file by repeatedly merging records in adjacent runs that are initially of length one.

An Example

Figure 13.6 demonstrates the workings of the algorithm on the standard data, that is,

$$27, 18, 29, 28, 39, 13, 16, 38, \text{ and } 53$$

The representation of the process is an inverted tree in which the branches between the nodes indicate the runs involved in the formation of a larger run. The example tree is *binary* since only two files are merged at a time.

Discussion

The advantage of the merge sort is its simplicity. Among its disadvantages is the fact that it cannot be done in place. For a file of *n* records, 2*n* storage locations are needed since on each pass, a record is copied from one location into another. The considerable

Algorithm 13.7
MERGE SORTING TWO FILES

proc   merge—files

    /*   The n elements of one file are indexed by the variable
    i and the m elements of the other file are indexed by
    j. k is the index for the output file.   */

```
1       i := 1; j := 1; k := 0
2       while i ≤ n and j ≤ m do
3             k := k + 1
4             if key(record[i]) < key(record[j]) then
5                   do
6                         record[k] := record[i]
7                         i := i + 1
8                   end
9             else
10                  do
11                        record[k] := record[j]
12                        j := j + 1
13                  end
14      end
```

    /*   Copy the records from the remaining list.   */

```
15      while i < n do
16            k := k + 1; record[k] := record[i]
17            i := i + 1
18      end
19      while j < m do
20            k := k + 1; record[k] := record[j]
21            j := j + 1
22      end
end merge—files
```

amount of data movement is another disadvantage. The computational complexity of the merge sort is O(n log$_2$ n) since log$_2$ n passes are needed and each pass handles (including copying) all n records. The merge sort is suitable only as an internal sort for it is unreasonable or inefficient, or both, on most computer systems to have n files open simultaneously. For an external sort, we then need to modify the merging process to limit the number of files that are dealt with at any one time. The balanced P-way merge is such a process.

External Balanced P-Way Merge

Since the primary obstacle to using the merge sort as an external sorting routine is the large number of files required initially, we modify the merging process to limit that number of open files. The original file is decomposed into runs by internally sorting

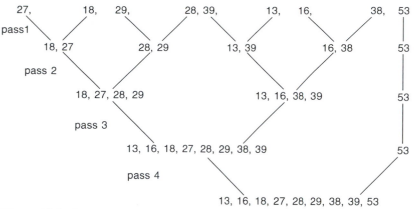

Figure 13.6 Merge sort example.

blocks of data as large as will fit into primary memory. The "balanced" in the name of this sort results from the fact that the resulting runs are evenly distributed on P external files. These files reside physically on either disks or tapes. If sufficient space is available on disks, they are preferable to tapes since they have faster access, do not require rewinding, and allow any record to be accessed directly. The initial runs on each of the P input files are merged using the merge sort. The merged output forms a larger run of P times the size of the original run; these resulting runs are evenly distributed on a second set of P output files. When all the runs on the original P files have been processed, one pass is completed. On the next pass, the output files from the previous pass become input files and the former input files become the output files. The next pass merges the files resulting from the first pass to form files P times their size. The merging process continues forming larger and larger runs until only a single run, the sorted file, exists. Since there are separate input and output files, a total of $2P$ files are needed. If tapes were being used, as might be necessary for quite large files, $2P$ tape drives would be necessary. A balanced two-way merge sort requires four files. If the number of available tape drives or files, F, is odd, then the number of input files alternate between

$$P = \left\lceil \frac{F}{2} \right\rceil \quad \text{and} \quad P = F - \left\lceil \frac{F}{2} \right\rceil$$

An Example

Assuming that we are using disk files with $P = 3$, let's process the standard data, that is,

27, 18, 29, 28, 39, 13, 16, 38, and 53

For illustrative purposes only, we suppose that the initial run size is one; in actual practice we would obviously have a much higher number of records in a run. The example data is then distributed as shown here.

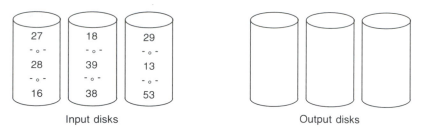

The data in the first run of each disk are merged to form a run that is placed on the first output disk. The data in the second runs are merged to form a run on the second output disk, and so forth. At the end of the first pass, the disks appear as

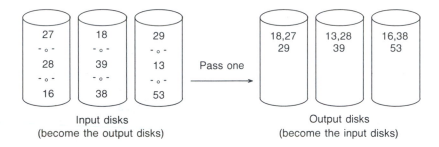

The disks that received the merged data from the first pass now become the input disks for the second pass, which yields the final sorted file, that is,

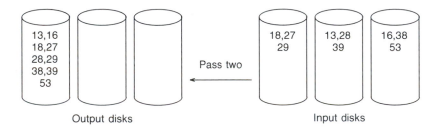

Discussion

The number of passes required in a balanced P-way merge sort provides an indication of its performance since all the records are processed on each pass and the input/output time required to access the records on auxiliary storage dominates the time for performing the operation. A balanced *two-way* merge sort requires

$$k = \lceil \log_2 r \rceil \text{ passes}$$

where r is the number of initial runs. In other terms, the range of runs r, requiring k passes, is $2^{k-1} < r \le 2^k$. Notice that this number of passes is the same as that for the internal merge sort when the runs are considered the same as the initial files. How many passes would have been necessary if the standard data were sorted with a balanced two-way merge with initial runs containing two records each? Since r would have been five in that case, three passes would have been needed. Verify this both through the formula and through processing of the data.

Many improvements have been made to this basic balanced P-way merge. Rather than investigating these variations here, since external sorting routines are needed less often than internal ones, we direct the reader to other sources on sorting, see, for example, [1] and [9]. We will, however, consider a variation of the merge sort introduced by Kernighan and Plauger [10] which is more adapted to disks in contrast to the balanced P-way merge, which was developed with tape files in mind. We refer to this sort as the disk sort.

DISK SORT

Instead of having two banks of tapes or disk files that alternate as input and output files, as in the balanced P-way merge sort, the disk sort merges runs from m input files onto a *single* output file. Initially, the runs from the original input file are placed into individual files. Unlike the balanced P-way merge, each file contains only a single run. After the initial distribution of the records, the first m files are merged onto a new output file and the m input files are removed since they are no longer needed. This new output file is positioned after the other files remaining to be processed. The procedure continues merging m files at a time until only a single file remains which contains the sorted data. The final merging may process fewer than m files. The *merge order, m,* usually ranges from three to seven. The process may be depicted as

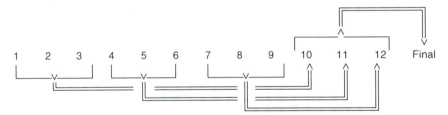

for a merge order of three. Files 1, 2, and 3 are merged to form file 10; files 4, 5, and 6 are then merged to form file 11, etc. Finally, the remaining three files, 10, 11, and 12 are merged to produce the final ordered file. The algorithm is described in Algorithm 13.8.

An Example

Let's work through an example of the disk sort again using the standard data of

$$27, 18, 29, 28, 39, 13, 16, 38, \text{ and } 53$$

with a run size of two and a merge order of three. The data is initially distributed into runs on five files as

Algorithm 13.8
DISK SORT

I. Internally sort the largest possible segments of the original data file and place the resulting runs into distinct output files. All the runs except the last will be the same size.

II. While the number of files remaining is greater than one
   A. Create a new file with an identifier one greater than the current last.
   B. Merge the records from the first *m* remaining files, or all the remaining files if fewer than *m,* onto the new file.
   C. Delete the input files for that merge.

III. Terminate with the sorted records in the one remaining file.

18, 27	28, 29,	13, 39,	16, 38	53
file 1	file 2	file 3	file 4	file 5

The first three files are merged into an output file six, and the three input files are removed. The data then appears as

16, 38,	53,	13, 18, 27, 28, 29, 39
file 4	file 5	file 6

The remaining three files are then merged to form the sorted output file of

13, 16, 18, 27, 28, 29, 38, 39, 53

and the input files are removed.

Discussion

The disk sort is conceptually simpler than even the balanced *P*-way merge. It requires the same number of run merges given the same data but contains more iterations or passes than a balanced *P*-way merge since the latter may perform several run merges in one pass. The primary requirement of using the disk sort is an operating system that allows the easy creation and deletion of files.

KEY TERMS

Balanced *P*-way merge
Disk sort
External sorting
Internal sorting
Merge order

Merge pass
Merge sort
Run
System sort

EXERCISES

1. **(a)** Perform a disk sort on the months of the year given in calendar order using a run size of two and a merge order of three. Show the contents of each file generated. You may abbreviate the names of the months. **(b)** How many initial files are needed? **(c)** How many additional files are created during the sorting process?

2. How many files are necessary to perform a disk sort on a file of 100,000 records using a run size of 1000 and a merge order of four?

3. Repeat Problem 2 with a run size of 2500.

4. How much internal memory is needed for record storage when sorting the records in Problem 2 using an internal merge sort if each record is 200 bytes?

5. What is the relationship between the depth d of a binary tree of n nodes and the number of passes k required to sort n records using a binary merge sort?

6. Other than conceptual simplicity, what are reasons for using a disk sort rather than a balanced P-way merge sort?

7. Prove the formula for the number of passes required to process r runs in a balanced two-way merge.

REFERENCES

1. Knuth, D. E., *Sorting and Searching,* Addison-Wesley, Reading, MA, 1973.
2. Bentley, Jon L., "How to Sort," *CACM,* Vol. 27, No. 4 (April 1984), pp. 287–291.
3. Hoare, C.A.R., "Quicksort," *Computer Journal,* Vol. 5, No. 1 (1962), pp. 10–15.
4. Singleton, R. C., "Algorithm 347: An Efficient Algorithm for Sorting with Minimal Storage," *CACM,* Vol. 12, No. 3 (March 1969), pp. 185–187.
5. Wainwright, Roger L., "A Class of Sorting Algorithms Based on Quicksort," *CACM,* Vol. 28, No. 4 (April 1985), pp.396–402.
6. Sedgewick, R., "Implementing Quicksort Programs," *CACM,* Vol. 21, No. 10 (October 1978), pp. 847–848.
7. Williams, J.W.J., "Heapsort," *CACM,* Vol. 7, No. 6 (June 1964), pp. 347–348.
8. Bentley, Jon, "Thanks, Heaps," *CACM,* Vol. 28, No. 3 (March 1985), pp. 245–250.
9. Martin, William A., "Sorting," *Computing Surveys,* Vol. 3, No. 4 (December 1971), pp. 147–174.
10. Kernighan, Brian W., and P. J. Plauger, *Software Tools,* Addison-Wesley, Reading, MA, 1976.

PART SUMMARY

Organizing data such that it is ordered according to the values in a field is a common requirement. In this part, we examined several of the principal routines for ordering data. Usually when you need an ordering, you will be able to invoke a system or built-in sort, but for those cases when that is not possible, you now have a variety of techniques from which to choose. When sufficient primary storage is available, an internal sort is much faster than an external one and is therefore preferable. To assist you in choosing from among the internal sorting routines, their average computational complexities and storage requirements are given in the following table:

PERFORMANCE CHARACTERISTICS OF INTERNAL
SORTING ROUTINES

Method	Average computational complexity	Storage requirements
Insertion sort	$O(n^2)$	n
Quicksort	$O(n \log_2 n)$	$n + \log_2 n$
Heapsort	$O(n \log_2 n)$	n
Merge sort	$O(n \log_2 n)$	$2n$

From this table and the previous discussions, rules of thumb for which internal sorting routine to select are

- For a small value of n, use an insertion sort.
- For the best average performance, if sufficient space is available, choose the quicksort.
- For the best worst case performance, choose the heapsort.

As noted previously, we introduced the merge sort primarily because it is a component of the external merge sort routines. If sufficient primary memory is not available, one must resort to an external sort. If the creation and deletion of files is easy and if sufficient disk space is available, the disk sort is preferred. Otherwise a balanced

P-way merge is appropriate. If insufficient disk space is available, tapes must be used and the choice for *P* is dependent upon the number of tape drives available. Since tapes require operator intervention, and tape drives are relatively expensive and unreliable components of a computer system, a tape sort is a last resort.

In addition to sorting procedures, this part also described the heap data structure, which is useful in implementing priority queues.

PART EXERCISES

1. Which internal sorting technique will give the best performance if the input data are already in sorted order?
2. Which internal sorting technique will give the best performance if the input data are in reverse sorted order?
3. Which sorting technique is preferable for a small number of records?

Applying File Structures

Winnie-the-Pooh sat down at the
foot of the tree, put his head
between his paws and began to
think.[1]

BACKGROUND

You have now learned about a variety of the "tools" of the software development trade.
You have discovered the advantages and disadvantages of the various data structures
presented in the preceding chapters. You have a grasp on the utility of the data
structures and the associated processing techniques. But what you really have not had
an opportunity to do yet is to *use* this knowledge to solve meaningful problems. The
goal of this chapter is to give you that opportunity. It will still be impossible to provide
you with the type of in-depth problems that you are likely to encounter after you
complete your studies, but this chapter will act as a bridge. Even if you do not take
more advanced computer science courses that build upon the material of this text, you
will still have the occasion to use the knowledge you have gained. This chapter provides
a means of exercising your newly acquired knowledge in a more controlled environment
than you may experience elsewhere. It will reinforce your learning just as the exercises
do at the end of each section. An intention of this chapter is to demonstrate to you that
you do have the ability to solve problems, that you can make improvements.

Each section of this chapter presents a situation that may be improved upon or
resolved with the development of software. Your task will be to take the information
presented, analyze it, and give your suggestions for solving the problem. The important
question is, now that you understand all these "tools," which one do you apply where?
Often, there will be not only *one* correct answer but rather a range of answers; in some
instances, the exact choice may be dependent upon information that is not presented.
In a realistic situation, you would have an opportunity to carry on a dialogue with an
individual from the organization requesting assistance. This individual would help you
make the proper interpretation of the constraints. Such communication is not uncomplicated
since the requester and the designer often have quite different experiences and
perspectives.

[1]Milne, A. A., *Winnie-the-Pooh,* Dell Publishing, 1954.

For most of the situations of this chapter, correct answers will not be presented. The solutions that you discover may be better than any of the common, known solutions. You may need to apply more than one data structure to resolve some situations. You will need to interact with others to determine the effectiveness of your solutions. In some ways, at this stage, the *process* that you go through to arrive at a solution is as important as the solution that you obtain. Questions are provided in the next section to guide you through your deliberations. In the early cases, you will need to refer back to these questions. As you gain more experience, you will begin to ask these questions instinctively. In some respects, you may consider this chapter one big examination prior to your graduation from the subject.

Considerations

- Analyze the situation.
  - What type of access is needed?
  - What are the retrieval time constraints placed upon the problem solution?
  - What are the space limitations placed upon the problem solution?
  - How frequently will the information be processed?
  - How much data will be processed?
  - How static is the quantity of the data?
  - How dynamic are the contents of the data?
  - How large are the records and associated keys?
- Ponder solutions.
  - Which data structures or processing techniques are most applicable?
  - What are the advantages and disadvantages of these potential solutions?
  - What are the priorities in making a selection?
  - What are the implementation costs?
  - How much time will it take to implement the solution?
  - How reliable is the implementation likely to be?
  - Does one solution do more than the others?
  - Is one candidate a natural solution? (The answer is usually "no," but it is a question still worth asking.)

Summary

The purposes of this chapter are to encourage you to

- Analyze problem situations,
- Develop software solutions based upon the data structures discussed in the text,
- Exercise your newly acquired knowledge,
- Build your confidence that you *can* solve problems.

You are now ready to do something.

This will be the really satisfying part!

SAVINGS AND LOAN

Problem Description

A savings and loan association has decided to add automatic teller machines (ATMs) to its branches as a service to its customers. Each customer requesting to use the machines will be given a special card to operate them. The card will contain a unique customer identifier. The savings and loan has not been able to estimate how many customers will request ATM cards, but it expects to issue a small number of cards initially and gradually to increase that number over time. It needs to modify its current computer system to handle this expansion of services. In particular, it needs a data structure for mapping the ATM customer ID to the existing customer information when the customer activates an ATM. It wants to use its existing computer system which has little primary memory available for this new application. It also realizes that speed in processing a customer request is important since it is a service organization. A software system already exists to process the customer account information; the purpose of this new application is to locate customer account information when given an ATM customer identifier.

A Solution

Let's go through an analysis of this problem to give you an impression of how to proceed when resolving the other situations described in this chapter. First, though, give thought to the type of data structure or structures that one might use for this application. After you have decided upon a solution or if you are not sure how to continue, read the following discussion. Basically, the savings and loan needs a technique for linking the ID on the customer's ATM card to his accessible accounts when he uses an ATM. After considering the questions in the previous section for helping to analyze a situation, you have probably concluded that:

- Direct access is essential. The system needs to retrieve the link information associated with a single customer ID.
- Fast response time is important. Since the savings and loan is a service organization, good service demands that the system respond in a reasonable amount of time after the customer has inserted a card into the ATM.
- Meager primary memory is available. The existing computer system is to be used to implement this new application. Since it appears that current applications use most of the existing primary memory and there is no mention of purchasing additional primary memory, the solution should not use much primary memory.
- Access to the link information would be frequent. An individual customer may only use an ATM a few times a week, but during the course of a day, many different customers would use an ATM.
- The amount of information to be processed is minimal. The purpose of this application is only to provide a link from the customer's ATM ID number to the account numbers that he might want to access; for example, a record might appear as

Customer ID	Account no.	. . .	Account no.

- The number of records will continue to grow. The savings and loan expects to provide ATM cards to a few customers initially, but over a period of time, it anticipates providing this service to many more customers.
- The contents of the data is stable but not static. Customers will occasionally want to add or delete the account numbers that they may wish to access through an ATM.
- The record size of the data is short. Since the record format as discussed previously contains only an ID number and account numbers, it does not require extensive storage. We assume fixed-length records, but depending upon the range of the number of possible accounts, a variable-length record may be preferable.

Having answered these questions should help you in narrowing the choices for potential data structures. Direct-access retrieval suggests a type of direct-access or hashed file organization. The fluctuation in the customer base probably rules out a scheme in which the customer ID is also the address in the table. Since the growing customer base with access to the ATMs suggests an expanding file, we should focus on those methods for expandable files, that is, extendible hashing, dynamic hashing, linear hashing, and trie hashing. Which of these four is most appropriate? To be able to answer that question, you must understand the differences, advantages, and disadvantages among them. One important piece of information given was that primary memory was limited. Which of these four hashing methods for expandable files requires the least amount of primary memory? Linear hashing, since it does not need space for a directory. If we were to use a method that requires a directory, the directory would most likely have to be stored on auxiliary storage, which would add another "$10,000" type of access to a retrieval and hence would degrade its performance. Since primary memory is limited, the choice of linear hashing is relatively clear. Linear hashing is fast and straightforward to implement, so those aspects of choosing a data structure should not pose a problem.

Notice that this part was titled "A Solution" rather than "The Solution." The presented solution was based upon the information given and an interpretation of it. Based upon your understanding and experience with the application described, you may have interpreted something differently or noted a potential problem not immediately obvious.

STUDENT TELEPHONE NUMBERS

Problem Description

A university wishes to provide a service for assisting someone who wishes to communicate with a student. Often a person such as a friend, relative, instructor, or prospective employer may wish to communicate with a student and does not know how to proceed. The university service will keep a file of all the current students and a local telephone number where each may be contacted. Since students are often quite mobile, one goal of this service is to keep the telephone numbers current. The service will be manned by somebody at the information desk at the student center. A person needing to contact a student may then, either in person or by calling the information desk, request the telephone number of the individual.

COMPUTER SCIENCE ALUMNI

Problem Description

A computer science department wishes to keep in touch with its alumni for various reasons, including feedback on curriculum development and program support. Depending upon the departmental need at hand, all or a selected group of alumni would be contacted. The alumni file should then have information on each individual pertaining both to his activities while a student and his current situation.

GRADUATE STUDENT RECRUITING

Problem Description

A computer science department wishes to establish a mailing list of colleges and universities to which it will send flyers each year for recruiting graduate students. It plans to use the mailing list without modification for a 10-year period. The department reasons that few new schools will appear during this period and that few existing schools will disappear. To assist in keeping the mailing list from becoming out of date, an address entry will use a generic title instead of an individual's name since people change positions frequently. Each year's mailing of the flyers will require address labels.

PATIENT RECORD SYSTEM

Problem Description

A doctor wishes to organize her patient information with the aid of a computer system. The purpose of the system is to aid the doctor in managing her practice and to provide better medical treatment for her patients. The system will be used to log information about each patient visit such as the day, diagnosis, charge, payment, medication prescribed, summary of treatment, subsequent action, and date of the next visit. The system will be used to retrieve patient information at the time of a visit and periodically to determine situations pending, such as payment due or a return visit that has not occurred. It will also be used for such activities as locating patients using a specific medication should a recall or special advisory be necessary, or locating patients having a certain diagnosis should a new treatment for it be discovered or released. This doctor is in practice by herself. The supporting staff consists of a bookkeeper-administrative assistant and a nurse.

PRODUCT ASSEMBLY

Problem Description

A manufacturer seeks to produce items such that each one is unique (e.g., the cabbage patch dolls). These products will be manufactured in a new facility in which the assembly line will be automated with the aid of robots to perform the assembly operations. Since each item will be different, the manufacturer realizes that he will need the help of a computer system to provide information to the robots on what component to add to each item as it moves down the line. Each item will have a bar code associated with it that will inform the robot what to add to the item. The robot will "read" the information from the bar code (using a scanner) and then retrieve the information needed for it to carry out its operation; for example, with cabbage patch dolls, it might need to know what type and color of hair to add. It is important to note that the assembly line moves continuously. The manufacturer plans to produce millions of these products, and no two are ever to be the same.

PURCHASE ORDER TRACKING SYSTEM

Problem Description

A large organization wishes to improve the management of its expenditures by installing an automated system to track its purchase orders. The organization wants to know daily how much money it has spent during the current budget year, how much it has committed which remains unpaid, and how much remains to be committed. This tracking system will be a centralized one that the many departments will use. The primary identifier for a purchase order will be a consecutive number generated and assigned by the system.

A purchase order may be updated until the requested item is delivered and the purchase order is paid; after that, a record may only be read. Only the department generating a purchase order may update it and only that department and the central office may read it. The organization plans to keep the information on the purchase orders on file for one year after the current budget year; after that, the information will be archived to tape. The system should also have the facility to list the outstanding and paid purchase orders by either department name or vendor name. Retrieval performance is a primary consideration.

UNIVERSITY PLACEMENT SERVICE

Problem Description

A university placement service wishes to assist both students who are seeking permanent employment after graduation *and* employers who are searching for prospective employees. The placement service wants to install an information retrieval system in which prospective employers would list their available positions with associated information such as the name of the organization, location, company strengths, company growth, job classification, minimum starting salary, maximum starting salary, and the requirements for each position, for example, knowledge of FORTRAN, experience with the UNIX[1] operating system, and previous employment in computer science. The information base would contain a record for each opening with a record's primary key being a placement office assigned sequence number. The retrieval system is intended to assist the student in locating a position compatible with her interests and abilities. A student would use the system by specifying a multiattribute query of up to 12 values for qualifiers that could be connected by Boolean operators. The system would respond with all job openings satisfying the entered requirements and student capabilities. The student could then make campus interview appointments with those companies having compatible openings. The employers would benefit from this system since their interviewing would be more confined to those students with backgrounds meeting their needs.

As is typical, the university's budget is limited, so that it may be necessary to sacrifice system response time to cut costs. If the response time is too slow, however, the students will not use the system. The university's existing computer system is to be used for running this placement retrieval system. Each year, usually about 400 companies interview on campus for about 3000 positions, and 2000 graduating students typically use the placement service.

GEOGRAPHY GURU

Problem Description

To encourage a greater knowledge of U.S. geography among students today, a national group wishes to develop a system containing facts on the subject. The geography system will be developed for use on low-cost microcomputers with minimal primary memory; it will be distributed at cost to interested school systems for use in elementary schools. Initially, the geography guru will contain a wealth of knowledge about each state, for example, the population, the area, the primary industry, the median income, the state capital, its average temperature by month, the annual rainfall, the longest river, the tallest peak, bordering states, the state insect, and much, much more. The primary key for each record will be the full name of the state, for example, California. To retrieve a field of information, the user will enter the field name and the state name. All the fixed-length records will contain the same fields. The retrieval time must be short to maintain the student's interest.

If time and resources permit, the organization wants to add the following capabilities to the system:

- Comparative capabilities, for example, the state with the greatest population or the one with the longest river.
- Secondary classifications, for example, the states with steel as the primary industry.

[1]UNIX is a trademark of AT&T Bell Laboratories.

● Range queries, for example, the states with capitals with a mean January temperature greater than 60°.

CATALOG MERCHANDISER

Problem Description

A fledgling catalog merchandiser needs a computerized information retrieval system to assist it in processing customer orders and managing its inventory. As an aid for controlling inventory, the system will automatically reorder an item when its inventory falls below a preset threshold, and on demand, it will print an ordered list of how much of each item has been sold over a given period. Knowing which items are selling well and those which are not will aid the merchandiser in deciding what should be included in future catalogs. Currently the business is just getting started and so it does not offer a diversity of products, but as it grows and becomes more successful, its number of offerings will increase. The name of a product will be the primary key for the record containing the associated information. These names vary from "ax" to "zymoscope with a power assisted adjustment." Primary memory is not extremely limited, but the data records *will* need to be stored on auxiliary storage.

COMPUTER FILE SYSTEM

In this text, we have considered a multitude of methods for managing a single file or related files. It is perhaps fitting then that we conclude with a problem that asks how to structure the information about the files themselves. This is just another information handling problem, but its solution is obviously necessary before a computer system can do anything with even one file. To have done anything with a computer, you must have had some experience with a file system. A *file system,* which is the major component of a computer's operating system, is the set of all the files within a computer system *plus* (and this is what we are interested in now) the information necessary to locate and process them. From the perspective of a file system, the organization of the specific files is immaterial, for that is determined by the individual applications. The focus of this text and the previous problems in this chapter has been on structuring and processing such files for user applications.

The components of a file system are

● The data within the files.
● The free space.
● The locator information.

The locator information in turn consists of:

● A directory.
● Data pointers.

Although the primary function of the directory, or index, is to locate the start of a file, it can also contain information associated with a file such as the creation date, creation time, length, and authorized users. The directory is then just another file but one that the system always needs to be able to find. Among the conventions used for locating a directory are to (1) place it in a known, fixed location, (2) place its address at a known, fixed address, and (3) place its address plus associated information in a known, fixed address. The directory points us to the beginning of a file and then the data pointers allow us to step through the contents of the file. The directory provides a structure for the file system.

The data pointers give us additional flexibility for handling the expansion and contraction of a file since they enable a file to be located in multiple areas that are noncontiguous. Even what is considered to be a sequentially organized file may then not be actually stored in consecutive locations. Since the pointer overhead would be excessive if files could expand in increments as small as a byte, space is usually allocated in segments (blocks or sectors) with 128, 256, 512 bytes, etc. What are the tradeoffs when using the different sizes? Some file systems allocate storage in terms of a data group that is a fixed multiple of contiguous data segments. Two common ways of implementing data pointers are to (1) provide a list of all data groups within a file or (2) build a table of all data groups within the file system and then chain together those for a particular file.

In the UNIX file system, there are three types of files: ordinary files, special files, and directories. The contents of the ordinary files are determined by the individual applications. Special files are used to handle input/output. Each input/output device has at least one special file associated with it, and when a request is made to read or write a special file, the associated device is activated. The directory is like any other file except that the system controls its contents.

Building a file system then is plainly another application of file structuring and processing techniques.

Problem Description

A start-up computer company needs a file system for a microcomputer that it is designing. The microcomputer system is intended to have a better price/performance ratio than any existing system aimed at small to midsized businesses. The computer company may ultimately decide to use a file system that is compatible with an existing one such as that used with PC-DOS or UNIX, but before making the final decision, it wants to consider alternatives. It wants a file system that has high performance and a "user-friendly" interface. For this problem, design a directory or index structure for the file system.

As you formulate a solution to this problem, be certain to review the questions given at the beginning of this chapter plus the knowledge that you have gained from being a user of one or more file systems. When you complete your solution, then (and only after you have at least made your own proposal) read the next section, which contains information on existing commercial systems. Compare your answer with those to determine if any of your ideas have advantages over existing ones. Also note the advantages that the existing ones may have over yours.

Stop! Think! Ponder! Then proceed.

More on File Systems

One of the most important questions that must be answered before choosing a file structure is what type of access is to be made of the information contained in it. In the case of a file system directory, what type of access is likely? Think back to your own experiences with file systems. You definitely want direct access since you may need to access a single file, for example, to edit it or to execute it. But you also need sequential access, for you may want to list all your files. So that suggests a data structure that will allow both sequential and direct access. What immediately comes to mind? That's correct, a tree structure. Most file systems therefore use a tree or hierarchical directory. The leaf nodes of the tree structure contain the pointers to individual files as shown in the example PC-DOS directory of Figure 14.1. The root node of the tree corresponds to what is referred to as the root directory. All the files that are contained within the computer system are descendants of this root directory; to locate any file within the system, the search must begin at the root directory. The other nonterminal nodes in the directory refer to subdirectories. In addition to being an efficient mechanism for computer retrieval, a hierarchical directory also assists the human users in conceptually organizing their information for subsequent retrieval.

The specification of a file or directory level (other than the root) in a hierarchical file system often takes the form:

delimiter node_identifer {delimiter node_identifier}*

where { }* means that the contents may be repeated zero or more times. The delimiters are typically a "/", a "\", or a ".". The node_identifier names the subdirectory or file. The root node is referenced by a single delimiter.

Both PC-DOS and UNIX employ a hierarchical file directory. It is also interest-

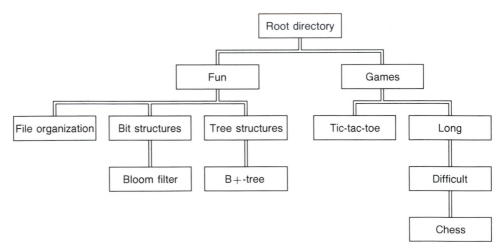

Figure 14.1 Hierarchical file directory.

ing to note that the file directory for VSAM is in fact a VSAM file; since it is just treated like any other file it contains an entry describing itself. In implementing a hierarchical file directory, what would you do to enable the files within a directory or subdirectory to be listed in lexicographic order? Also, can you think of any disadvantages of a hierarchical structure?

The UNIX file system has a linking feature that allows a nondirectory file to appear in more than one directory under possibly different names. Unlike other systems that allow linking, UNIX treats all links to a file with equal status; that is, a file does not exist within a particular directory, but the directory entry consists of its name and a pointer to the information actually describing the file.

As an aside, UNIX uses a noteworthy variation of the table technique for handling its data pointers for defining a file. The table has 13 entries with the first 10 each containing a pointer to a single 512-byte data block. For files greater than 5120 bytes, the eleventh entry is used to point to an area that contains up to 128 pointers for additional data blocks. The twelfth entry in the table points to 128 areas each of which contains pointers for up to 128 additional data blocks. And finally, the thirteenth entry contains three levels of additional pointers: 128 on the first level, 128^2 for the second, and 128^3 for the third. In this way, a file may contain up to 1,082,201,088 bytes, that is, $[(10 + 128 + 128^2 + 128^3) \times 512]$, truly a large file. The smaller files or the beginning portions of larger ones require only a single access of the pointer table, whereas the end of the largest file requires four accesses of the table.

Typical operations associated with a file system for performing the input/output of files include:

Create   Establishes the system parameters needed to process the file; it may also open the file.

Open   Prepares the file for reading, writing, or updating (both).

Close   Terminates use of the file; the file may then be available for other users.

Delete   Removes the file and associated information from the system.

Write   Transfers information from main memory to a system device.

Read   Transfers information from a system device to main memory.

Location   Positions a pointer for a subsequent direct-access read or write.

Since the implementation specifics of file systems vary greatly, we will not discuss them in detail. Golden and Pechura [1] give additional details on PC-DOS and Ritchie and Thompson [2] describe the UNIX file system; more information on UNIX is given in [3]. For more information on file systems in general, consult [4].

CHAPTER REVIEW

Unlike the previous chapters, this one has not presented new data structures, algorithms, principles, or techniques. Instead it has given the reader an opportunity to apply the ones presented in the prior chapters. The real value of the tools described in this text is in using them. They are not concepts and ideas to be tucked back into the far reaches of one's memory, but they are to be used for designing the necessary and innovative software of the future.

Having reached this point in the text, you now have

- An understanding of the basic mechanisms for storing and retrieving information.
- A knowledge of the differences of the various file processing techniques.
- Additional confidence in applying your problem solving skills in new circumstances.

> "Is that all?"
>   Alice timidly asked.
> "That's all,"
>   said Humpty Dumpty.
> "Good-bye." [1]

REFERENCES

1. Golden, Donald, and Michael Pechura, "The Structure of Microcomputer File Systems," *CACM,* Vol. 29, No. 3 (March 1986), pp. 222–230.
2. Ritchie, D. M., and K. Thompson, "The UNIX Time-Sharing System," *Bell System Technical Journal,* Vol. 10, No. 2 (July–August 1978), pp. 1905–1929.
3. The UNIX System, *AT&T Bell Laboratories Technical Journal,* Vol. 63, No. 8, Part 2 (October 1984), pp. 1571–1910.
4. Grosshans, Daniel, *File Systems: Design and Implementation,* Prentice-Hall, Englewood Cliffs, NJ, 1986.

[1]Carroll, Lewis, *Through the Looking Glass,* Simon and Schuster, Inc., 1951.

Answers to Selected Exercises

CHAPTER 1

Page 11, Prob. 5

six

Page 22, Prob. 1

572070 bytes

Page 22, Prob. 6

2.3 days

CHAPTER 2

Page 35, Prob. 1

(a) 10; (b) 11; (c) 3

Page 35, Prob. 2

(a) 10; (b) 1; (c) 3

Page 35, Prob. 8

(a) a b c d e i g f h j l k m n o p r q t s u v w x z y
(b) The reorganization of the list changes much more slowly with Transpose; n and o traded places several times; the occurrence of z did not affect the retrieval of other records.

Page 35, Prob. 10

The link fields used in implementing the Move—to—front algorithm require significant storage.

CHAPTER 3

Page 45, Prob. 4

(a) 456; (b) 852

Page 56, Prob. 1

Key

0	44	6
1		
2		
3		
4	15	10
5	49	
R → 6	99	
7	73	
8	30	
9	59	
10	37	9

Page 56, Prob. 2

1.5 probes

Page 56, Prob. 4

(a)

Key

0	49	
1	15	7
2	44	10
3	73	8
4		
5		
6		
R → 7	99	
8	59	Cellar
9	30	
10	37	9

Page 56, Prob. 4

(b) 1.6 probes

Page 60, Prob. 1

	Key
0	44
1	99
2	
3	
4	15
5	37
6	59
7	73
8	30
9	49
10	

Page 60, Prob. 2

2.0 probes

Page 68, Prob. 1

	Key
0	44
1	
2	
3	99
4	15
5	49
6	
7	73
8	30
9	59
10	37

Page 68, Prob. 2

1.9 probes

Page 69, Prob. 8

If an algorithm works properly with one set of data, that does not imply that the algorithm is correct.

Page 76, Prob. 1

	Key
0	44
1	
2	
3	59
4	37
5	49
6	15
7	73
8	30
9	99
10	

Page 76, Prob. 2

1.6 probes

Page 86, Prob. 1

	Key
0	44
1	
2	
3	59
4	37
5	49
6	15
7	73
8	30
9	99
10	

Page 86, Prob. 2

1.6 probes

Page 98, Prob. 1

	Key	nof
0	44	3
1	99	
2		
3		
4	15	2
5	49	
6	37	1
7	73	
8	30	
9	59	
10		

Page 98, Prob. 2

1.5 probes

Page 98, Prob. 3

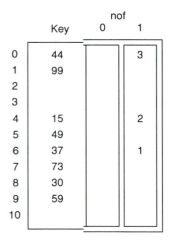

	Key	nof 0	1
0	44		3
1	99		
2			
3			
4	15		2
5	49		
6	37		1
7	73		
8	30		
9	59		
10			

Page 98, Prob. 4

1.5 probes

Page 98, Prob. 5

```
                    nof
            Key    0     1
     0      44           3
     1      99
     2
     3
     4      15     2     5
     5      49
     6      48
     7      73
     8      30
     9      37
    10
```

Page 98, Prob. 6

1.4 probes

Page 110, Prob. 2

$a = 0; b = 2; d = 0; f = 5$

CHAPTER 4

Page 123, Prob. 1

(a) track index *(for cylinder 1)*

```
0   10      1      13    9–2    41     2     57    9–3   103    3    103    Λ
```

(b) two

CHAPTER 5

Page 136, Prob. 3

(a) Peanut-fudge pudding cake; **(b)** no false drops

Page 145, Prob. 2

(a) 8 bytes: 36, 33, 9, 12, 33(collision)
(b) 16 bytes: 92, 53, 108, 109, 111

Page 145, Prob. 8

(a) Searching for a pattern of length less than three would be impossible.
(b) There would be a large number of false drops.

CHAPTER 6

Page 152, Prob. 5

four

Page 159, Prob. 1

Records 1, 5, and 24.

Page 166, Prob. 1

(a) 4, 5, 6, 7, 12, 13, 14, 15; (b) three records

CHAPTER 7

Page 177, Prob. 2

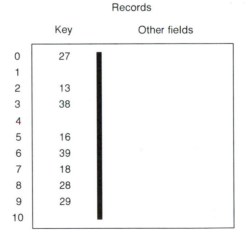

Page 187, Prob. 1

(a) Set bits 3, 5, 6, 10, 11, 14; (b) 46

Page 190, Prob. 1

(a) 66

Page 194, Prob. 1

(a)

	Check hash	Data (optional)
0		
1		
2		
3	1	(24)
4	3	(32)
5	3	(17)
6		

(b) Two

CHAPTER 8

Page 202, Prob. 2

(a) 3.5; **(b)** six; **(c)** six; **(d)** four

Page 209, Prob. 1

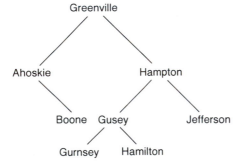

Page 209, Prob. 2

Page 209, Prob. 3

Page 219, Prob. 1

Page 219, Prob. 2

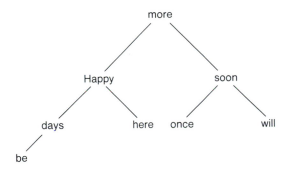

Page 220, Prob. 11

(a) Sort the data records. Insert the middle record into the root node. Perform an operation analogous to a binary search to insert the remaining records.

CHAPTER 9

Page 232, Prob. 1

Page 232, Prob. 2

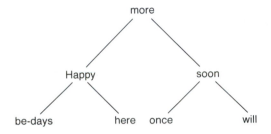

Page 232, Prob. 4

(a) 50 percent; (b) 57 percent

Page 232, Prob. 10

(a) 3; (b) 2

Page 233, Prob. 17

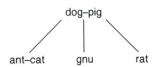

Page 243, Prob. 1

Page 243, Prob. 2

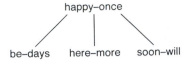

Page 243, Prob. 3

(a) 87.5 percent; (b) 100 percent

Page 243, Prob. 6

(a) 2.65; (b) 1.76

Page 243, Prob. 7

Page 252, Prob. 1

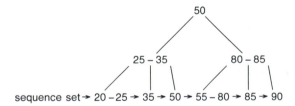

Page 252, Prob. 2

(a) 1; (b) 5

Page 253, Prob. 6

(a) 3; (b) 3

CHAPTER 10

Page 266, Prob. 1

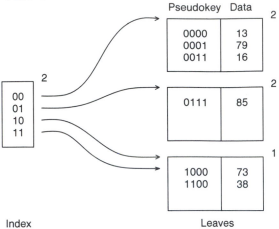

Index   Leaves

Page 276, Prob. 1

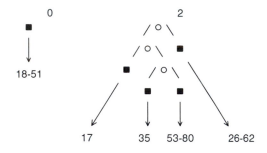

Page 276, Prob. 9

$$\text{splittings} = \left\lfloor \frac{r - 1}{\tau \times bs} \right\rfloor$$

Page 284, Prob. 1

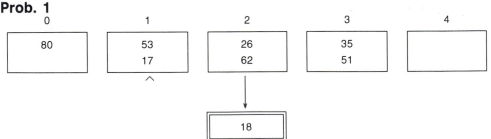

CHAPTER 11

Page 298, Prob. 2

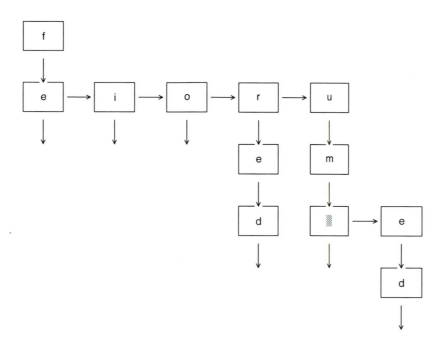

Page 298, Prob. 3

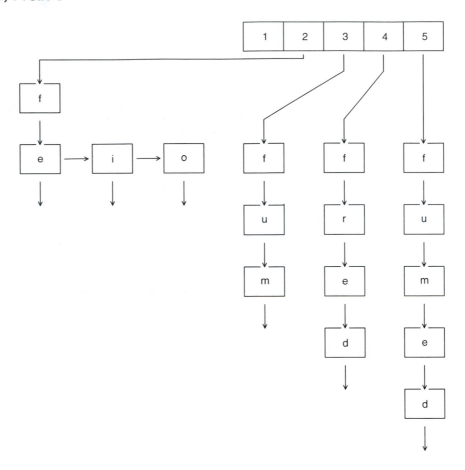

Page 304, Prob. 1

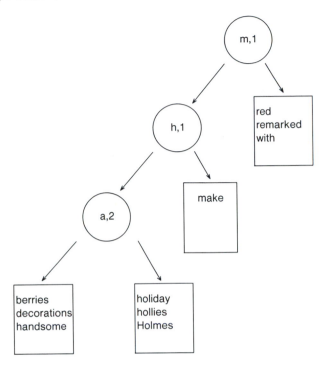

Page 312, Prob. 1

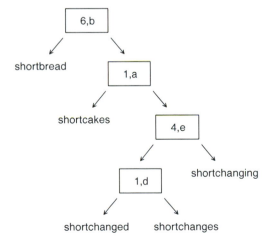

Page 312, Prob. 3

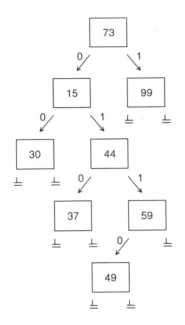

CHAPTER 12

Page 320, Prob. 1

(a) 9; (b) 5

Page 320, Prob. 3

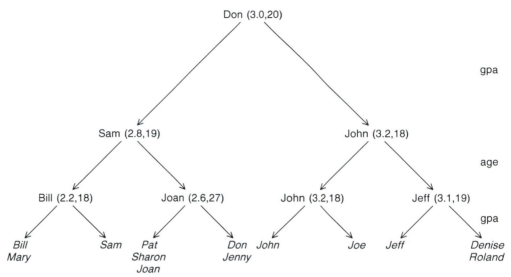

(Other answers are possible depending upon the choice of comparison records at levels two and three—the goal is to fill each bucket with approximately the same number of records.)

Page 333, Prob. 8

The gpa linear scale is [2.0,1; 2.5,2; 2.75,3; 3.0,4; 3.5,5; 4.0,6]. The age linear scale is [15,1; 17,2; 18,3; 19,4; 20,5; 30,6]. Twelve buckets are needed; only one bucket has multiple records, the one with the records for Don and Jenny.

CHAPTER 13

Page 342, Prob. 2

(a) 23 comparisons (including the 7 needed to find the record with the smallest key); (b) 10 swaps

Page 346, Prob. 2

16 exchanges

Page 352, Prob. 1

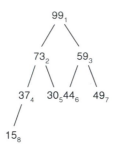

Page 359, Prob. 1

(b) Six initial files; (c) three additional files

Page 359, Prob. 2

133 files total

CHAPTER 14

Page 370

The primary data structure for this application needs to be one that will allow for both sequential and direct access plus file expansion without reorganization. The wide range of key lengths narrows the number of acceptable choices. Trie hashing is a good choice; if no two of the product names were similar, a prefix B+-tree would be acceptable.

Index